# Springer Series in Statistics

For other titles published in this series, go to
http://www.springer.com/series/692

Olivier Thas

# Comparing Distributions

Olivier Thas
Department of Applied Mathematics
Biometrics, and Process Control
Ghent University
Coupure Links 653
B-9000 Gent
Belgium
olivier.thas@UGent.be

ISSN: 0172-7397
ISBN: 978-1-4614-2449-9 e-ISBN: 978-0-387-92710-7
DOI: 10.1007/978-0-387-92710-7


Softcover reprint of the hardcover 1st edition 2010


Printed on acid-free paper

springer.com

To Ingeborg and my parents

# Preface

This book is mainly about goodness-of-fit testing, particularly about tests for the one- and the two- and $K$-sample problems. In the one-sample problem we need to test the hypothesis that the sample observations have a hypothesised distribution, whereas the two-sample problem is concerned with testing the equality of the distributions of two independent samples. Both testing problems are almost as old as statistical science itself. For instance, the well-known Pearson chi-squared test for testing goodness-of-fit to a discrete multinomial distribution, was proposed back in 1900 by Karl Pearson, who is generally recognised as one of the fathers of statistics. Another important test is the smooth test for testing uniformity which was proposed in 1937 by Jerzy Neyman, another founder of modern statistics. The Kolmogorov–Smirnov test dates from the same period, and in the middle of the century the Anderson–Darling and Cramér–von Mises tests were published. The roots of the two-sample problem also date back to the first half of the twentieth century. Frank Wilcoxon published his nonparametric rank test in 1945, and if we consider the Student-$t$ test also as a two-sample test, though under very restrictive parametric assumptions, then we even have to go back to 1908. Despite the age of many of these methods, they are still very often used in daily statistical practice, and they are taught in almost any basic statistics course. These older methods are also frequently referred to in the contemporary statistical literature. Moreover, goodness-of-fit is still a very active research domain, and many of the newer techniques are based on these older tests. I give a few examples. In the 1980s Neyman's smooth test was extended to more complex testing situations, and in the 1990s the method was further improved so that the user does not have to make any arbitrary choices anymore of the order of the test. These tests are now known as data-driven smooth tests. In 1988 Read and Cressie dedicated a whole book to generalisations of the Pearson chi-squared test. Thanks to the advances made in the theory of stochastic and empirical processes, the distribution theory of tests like the Anderson–Darling and Cramér–von Mises are nowadays much easier to tackle. Because of their relation to the empirical distribution function

(EDF), these tests are sometimes referred to as EDF tests. It is now also known that these latter test statistics can be represented in their "principal component" representation, and their "principal components" are now recognised as the components of Neyman's smooth test statistic. Also many tests for the two-sample problem belong to the class of EDF tests, and their theory is thus quite similar. This is, however, not the only relation between tests for the one- and the two-sample problems. Although they are maybe not generally known, there is a group of smooth tests for the two-sample problem, and, interestingly, their lower-order components are related to the Wilcoxon rank sum statistic, and generalisations thereof.

In the previous paragraph I have very briefly illustrated that there are many old and new tests for both the one- and the two-sample problems, and many of these methods are related to one another. It is one of the objectives of this book to give an overview of the several different classes of methods, and how they interrelate.

In the beginning of this preface I said that this book is about goodness-of-fit testing, but the title of the book is *Comparing Distributions*. This asks for an explanation. It is indeed true that most of the goodness-of-fit techniques are essentially statistical hypothesis tests, but testing does not give the whole answer to the question. A statistical test is just a formal way to make a decision between two mutually exclusive hypotheses: the null hypothesis is very clear-cut and states the hypothesised distribution, whereas usually the alternative hypothesis is very broad, and, as is the case of omnibus tests, it is just the negation of the null hypothesis. Thus, when the null hypothesis is rejected, many tests do not give any information about what the true distribution might look like, or how the true distribution differs from the hypothesised. The same reasoning holds for the two-sample problem: when the null hypothesis of equality of the two populations is rejected, there is often no information about how the two distributions disagree. First note that in the description of the testing problems just given, it might have become clear that it is basically a question of *comparing distributions*: comparing the true distribution of the sample with an hypothesised distribution, and comparing two unspecified distributions in the two-sample problem. Hypothesis testing is just one, but very popular method to compare distributions. In this book I look for more *informative* statistical analyses that provide useful information about the differences between distributions. Although crude application of goodness-of-fit tests is not informative in the sense just explained, a complementary analysis, sometimes even very closely related to the test statistic, may shed some more light on the comparison question. For instance, the goodness-of-fit test statistic is often an estimator of a distance measure between two distributions. Thus if this distance measure is well understood, the rejection of the null hypothesis suggests in what "direction" the distributions differ. Another example is the decomposition of a test statistic: the smooth and EDF test statistics can often be decomposed into component statistics, and each of them reflects another aspect of the difference between the distributions. This may help in, for instance, concluding that two distributions only differ in scale, and not in skewness. This book stresses such informative statistical analyses. Particularly for the two-sample problem, these informative analyses can give a deep understanding and a very relevant answer to the comparison question. Because I have the impression that, for instance, the conclusions from a nonparametric Wilcoxon rank sum test are far too often misunderstood, because it is only used as a nonparametric counterpart of the parametric $t$-test, I spend much time on explaining the correct, but informative, application and interpretation of goodness-of-fit tests.

Statistical hypothesis tests are not the only solutions to answer questions about the comparison of distributions. Graphs are also most helpful in understanding what is going on. I discuss some graphical tools, and I particularly focus on those graphs that are closely related to statistical tests. When a graph is a visual representation of the information that the statistical test uses in making its decision between the null and the alternative hypothesis, it is unlikely that the graph suggests a different answer and confuses the analyst. Examples of such graphs include the PP plot and the plot of the comparison distribution.

The book is written at an intermediate level. I have tried to provide the reader with some of the basic theory which is needed to understand the techniques, but some of the more technical issues are ignored. For instance, I give a very brief introduction to empirical processes, but I do not say anything about the measure-theoretical aspects. I think that our introduction to this theory is sufficient for the reader to understand the rationale of the methods. For more details I refer to the literature. The text is aimed for two groups: first, for researchers and for master or graduate students in statistics. To understand all the theory in the book, I assume that the reader is familiar with matrix algebra, calculus, and asymptotic statistical inference. Although some theory is given, I also hope that the book may be useful for applied statisticians and for practitioners who have to do statistical analyses involving the comparison of distributions. Particularly because the problems treated here are very important and so widespread in daily statistical practice, I feel that this book may be helpful for many practitioners of statistical methods. Throughout the book many of the statistical methods are applied to example datasets, and a detailed interpretation and discussion is given. All methods that I used are collected in an R-package that is available at the website accompanying this book: the cd package. The website is http://biomath.ugent.be/~othas/cd. The R-code is provided for most examples. Also, at the end of each chapter I give a summary from a purely practical point of view. The examples, together with these practical guidelines should be enough for a nonstatistician to help him or her in statistical analyses.

The book is organised as follows. The text is divided into two parts. The first part concerns the one-sample problem, and the two- and $K$-sample problems are discussed in Part Two.

The first part starts with an introduction in which a brief historical overview is given of some of the main early contributions to the methods for goodness-of-fit. In that same chapter, the Pearson chi-squared test is reviewed. The second chapter provides some essential theory and methodology that is used in further chapters. For instance, some very basic concepts, such as the empirical distribution function, are introduced, but also some more advanced topics such as empirical processes and Hilbert spaces are discussed. This chapter may be skipped at first reading, and depending on the background of the reader, one or more sections may be of interest to understand the theory in later chapters. In Chapter I introduce some graphical exploration tools that are useful in assessing goodness-of-fit. As I mentioned earlier, I prefer to focus on graphical aids that are in some way related to the more formal goodness-of-fit tests. For instance, a PP plot can be used in conjunction with a Kolmogorov–Smirnov test, as the Kolmogorov–Smirnov statistic is defined as the maximal deviation between the sample PP plot and the diagonal reference line. Chapter 4 is completely devoted to the important class of smooth tests, and in Chapter 5 I discuss the class of EDF tests (e.g., Kolmogorov–Smirnov, Anderson–Darling, but also generalisations and some more recent tests). The stress is on the relation among all methods, and on how they can be applied in an informative manner. An important example of an informative analysis is the diagnostic property and interpretation of the components of smooth test statistics, which also appear as the components of some EDF statistics. In particular, sometimes these components may be helpful in understanding in which moments the true and the hypothesised distributions are different.

In Part Two I treat the methods for the two-sample problem. In Chapter 6 I define the two- and $K$-sample problems, and I introduce the example datasets. Chapter 7 provides some more concepts and building blocks that are particularly useful for understanding the theories and concepts discussed in the chapters following. For instance, the basic theory of rank tests and exact permutation tests are introduced here. Graphical exploration tools are the topic of Chapter 8. It includes PP and QQ plots, as well as graphs of the comparison distribution.

In Chapter 9 some important statistical tests for the two- and $K$-sample problem are discussed in detail: $t$-tests for comparisons of means and Wilcoxon and other rank tests for nonparametric testing. I stress that all tests test different hypotheses, and that comparing means, as does the $t$-test, is not always the most relevant question, and that the Wilcoxon rank sum test does not necessarily test for equality of means. For the two next chapters I keep the same classification of the methods as in Part I: smooth tests in Chapter 10 and EDF tests in Chapter 11. The analogy with the methods and concepts from the one-sample problem in Part I are stressed, so that it provides us with a better understanding, and more generalisations arise easily. For instance, when a smooth test statistic for the two-sample is constructed in a particular way, its first component is the Wilcoxon rank sum statistic, its second component is the Mood rank statistic for comparing dispersions, and the third component is a rank statistic that may be used to detect differences in skewness, at least under some distributional assumptions. Throughout these two chapters I always illustrate how the methods should be used to get the most information out of the data. Some of the techniques are well known to most statisticians, but I try to make clear that even these methods can be used in a more informative and correct manner. Other methods are not very common, but I aim at showing that they are just as simple as many other popular tests, and that some of them can guide very well in understanding how the two populations differ. I always focus on the interpretation of the tests so that eventually a very informative statistical analysis may be obtained. The R-package helps using the methods in a flexible way.

I did, however, not aim at writing an encyclopedic work on goodness-of-fit tests. Writing a book that describes all tests for goodness-of-fit might have resulted in two volumes of about 500 pages. For this reason I had to make some choices along the way, so that I could focus on the relations among various types of tests and methods, and on how they may be used for informative statistical analyses. As a result, I did not give as many details on tests for discrete distributions as I did for tests for continuous distributions, and I did not thoroughly discuss rank tests in the presence of ties.

Finally I want to thank some people without whom this book could never have been at all possible. First I want to thank John Kimmel from Springer, who kept believing in the project and whose endless patience I really appreciate. This book could never have existed without the many scientific discussions I had with my Australian colleagues and friends, John Rayner and John Best. They are well known for their work on smooth tests of goodness-of-fit for the one-sample problem, and they are the founders of the *contingency table approach.* Both ideas are very central in this book. We share the idea that statistical hypothesis testing should result in an informative analysis, not necessarily focussing on the mean. I therefore want to thank them deeply, and I hope that we can work further on these ideas in the future. Writing a book takes time, a lot of time. It was not always straightforward to find quality time during the "usual" office hours when I was also supposed to teach, advise PhD students, and be absorbed with administrative jobs. Every now and then I needed "time off" so that I could work intensively on the book. Well, actually it was rather "extra time", that is, time that I would have loved to spend with my wife. I therefore thank Ingeborg; without her continuing support I would have given up the project long before. Thanks.

I hope that this book may be stimulating for researchers in the field, and that it may be helpful to practitioners, and I particularly hope that a more informative statistical practice gets promoted.

Gent, Belgium *Olivier Thas*
April 2009

# Contents

# Part I
# One-Sample Problems

# Chapter 1
# Introduction

In this introductory chapter we start with a brief historical note on the one-sample problem (Section 1.1). A first step in a data analysis is often the graphical exploration of the data. In Section 1.2 we give some graphical techniques which may be very useful in assessing the goodness-of-fit. In this section also most of the example datasets are introduced which are further used to illustrate methods in the remainder of the first part of the book. One of the earliest goodness-of-fit tests is the Pearson chi-squared test. Although it is definitely not the best choice in many situations, it is still often applied. It also often serves as a cornerstone in the construction of other goodness-of-fit tests. We give an overview of the most important issues in applying the Pearson test in Section 1.3. Moreover, many of the more recent methods still rely on the intuition of this test.

## 1.1 The History of the One-Sample GOF Problem

Probably the oldest and best known goodness-of-fit test is the Pearson $\chi^2$ test (Pearson (1900)). The test was originally constructed for testing a simple null hypothesis in a multinomial distribution. For many years it was the only GOF test, so that when other types of goodness-of-fit problems had to be solved, statisticians tried to adapt the Pearson test to these new problems. For instance, in the first half of the twentieth century, the Pearson test was frequently used to test the goodness-of-fit of continuous distributions. Because the Pearson test actually works on multinomial data, the continuous data had first to be grouped or categorised. It is even intuitively already clear that this categorisation results in information loss, and consequently in a less powerful testing method. Nowadays we have many GOF tests available which are constructed particularly for continuous data, and it is only in very exceptional cases that one still chooses to apply the Pearson test to grouped continuous data. However, because of the historical importance and

O. Thas, *Comparing Distributions,* Springer Series in Statistics,

DOI 10.1007/978-0-387-92710-7_1, 

because some of the more modern methods described later in this book rely on the Pearson $\chi^2$ test, we give a brief overview of its history and theory. For a more detailed account, we refer to, e.g., Bishop et al. (1975) and Read and Cressie (1988).

## 1.2 Example Datasets

### *1.2.1 Pseudo-Random Generator Data*

The generation of random numbers is important in many areas. For instance, in modern cryptographic algorithms 'good' random numbers are needed. Good random number generators are also essential in many sciences, e.g., in physics and, of course, in statistics, where it is common practice today to assess empirically the validity of theoretical distribution theory by means of a simulation experiment in which statistics are calculated on repeatedly generated random samples from a given distribution.

A device that generates true random numbers is hard to achieve. A true random generator is, for instance, based on a radioactive source, but it is unrealistic to have this built into every computer. Therefore, computer scientists, mathematicians, and engineers have created algorithms that generate *pseudo-random numbers*. These algorithms are based on a sound mathematical theory, and despite their deterministic nature they generate sequences of numbers that come close to true random number sequences. Apart from having as much randomness in the sequence as possible, pseudo-random generators 'sample' the numbers from a particular distribution. Often this is the uniform distribution over $[0,1]$. Whenever a new pseudo-random generator is developed, it should be tested. Using the terminology of Knuth (1969), two types of tests exist: theoretical and empirical tests. The former are based on algorithmic properties and their application does not need to let the algorithm generate sequences of pseudo-random numbers. The result of the test is a score of the randomness. The empirical tests, on the other hand, are basically statistical goodness-of-fit tests that should be applied to a generated sequence. These tests are used to test the null hypothesis that the generated numbers are indeed sampled from a uniform distribution over $[0,1]$. Atkinson (1980) is a reference in the statistical literature describing the problem. A nice reference in the computer science literature in which goodness-of-fit tests are applied to several pseudo-random generators, is Entacher and Leeb (1995).

As an example we examine the quality of the uniform pseudo-random generator in the R software, i.e., the `runif` function. We have generated 100,000 numbers. Because it would be quite useless to list all 100,000 numbers, we only present the histogram and the boxplot in Figure 1.1. The dataset (`PRG`) is available at the website accompanying the book.

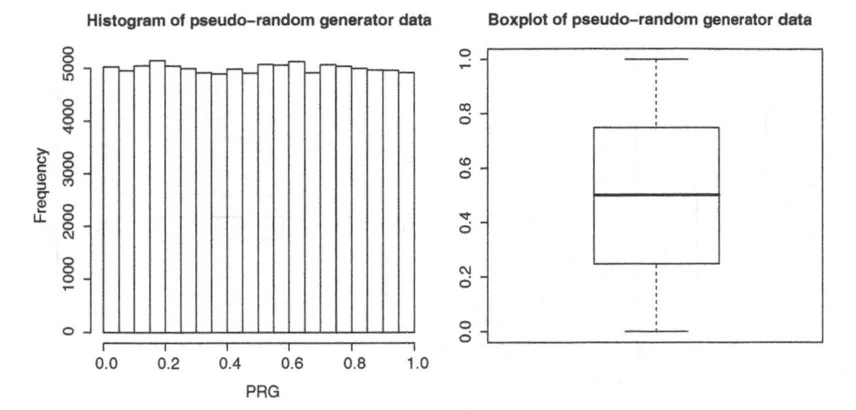


**Fig. 1.1** The histogram (left) and the boxplot (right) of the pseudo-random generator data

**Table 1.1** Concentration of PCB in the yolk lipids of 65 pelican eggs

| Concentration | | | | | | | | | | | | |
|---|---|---|---|---|---|---|---|---|---|---|---|---|
| 452 | 184 | 115 | 315 | 139 | 177 | 214 | 356 | 166 | 246 | 177 | 289 | 175 |
| 324 | 260 | 188 | 208 | 109 | 204 | 89 | 320 | 256 | 138 | 198 | 191 | 193 |
| 305 | 203 | 396 | 250 | 230 | 214 | 46 | 256 | 204 | 150 | 218 | 261 | 143 |
| 132 | 175 | 236 | 220 | 212 | 119 | 144 | 147 | 171 | 216 | 232 | 216 | 164 |
| 199 | 236 | 237 | 206 | 87 | 205 | 122 | 173 | 216 | 296 | 316 | 229 | 185 |

### *1.2.2 PCB Concentration Data*

In a study on the effect of environmental pollutants on animals, Risebrough (1972) gives data on the concentration of several chemicals in the yolk lipids of pelican eggs. The data considered here are the PCB (polychlorinated biphenyl) concentrations for 65 Anacapa birds. The complete dataset is presented in Table 1.1. The example is referred to as the *PCB concentration data.*

In the original study the mean PCB concentration in Anacapa eggs was compared to the mean concentration in eggs of other birds. Here we concentrate on the Anacapa eggs. A histogram and a boxplot are presented in Figure 1.2.

### *1.2.3 Pulse Rate Data*

At a hospital the pulse rates of 50 patients were measured in beats per minute. The data are presented in Table 1.2 and are taken from Hand et al. (1994) (dataset 416). Figure 1.3 shows the histogram and the boxplot of the data. The example is referred to as the *pulse rate data.*

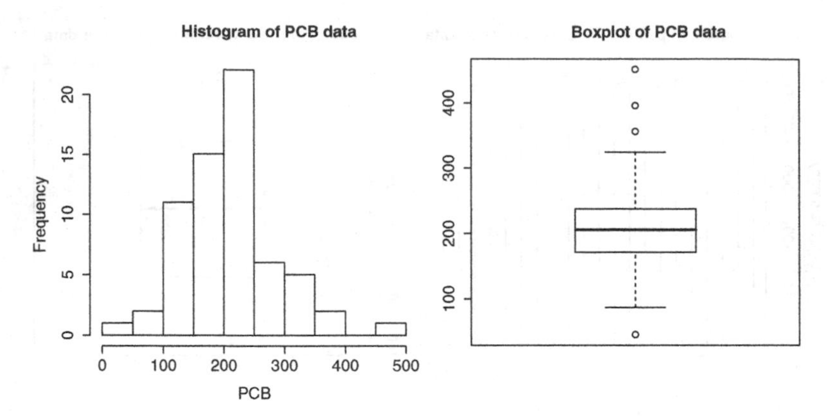


**Fig. 1.2** The histogram (left) and the boxplot (right) of the PCB data

**Table 1.2** Pulse rate of 50 patients

| Pulse rate (beats per minute) | | | | | | | | | |
|---|---|---|---|---|---|---|---|---|---|
| 68 | 80 | 84 | 80 | 80 | 80 | 92 | 92 | 80 | 80 |
| 80 | 80 | 80 | 78 | 90 | 80 | 72 | 80 | 82 | 76 |
| 84 | 70 | 80 | 82 | 84 | 116 | 80 | 95 | 80 | 76 |
| 100 | 88 | 90 | 90 | 90 | 80 | 76 | 80 | 84 | 80 |
| 80 | 80 | 80 | 104 | 80 | 68 | 84 | 64 | 84 | 72 |

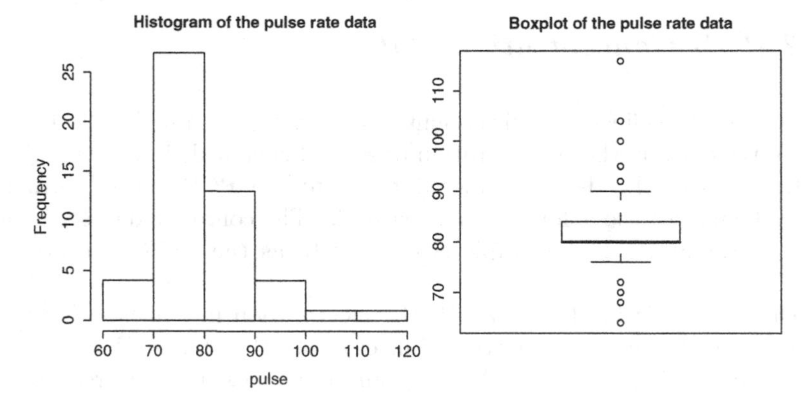


**Fig. 1.3** The histogram (left) and the boxplot (right) of the pulse rate data

### *1.2.4 Cultivars Data*

The cultivars dataset is taken from Karpenstein-Machen et al. (1994) and Karpenstein-Machan and Maschka (1996). It has also been analyzed by Piepho (2000). This example is referred to as the *cultivars data.*

**Table 1.3** Yields (in tons per hectare) of two cultivars, and the fertility score (AZ) from 19 environments

| Environment | Yield | | AZ |
|---|---|---|---|
| | Alamo | Modus | |
| 1 | 98.250 | 96.200 | 61 |
| 2 | 112.950 | 115.400 | 60 |
| 3 | 66.875 | 69.175 | 39 |
| 4 | 106.500 | 123.900 | 82 |
| 5 | 64.800 | 53.750 | 30 |
| 6 | 82.900 | 88.350 | 55 |
| 7 | 96.433 | 101.033 | 35 |
| 8 | 78.950 | 82.650 | 75 |
| 9 | 74.200 | 80.000 | 28 |
| 10 | 71.600 | 79.300 | 42 |
| 11 | 88.550 | 86.250 | 28 |
| 12 | 93.650 | 95.550 | 42 |
| 13 | 75.000 | 71.300 | 54 |
| 14 | 94.450 | 100.450 | 80 |
| 15 | 95.033 | 98.067 | 85 |
| 16 | 84.150 | 80.150 | 33 |
| 17 | 93.350 | 97.200 | 50 |
| 18 | 64.650 | 60.000 | 24 |
| 19 | 67.750 | 70.600 | 45 |

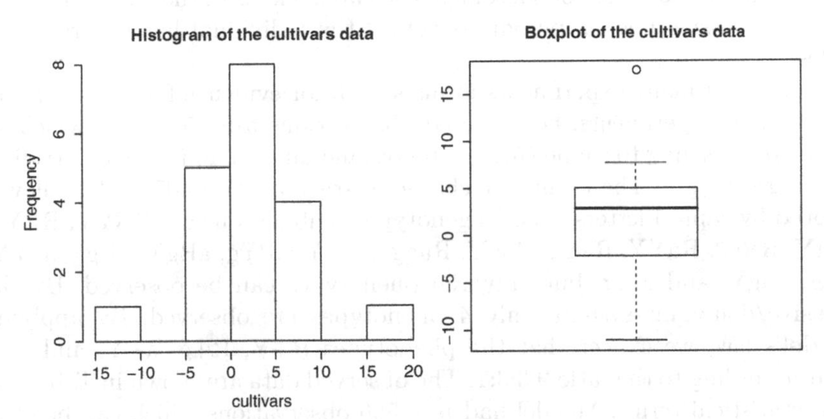


**Fig. 1.4** The histogram (left) and the boxplot (right) of the cultivars data

The dataset contains the yields (in tons per hectare) of two triticale cultivars: Alamo and Modus. Yields on both cultivars are obtained in 19 different environments. For each environment, a fertility score ("Ackerzahl" (AZ)) was recorded. The data are presented in Table 1.3 and a histogram and boxplot are shown in Figure 1.4.

One of the aims of the study was to assess if a difference existed between both cultivars in terms of the average yield. This question may be solved by performing a paired $t$-test on the paired data. An assumption underlying a paired $t$-test is that the difference between the yields of the cultivars is normally distributed. This is a classical one-sample goodness-of-fit problem.

## 1.3 The Pearson Chi-Squared Test

### *1.3.1 Pearson Chi-Squared Test for the Multinomial Distribution*

#### 1.3.1.1 The Simple Null Hypothesis Case

To introduce the original Pearson test, we consider a typical example dataset from the time that Karl Pearson developed his test. In the late 1800s and the early 1900s, there was a heavy discussion going on among scientists about the correctness of the theory of Mendel about inheritance. Mendelian law can be summarized as follows. (1) The two members of a gene pair (alleles) segregate (separate) from each other in the formation of gametes for the offspring. Half the gametes carry one allele, and the other half carry the other allele. (2) Genes for different traits assort independently of one another in the formation of gametes. The genotype of an individual is determined by the two alleles. Meldel further assumed that each gene consists of one of two possible alleles: a recessive and a dominant allele. The individual's phenotype, which is the characteristic that is expressed, corresponds to that of the dominant allele as soon as one of the two alleles is of the dominant type. Thus, a recessive phenotype only occurs if the individual has two recessive alleles.

Mendel did many experiments in his search for evidence for his theory. In one of these experiments, he collected observations from 556 peas which he classified according to shape (round (R) or angular (a)) and to colour (yellow (Y) or green (g)). The dominant characteristics are "round" and "yellow", denoted by capital letters. The 16 genotype combinations are RRYY, RRYg, RRgY, RRgg, RaYY, RaYg, RagY, Ragg, aRYY, aRYg, aRgY, aRgg, aaYY, aaYg, aagY, and aagg, but only the phenotypes can be observed. By the recessive/dominant system, only 4 phenotypes are observed. By applying Mendel's law, we expect that the phenotypes R+Y, R+g, a+Y, and a+g occur according to the ratio 9:3:3:1. The observed data are shown in Table 1.4.

In statistical terms, Mendel had $n = 556$ observations which can be classified into four classes. The observations can be denoted by $Y_i$ $(i = 1, \ldots, n)$ which can take one of the values in $\{1, \ldots, 4\}$, but usually the data are represented as counts. Let $N_j$ denote the count of observations in class $j$ $(j = 1, \ldots, 4)$. Clearly, $n = \sum_{j=1}^{4} N_j$. Note that we use capital letters for random variables, and their lowercase versions for their realisations or observed

**Table 1.4** Counts of the phenotypes of 556 peas

| | Phenotype | | |
|---|---|---|---|
| R+Y | R+g | a+Y | a+g |
| $n_1 = 315$ | $n_2 = 108$ | $n_3 = 101$ | $n_4 = 32$ |

values. The vector $\boldsymbol{N}^t = (N_1, N_2, N_3, N_4)$ has thus a multinomial distribution with parameters $n$ and $\boldsymbol{\pi}^t = (\pi_1, \pi_2, \pi_3, \pi_4)$, where $\pi_j = \Pr\{Y = j\}$. This is denoted by $\boldsymbol{N} \sim \text{Mult}(n, \boldsymbol{\pi})$. If Mendel's law is correct, the probabilities are equal to $\boldsymbol{\pi}_0^t = (\pi_{01}, \pi_{02}, \pi_{03}, \pi_{04}) = (\frac{9}{16}, \frac{3}{16}, \frac{3}{16}, \frac{1}{16})$. Thus, we are interested in testing the simple null hypothesis

$$H_0 : \boldsymbol{\pi} = \boldsymbol{\pi}_0$$

versus the alternative hypothesis that the Mendelian law is incorrect,

$$H_1 : \boldsymbol{\pi} \neq \boldsymbol{\pi}_0.$$

For the general case where there are $k$ classes, Pearson's test statistic is given by

$$X_n^2 = \sum_{j=1}^{k} \frac{(N_j - n\pi_{0j})^2}{n\pi_{0j}}, \tag{1.1}$$

which is often written as

$$\sum_{j=1}^{k} \frac{(O_j - E_j)^2}{E_j},$$

where $O_j$ and $E_j$ refer to the observed and the expected counts or frequencies, respectively. Pearson proved the next theorem, the proof of which is provided in Appendix A.1.

**Theorem 1.1.** *Suppose $H_0$ holds true. Then, as $n \to \infty$,*

$$X_n^2 \xrightarrow{d} \chi_{k-1}^2.$$

Because the Pearson $\chi^2$ test relies on asymptotic theory, it is important to understand under which finite sample size conditions the asymptotic $\chi^2$ null distribution is a good approximation to the exact null distribution. This has been studied by many authors before. See, e.g., Lancaster (1969) or Bishop et al. (1975). A very often applied rule of thumb is that all expected number of counts $E_j$ should be at least equal to 5 to have a good approximation. Note that under the simple null hypothesis, however, the exact null distribution of $X_n^2$ may also easily be approximated by simulation.

*Example 1.1 (peas).* Under the null hypothesis that the phenotypes R+Y, R+g, a+Y, and a+g occur with probabilities $\frac{9}{16}, \frac{3}{16}, \frac{3}{16}$ and $\frac{1}{16}$, respectively, the expected frequencies of the $n = 556$ are computed as

$$E_1 = n\pi_{01} = 556 \times \frac{9}{16} = 312.75\ , \quad E_2 = n\pi_{02} = 556 \times \frac{3}{16} = 104.25$$
$$E_3 = n\pi_{03} = 556 \times \frac{3}{16} = 104.25\ , \quad E_4 = n\pi_{04} = 556 \times \frac{1}{16} = 34.75.$$

This gives $X_n^2 = 0.47$. Under the null hypothesis $X_n^2$ is asymptotically distributed as $\chi_{4-1}^2 = \chi_3^2$. If $X^2$ denotes a $\chi_3^2$ random variable, the $p$-value is given by $\Pr\{X^2 \geq 0.47\} = 0.9254$. Thus the null hypothesis is accepted at the 5% level of significance.

The same analysis in R gives the following output.

```
> peas<-c(315,108,101,32)
> chisq.test(peas,p=c(9/16,3/16,3/16,1/16))

        Chi-squared test for given probabilities

data:  peas
X-squared = 0.47, df = 3, p-value = 0.9254
```

Note that the observed test statistic ($X_n^2 = 0.47$) is extremely small. Even if the null hypothesis is true there is only a probability of $\Pr\{X^2 < 0.47\} = 0.0746$ that a smaller value is observed. In the early 1900s, this led to the suspicion that Mendel (or his coworkers) might have cheated with the data to make them more supportive for his theory.

#### 1.3.1.2 The Composite Null Hypothesis Case

We first introduce an example in which the multinomial distribution probability parameter $\boldsymbol{\pi}$ depends on an $m$-dimensional nuisance parameter $\boldsymbol{\beta}$.

*Example 1.2 (Hardy–Weinberg equilibrium).* In population genetics the *Hardy–Weinberg equilibrium* is a model which predicts genotype and allele frequencies in stationary populations. There are five assumptions underlying the model: (1) the population is large; (2) there is no gene flow between different populations; (3) the number of mutations is negligible; (4) individuals mate randomly; and (5) natural selection is not operating on the population. For one single gene, we consider two types of alleles which are denoted by a and A. Let $p$ denote the probability in the population of the occurrence of A; i.e., $p = \Pr\{A\}$. Because there are only two alleles, we have $q = \Pr\{a\} = 1 - p$. Under the conditions of the Hardy–Weinberg model, the probabilities of the three possible genotypes AA, aA, and aa are given by $p^2$, $2pq$, and $q^2$, respectively. Note that $p^2 + 2pq + q^2 \equiv 1$. Thus, if $\boldsymbol{N}^t = (N_1, N_2, N_3)$ denotes the vector of counts of the three genotypes in a random sample of size $n = N_1 + N_2 + N_3$, and if the Hardy–Weinberg equilibrium applies, the probabilities of the multinomial distribution of $\boldsymbol{N}$ are given by $\boldsymbol{\pi}_0^t = (\pi_{01}, \pi_{02}, \pi_{03})$, where

$$\pi_{01} = p^2 \quad \pi_{02} = 2pq \quad \pi_{03} = q^2.$$

These three probability parameters depend on the nuisance parameter $\boldsymbol{\beta} = p$. Therefore, $\boldsymbol{\pi}_0$ is actually a function that maps the $m$-dimensional parameter $\boldsymbol{\beta} \in B$ into a subset of $\Pi = \{(p_1, \ldots, p_k); p_i \geq 0, i = 1, \ldots, k; \sum_{i=1}^k p_i = 1\}$, and $B$ is typically a subset of $\mathbb{R}^p$. We write $\boldsymbol{\pi}_0(\boldsymbol{\beta})$ to stress that $\boldsymbol{\pi}_0$ is a function. It is convenient to assume that $m < k$.

**Table 1.5** Counts of individuals with a given genotype for three different loci

| Locus | | EST | |
|---|---|---|---|
| Genotype | SS | SF | FF |
| Counts | 37 | 20 | 7 |
| Locus | | ICD | |
| Genotype | SS | SF | FF |
| Counts | 48 | 4 | 3 |
| Locus | | LA | |
| Genotype | SS | SF | FF |
| Counts | 20 | 11 | 2 |

Table 1.5 is taken from Lidicker and McCollum (1997). They studied one single isolated population of sea otters in California. When in 1911 this population became protected by law, it contained only 50 sea otters. It was the only group of sea otters left along the central California coast. Fur hunting had nearly led to their complete extinction. When Lidicker and McCollum studied the population, the population had grown to about 1500 sea otters. At six different loci (here we present only the results for three loci), they counted the number of sea otters with a given genotype. The genes they studied code for allozymes, which are a particular type of enzymes that can be easily genotyped. The alleles are represented by the letters S and F. The researchers were interested in testing the null hypothesis that the population of sea otters is in Hardy–Weinberg equilibrium, which implies that there is sufficient genetic variation in the population. Table 1.5 contains the data.

In statistical terms the null hypothesis is

$$H_0 : \boldsymbol{\pi} = \boldsymbol{\pi}_0(\boldsymbol{\beta}) \text{ for some } \boldsymbol{\beta} \in B. \tag{1.2}$$

We want a test to test $H_0$ against the alternative hypothesis $H_1$ : not $H_0$. A natural procedure is to first estimate the nuisance parameter, plugin this estimate in $\boldsymbol{\pi}_0$, and compute the Pearson $X^2$ test statistic as before. We first give some more details on how the nuisance parameter is estimated, and then, in Theorem 1.2 the asymptotic null distribution of the test statistic is established.

From the Hardy–Weinberg example it is clear that the nuisance parameter appears in the probabilistic model as specified under the null hypothesis. It is therefore obvious that it must be estimated under the assumption that the null hypothesis holds true. Because the null hypothesis in (1.2) completely specifies a multinomial distribution for $\boldsymbol{N}$, the maximum likelihood method is available. Let $\hat{\boldsymbol{\beta}}$ denote the MLE of $\boldsymbol{\beta}$. Pearson's $X^2$ statistic then becomes

$$\hat{X}_n^2 = X_n^2(\hat{\boldsymbol{\beta}}) = \sum_{j=1}^{k} \frac{\left(N_j - n\pi_{0j}(\hat{\boldsymbol{\beta}})\right)^2}{n\pi_{0j}(\hat{\boldsymbol{\beta}})}.$$

Theorem 1.2 gives the asymptotic null distribution of $X_n^2$ under slightly more general conditions on the estimator of $\boldsymbol{\beta}$. In particular, the theorem holds under the assumption that the estimator $\hat{\boldsymbol{\beta}}$ is *best asymptotically normal* (BAN). An estimator is BAN if (1) it is a consistent estimator, (2) it is asymptotically normally distributed, and (3) it is asymptotically efficient. Birch (1964) gives six regularity conditions for the function $\boldsymbol{\pi}_0(\boldsymbol{\beta})$. See also Bishop et al. (1975) for a detailed discussion. The proof of the theorem is given in Section A.2.

**Theorem 1.2.** *Suppose that $\hat{\boldsymbol{\beta}}$ is a BAN estimator of the p-dimensional parameter $\boldsymbol{\beta}$. Then, as $n \to \infty$,*

$$\hat{X}_n^2 \xrightarrow{d} \chi^2_{k-p-1}.$$

The test based on $\hat{X}_n^2$ is referred to as the Pearson–Fisher test because it was Sir Ronald Fisher who correctly proved that the number of degrees of freedom of the $\chi^2$ distribution should take the number of estimated nuisance parameters into account. Karl Pearson, on the other hand, was convinced that the correct number of degrees of freedom was still $k-1$. This famous controversy between Pearson and Fisher is told in a lively manner by Box (1978).

*Example 1.3 (Hardy–Weinberg equilibrium).* First the MLE of the parameter $p = \Pr\{\text{S}\}$ must be found. Let $N_0$, $N_1$, and $N_2$ denote the counts of genotypes SS, SF, and FF, respectively. The log-likelihood is

$$l(p) = 2N_0 \ln p + N_1 \ln p(1-p) + 2N_2 \ln(1-p).$$

Hence, the MLE is given by $\hat{p} = \frac{1}{2}((N_1+2N_2)/n)$. The results of the Pearson–Fisher tests on the three loci are presented in Table 1.6. We may conclude that at the 5% level of significance, only the gene at the ICD locus is not

**Table 1.6** Results of the Pearson–Fisher tests on the Hardy–Weinberg equilibrium example dataset

| Locus | EST | | | $\hat{p}$ | $\hat{X}_n^2$ | $p$-value |
|---|---|---|---|---|---|---|
| Genotype | SS | SF | FF | | | |
| Counts | 37 | 20 | 7 | | | |
| Expected counts | 34.5 | 25 | 4.5 | 0.266 | 2.53 | 0.111 |
| Locus | ICD | | | | | |
| Genotype | SS | SF | FF | | | |
| Counts | 48 | 4 | 3 | | | |
| Expected counts | 45.5 | 9.1 | 0.4 | 0.091 | 17.2 | $< 0.001$ |
| Locus | LA | | | | | |
| Genotype | SS | SF | FF | | | |
| Counts | 20 | 11 | 2 | | | |
| Expected Counts | 19.7 | 11.6 | 1.7 | 0.227 | 0.086 | 0.770 |

at the Hardy–Weinberg equilibrium. The sea otter population thus shows insufficient genetic variation, particularly at the ICD locus where the SS genotype is overrepresented. Thus at least one of the five assumptions underlying the Hardy–Weinberg model must be violated (e.g., no random mating, natural selection may have interfered, mutations may have changed the gene pool).

### *1.3.2 Generalisations of the Pearson $\chi^2$ Test*

In the previous section we relied on likelihood theory to obtain the Pearson–Fisher test. Inasmuch as likelihood theory is basically an asymptotic theory we can approximate the multinomial distribution of $\boldsymbol{N}$ by $k$ Poisson distributions for the $N_i$ ($i = 1, \ldots, k$) with means given by $n\pi_i$ (or $n\pi_{0i}$ under $H_0$). Using this Poisson model, the Pearson–Fisher test arises naturally as the score test. In a similar way, the Wald test statistic turns out to be Neyman's modified $X^2$ statistic (Neyman (1949)),

$$\widehat{NM}_n = \sum_{j=1}^{k} \frac{\left(N_j - n\pi_{0j}(\hat{\boldsymbol{\beta}})\right)^2}{N_j}.$$

The Wald statistic is a quadratic approximation of the likelihood ratio test statistic,

$$\widehat{LR}_n = 2\sum_{j=1}^{k} N_j \ln \frac{N_j}{n\pi_{0j}(\hat{\boldsymbol{\beta}})}. \tag{1.3}$$

Based on the likelihood theory, all three test statistics have the same asymptotic null distribution.

The three likelihood-based statistics are not the only ones used for goodness-of-fit testing in a multinomial distribution. For instance, the Freeman–Tukey statistic is derived independently from the likelihood, but it also has the same asymptotic $\chi^2$ null distribution. Cressie and Read (1984) introduced a generalisation of the above-mentioned statistics. They found a family of statistics indexed by a real-valued parameter $\lambda$. The family is called the family of power divergence statistics and it is given by

$$2nI^{\lambda}(\boldsymbol{N}; \hat{\boldsymbol{\beta}}) = \frac{2}{\lambda(\lambda+1)} \sum_{j=1}^{k} N_j \left\{ \left( \frac{N_j}{n\pi_{0j}(\hat{\boldsymbol{\beta}})} \right)^{\lambda} - 1 \right\}.$$

For $\lambda = 0$ and $\lambda = -1$, the corresponding statistics are defined by continuity. Then, $\lambda = 0$, $\lambda = 1$, $\lambda = -2$, and $\lambda = -\frac{1}{2}$ give the likelihood ratio, Pearson, Neyman's modified, and the Freeman–Tukey statistic, respectively.

We summarise the main results of Cressie and Read (1984) in the next theorem. For more details on the power divergence statistics, we refer to a monograph on these statistics by Read and Cressie (1988).

**Theorem 1.3.** *(1) Suppose that* $\boldsymbol{N}$ *is a multinomial random vector with probability vector* $\boldsymbol{\pi}$ *and let* $\hat{\boldsymbol{\pi}}$ *denote any* $\sqrt{n}$*-consistent estimator of* $\boldsymbol{\pi}$*. Then, as* $n \to \infty$,

$$2nI^{\lambda}(\boldsymbol{N};\hat{\boldsymbol{\beta}}) - 2nI^{1}(\boldsymbol{N};\hat{\boldsymbol{\beta}}) \xrightarrow{p} 0 \qquad -\infty < \lambda < +\infty.$$

*(2) Suppose* $H_0$ *is true, and* $\hat{\boldsymbol{\beta}}$ *is a BAN estimator of* $\boldsymbol{\beta}$*. Then, as* $n \to \infty$,

$$2nI^{\lambda}(\boldsymbol{N};\hat{\boldsymbol{\beta}}) \xrightarrow{d} \chi^2_{k-p-1} \qquad -\infty < \lambda < +\infty.$$

Cressie and Read (1984) thoroughly studied the large and small sample properties of goodness-of-fit tests based on their power divergence statistics. They concluded that the Pearson test ($\lambda = 1$) is good in the sense that its null distribution is well approximated by the $\chi^2$ distribution in small samples, and it has quite good power against many alternatives. From many perspectives ($\chi^2$ approximation and power) they generally recommend $\lambda \geq 0$, and because the likelihood ratio test is at the boundary of this recommendation ($\lambda = 0$), they suggest not to use this test. Based on their study, they eventually proposed a new test with overall good properties in terms of $\chi^2$ approximation and power,

$$2nI^{2/3}(\boldsymbol{N};\hat{\boldsymbol{\beta}}) = \frac{9}{5}\sum_{j=1}^{k} N_j \left[\left(\frac{N_j}{n\pi_{0j}(\hat{\boldsymbol{\beta}})}\right)^{2/3} - 1\right].$$

### *1.3.3 A Note on the Nuisance Parameter Estimation*

Theorems 1.2 and 1.3 rely on the condition that $\hat{\boldsymbol{\beta}}$ is a BAN estimator. Although in many situations this will be the MLE, there is a more general class of estimators that possess this property. Holland (1967), for instance, showed that the *minimum chi-square estimator* is also BAN. The latter is defined as

$$\hat{\boldsymbol{\beta}} = \text{ArgMin}_{\boldsymbol{\beta}\in B} X_n^2(\boldsymbol{\beta}).$$

If one believes that the Pearson $X^2$ statistic measures the discrepancy between the observed data and the hypothesised model at the right scale, it seems indeed meaningful to replace $\boldsymbol{\beta}$ by that value of the nuisance parameter that moves the hypothesised model as close as possible to the observed data. Note that the MLE minimises the likelihood ratio statistic of Equation (1.3). Because the Pearson $X^2$ and the likelihood ratio statistics are no true distance measures, these estimators are generally referred to as *minimum discrepancy estimators.* Cressie and Read (1984) introduced a large class of such estimators. They defined

$$\hat{\boldsymbol{\beta}} = \text{ArgMin}_{\boldsymbol{\beta} \in B} I^\lambda(\boldsymbol{N}; \boldsymbol{\beta})$$

as the minimum $I^\lambda$-discrepancy estimator. They showed that this $\hat{\boldsymbol{\beta}}$ is a BAN estimator, provided that the function $\boldsymbol{\pi}_0(\boldsymbol{\beta})$ satisfies the six Birch regularity conditions.

A BAN estimator is also a *locally asymptotically linear* estimator. See Section 2.7 for more details.

## 1.4 Pearson $X^2$ Tests for Continuous Distributions

Although the Pearson $X^2$ test is clearly developed for testing goodness-of-fit for a multinomial distribution, it is also a popular test to use with continuous distributions. When put into an historical perspective, it is easy to understand: in the early years of statistics, say in the first few decades of the 1900s, the Pearson $X^2$ test was the only goodness-of-fit test available. Thus when testing goodness-of-fit for a continuous distribution, the data were first grouped into $k$ groups or cells so that counts from a multinomial distribution were obtained. In this section, we give only a few general comments on this procedure, but no details are given because we believe that nowadays one should use other types of goodness-of-fit tests for continuous distributions. This does, however, not mean that there is no active research going on anymore. For instance, Aguirre and Nikulin (1994) and Pya (2004) constructed Pearson-type tests for the logistic distribution. The book of Greenwood and Nikulin (1996) is completely devoted to this class of tests.

Let $\mathcal{S}_n$ and $\mathcal{S}$ denote the sample and the sample space (i.e., the support of the hypothesised distribution $G$). Consider a partition of the sample space, say $\mathcal{S} = \cup_{j=1}^k \mathcal{S}_j$, where the $\mathcal{S}_j$ are disjoint subsets of $\mathcal{S}$. Usually, the partition is of the form $\mathcal{S} = \{]-\infty, c_1[, [c_1, c_2[, \ldots, [c_{k-1}, +\infty[\}$, where the constants $c_j$ are called the *cell boundaries.* The multinomial counts are then computed as $N_j = \#\{1 \leq i \leq n : X_i \in \mathcal{S}_j\}$, and their corresponding probabilities under the goodness-of-fit null hypothesis are given by

$$\pi_{0j}(\boldsymbol{\beta}) = \int_{\mathcal{S}_j} g(x; \boldsymbol{\beta}) dx,$$

$(j = 1, \ldots, k)$ where the nuisance parameter may be replaced by an estimator (see later).

When testing a simple null hypothesis, no nuisance parameter estimation is needed, and the Pearson $X^2$ statistic of Equation (1.1) has asymptotically a $\chi^2_{k-1}$ null distribution, as before. Despite the apparent simplicity of the procedure, there are some important issues left unanswered: how to choose the number of partitions ($k$) and where to place the cell boundaries ($c_j$). The optimal grouping depends on the alternative against which a large power is desirable, but in many realistic situations there is no clear idea about the alternative. Many of the theoretical studies about these issues are asymptotic in nature, and some of these even suggest to let the number of cells ($k$) grow with the sample size $n$. Among all grouping schemes that have been suggested, we only mention a simple but popular solution which says that the cell boundaries must be chosen so that equiprobable classes are obtained (Mann and Wald (1942)), i.e., $\pi_{01} = \cdots = \pi_{0k} = 1/k$. Under these conditions Mann and Wald (1942) showed that the Pearson test is unbiased. Later, Cohen and Sackrowitz (1975) and Bednarski and Ledwina (1978) showed that in most cases Pearson's test is biased when applied to an unequiprobable grouping. Mann and Wald also give a formula to determine an appropriate number of equiprobable cells based on some minimum power requirement. The intuitively appealing reasoning that the more cells are constructed the more information from the original sample of continuous data is retained and the higher the power will be is, however, not always correct (Oosterhoff (1985)) because the increase of the number of cells implies both an increase in the noncentrality parameter of the noncentral $\chi^2$-distribution of the test statistic under a local alternative hypothesis, and an increase of the variance of the limiting central $\chi^2$-distribution under $H_0$. Whenever the second implication beats the first, an increase in power under partition refinements is not guaranteed anymore. Since the publication of the Mann and Wald paper, many more papers on the choice and the number of cells have appeared. In general it is concluded that the Mann–Wald number of cells is too high (see, e.g., Quine and Robinson (1985)) and may even reduce the power for some specific alternatives. A comprehensive and practical oriented summary can be found in Moore (1986). A more modern approach to the problem of choosing $k$ is to make this choice data dependent (see, e.g., Bogdan (1995) and Inglot and Janic-Wróblewska (2003)). In most of the papers on the subject the authors agree with the initial recommendation of equiprobable cells; still it is important to recognise that some others have other recommendations. Kallenberg et al. (1985), for instance, suggest that for heavy-tailed distributions under the alternative hypothesis, smaller cells in the tails may result in better power characteristics.

When there are nuisance parameters involved the problem becomes even more complex. Only when the nuisance parameters are estimated as a BAN estimator (e.g., the MLE) based on the grouped data $\boldsymbol{N}$, the theory of Section 1.3.1 applies, and thus the Pearson–Fisher $\hat{X}_n^2$ statistic has a limiting $\chi^2_{k-p-1}$ distribution under the null hypothesis. However, when the original ungrouped observations $X_1, \ldots, X_n$ are available, it may seem more appropriate to use them directly for estimating the nuisance parameter, for example, the (ungrouped) MLE defined as $\text{ArgMax}_{\boldsymbol{\beta} \in B} \sum_{i=1}^n \ln g(X_i; \boldsymbol{\beta})$. When this ungrouped MLE is plugged into the Pearson $X^2$ statistic, the asymptotic null distribution has no simple expression anymore (it is a weighted sum of $\chi_1^2$ variates, but the weights may depend on $G$ and on the unknown nuisance parameter $\boldsymbol{\beta}$). This test, which is known as the Chernoff–Lehmann test (Chernoff and Lehmann (1954)), has thus little value in practice. Rao and Robson (1974) showed that if the $X_n^2$ test statistic is changed by replacing the variance–covariance matrix $\boldsymbol{\Sigma}$ in Equation (A.1) by a more complicated form (not shown), the resulting test statistic has asymptotically a $\chi_{k-1}^2$ null distribution, which does not depend on the number of nuisance parameters! This test is known as the Rao–Robson test. Numerous simulation studies have indicated that the Rao–Robson test has the largest power for many different alternatives (see, e.g., Moore (1986)). As in the no-nuisance parameter case, here also the issues related to the choice of the number and position of the cell boundaries are important. Many systems of choosing the cell boundaries (e.g., equiprobable cells) now result in random cell boundaries because of the dependence on the same data through $G(.;\hat{\boldsymbol{\beta}})$, which further complicates the theory.

Finally, we refer the interested reader to D'Agostino and Stephens (1986), Drost (1988), Rayner et al. (2009), and Greenwood and Nikulin (1996) for more detailed discussions on the issues briefly introduced here.

# Chapter 2
# Preliminaries (Building Blocks)

This chapter provides an introduction to some methods and concepts on which many of the goodness-of-fit methods are based. For instance, the empirical distribution function (EDF) plays a central role in many GOF techniques. Instead of introducing and discussing the EDF in the section where it is used for the first time, we have chosen to isolate it and put it into this chapter. When in further chapters a method is described which relies heavily on the EDF, the reader is referred to this introductory chapter. Other concepts treated in this way are empirical processes, comparison distributions, Hilbert spaces, parameter estimation, and nonparametric density estimation. Some of the topics are quite technical, but we have tried to focus on the rationale and intuition behind them, rather than providing all the technical details.

Despite the heterogeneity of topics included here, we have tried writing this chapter so that it can also be read as an introduction which demonstrates that GOF can be viewed from many different angles.

## 2.1 The Empirical Distribution Function

### *2.1.1 Definition and Construction*

The empirical distribution function (EDF) is basically an estimator of the distribution function $F$ of a random variable $X$, and it is directly based on the probability interpretation of $F$. In particular, for each $x$,

$$F(x) = \Pr\{X \leq x\}.$$

O. Thas, *Comparing Distributions*, Springer Series in Statistics,
DOI 10.1007/978-0-387-92710-7_2, © Springer Science+Business Media, LLC 2010

Thus, for each $x$, $F(x)$ is a probability, and because probabilities are easy to estimate, $F(x)$ also has a simple estimator. In particular, let $\mathcal{S}_n = \{X_1, \ldots, X_n\}$ denote a sample of $n$ i.i.d. observations; then $F(x)$ is consistently estimated as

$$\hat{F}_n(x) = \frac{1}{n}\#\{X_i \in \mathcal{S}_n : X_i \leq x\} = \frac{1}{n}\sum_{i=1}^{n} \mathrm{I}(X_i \leq x); \tag{2.1}$$

i.e., $\hat{F}_n(x)$ equals the number of sample observations not larger than $x$, divided by the sample size $n$. From this construction, it is clear that $\hat{F}_n(x)$ is a nondecreasing step function, with steps at the sample observations $x = X_i$. Each step is a multiple of $1/n$.

The EDF may also be constructed by using the order statistics. Suppose that no ties occur in the sample: i.e., all sample observations are different (this happens with probability one when $F$ is continuous). Then the $n$ observations can be ordered so that $X_1 < X_2 < \cdots < X_n$. For this ordering, the $i$th order statistic, denoted by $X_{(i)}$, equals $X_i$ $(i = 1, \ldots, n)$. Using the order statistics, the EDF may be defined as

$$\begin{cases} \hat{F}_n(x) = 0 & \text{if } x < X_{(1)} \\ \hat{F}_n(x) = \frac{i}{n} & \text{if } X_{(i)} \leq x < X_{(i+1)}, \quad i = 1, \ldots, n-1 \\ \hat{F}_n(x) = 1 & \text{if } X_{(n)} \leq x \,. \end{cases}$$

*Example 2.1 (PCB concentration).* Figure 2.1 shows the EDF of the PCB data. The R-code is given below.

```
> PCB.edf<-ecdf(PCB)
> plot(PCB.edf,verticals= TRUE, do.p = FALSE,
+ main="EDF of PCB data")
```

The EDF is closely related to the binomial distribution. From its definition in (2.1), we may see that, for each $x$, $n\hat{F}_n(x)$ is binomially distributed with parameters $n$ and $F(x)$. Thus, for every $x$ the exact distribution of $\hat{F}_n(x)$ is known. Many of the results presented later in this chapter, however, are based on asymptotic properties of the EDF. For instance, the next three properties follow immediately from the binomial distribution of $n\hat{F}_n(x)$.

1. $\hat{F}_n(x)$ is an unbiased estimator of $F(x)$; i.e.,

$$\mathrm{E}\left\{\hat{F}_n(x)\right\} = F(x) \text{ for every } x \text{ and every } n.$$

2. By the strong law of large numbers, $\hat{F}_n(x)$ is consistent; i.e., as $n \to \infty$,

$$\hat{F}_n(x) \xrightarrow{\text{a.s.}} F(x) \text{ for every } x.$$

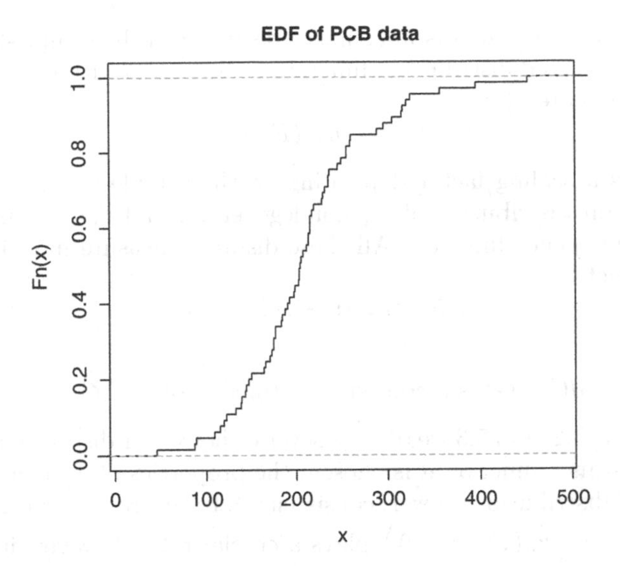


**Fig. 2.1** The EDF of the PCB data

3. By the central limit theorem (CLT), the asymptotic normality of $\sqrt{n}\hat{F}_n(x)$ follows; i.e., as $n \to \infty$,

$$\sqrt{n}\left(\hat{F}_n(x) - F(x)\right) \xrightarrow{d} N\left(0, F(x)\left(1 - F(x)\right)\right) \text{ for every } x.$$

Note that these are all pointwise convergences. Property (2) may even be extended,

$$\sup_x |\hat{F}_n(x) - F(x)| \xrightarrow{\text{a.s.}} 0.$$

This result is known as the *Glivenko–Cantelli theorem* . The estimation error of $\hat{F}_n$ is controlled by the *Dvoretzky–Kiefer–Wolfowitz inequality*, which says that for any $\epsilon > 0$,

$$\Pr\left\{\sup_x |\hat{F}_n(x) - F(x)| > \epsilon\right\} \le 2\exp(-2n\epsilon^2).$$

### *2.1.2 Rationale for Using the EDF*

All these properties essentially say that $\hat{F}_n$ is close to $F$ for large sample sizes. Thus, when the interest is in testing the GOF null hypothesis $H_0 : F(x) = G(x)$, it is sensible to measure in some sense how different the EDF is from the hypothesised distribution $G$. This is exactly what EDF test statistics do (see Chapter 5). They are distance measures between the sample-based EDF and the hypothesised distribution function $G$. Statistics within this class may be generally denoted by

$$T_n = c(n)d(\hat{F}_n, G), \tag{2.2}$$

where $c(n)$ is a scaling factor depending on the sample size $n$ to make the asymptotic null distribution of $T_n$ nondegenerate, and $d(.,.)$ denotes a distance or a divergence function. All these distance measure have in common that they satisfy

$$d(F,G) = 0 \Leftrightarrow H_0 \text{ is true},$$

and

$$d(\hat{F}_n, G) \text{ is a consistent estimator of } d(F,G).$$

In Sections 5.1, 5.2, and 5.3 we discuss several choices for distance functions $d$.

Once a distance function $d$ is chosen, the properties of $T_n$ can be studied (e.g., the null distribution, power, consistency). In this respect the asymptotic normality of $\sqrt{n}\left(\hat{F}_n(x) - F(x)\right)$ plays a crucial role. However, it turns out that pointwise convergence is not sufficient for obtaining the null distribution of most $T_n$. Intuitively, this may be seen from (2.2), where $d$ is a distance between two *functions*. Hence, some *functional* central limit theorem is needed. This is the topic of Section 2.2.

## 2.2 Empirical Processes

### 2.2.1 Definition

In the previous section the EDF was introduced, and some of its asymptotic properties were given. It is important to see that all these properties only hold in a pointwise fashion; i.e., they are statements about $\hat{F}_n(x)$ for a given $x$. On the other hand, $\hat{F}_n(x)$ is clearly a function of $x$, and it is in this sense that the EDF is used in the distance function in (2.2). Although our focus is on $\hat{F}_n(x)$, it turns out that it is more convenient to work with the empirical process

$$\mathbb{B}_n(x) = \sqrt{n}\left(\hat{F}_n(x) - F(x)\right).$$

When $F$ is the uniform distribution, $\mathbb{B}_n$ is sometimes referred to as the uniform empirical process. Because $\mathbb{B}_n$ depends on the sample observations it is a random function. Figure 2.2 illustrates the concept of a random function by showing some realisations of the uniform empirical process $\mathbb{B}_n$. A realisation of an empirical process is called a *sample path*.

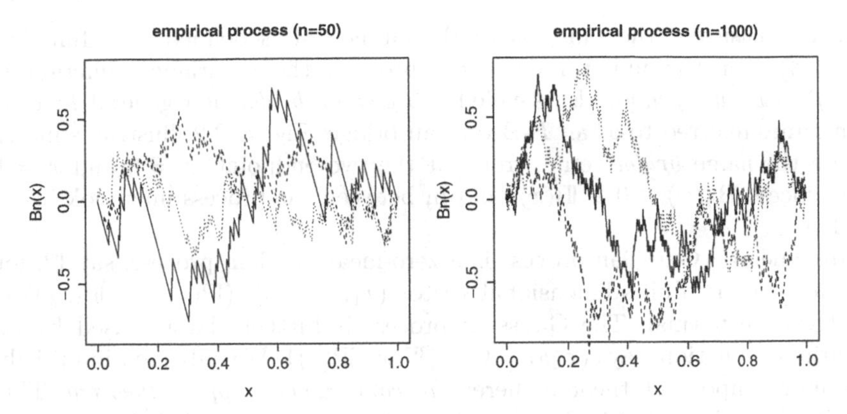


**Fig. 2.2** Realisations of a uniform empirical process with $n = 50$ (left), and with $n = 1000$ (right)

## 2.2.2 *Weak Convergence*

When going from pointwise to *functional* properties of $\mathbb{B}_n(x)$, it is a natural step to first have a look at the properties of the finite-dimensional random vector

$$(\mathbb{B}_n(x_1), \ldots, \mathbb{B}_n(x_k)), \tag{2.3}$$

for any $x_1, \ldots, x_k$ in the support of $F$. In particular, it is the multivariate CLT that gives, for any $x_1, \ldots, x_k$,

$$(\mathbb{B}_n(x_1), \ldots, \mathbb{B}_n(x_k)) \xrightarrow{d} (\mathbb{B}(x_1), \ldots, \mathbb{B}(x_k)),$$

where the vector on the right has a multivariate normal distribution with zero mean and a variance–covariance matrix with the $(i, j)$th element given by

$$\operatorname{Cov}\{\mathbb{B}(x_i), \mathbb{B}(x_j)\} = F(x_i \wedge x_j) - F(x_i)F(x_j). \tag{2.4}$$

As the dimension $k$ grows, the vector (2.3) becomes a better approximation of the function $\mathbb{B}_n$. To move further on to a functional CLT, however, it is not sufficient to let $k$ grow infinitely large. A more technical condition (*tightness*) is needed. Nevertheless, for most results in this book, it is sufficient to think of a functional CLT as the limit of a multivariate CLT. We say that the empirical process $\mathbb{B}_n$ converges weakly to a limiting process $\mathbb{B}$, which is denoted by

$$\mathbb{B}_n \xrightarrow{w} \mathbb{B}.$$

Here, the limiting process is a zero-mean *Gaussian process* with covariance function

$$c(x, y) = \operatorname{Cov}\{\mathbb{B}(x), \mathbb{B}(y)\} = F(x \wedge y) - F(x)F(y), \tag{2.5}$$

which is basically the same as (2.4), but now it is defined as a function of $(x, y)$. For the uniform empirical process, the covariance function is $c(x, y) = x \wedge y - xy$, and $\mathbb{B}$ is called a *Brownian bridge.* For general $F$, $\mathbb{B}$ is sometimes referred to as an $F$-Brownian bridge. Figure 2.2 illustrates nicely where the name *bridge* comes from: at the two endpoints $x = 0$ and $x = 1$ the process $\mathbb{B}_n(x) \equiv 0 \equiv \mathbb{B}(x)$, and in between the process may look like a bridge.

In general a Gaussian process is a zero-mean random process, say $\mathbb{P}$, for which for every finite-dimensional vector $(x_1, \ldots, x_k)$, $(\mathbb{P}(x_1), \ldots \mathbb{P}(x_k))$ is multivariate normal. The Gaussian process is further characterised by its covariance function, say $c(x, y) = \text{Cov}\{\mathbb{P}(x), \mathbb{P}(y)\}$. We only mention briefly one more important theorem here: *the continuous mapping theorem.* This result is loosely stated in Theorem 2.1 in which we discarded the measurability conditions.

**Theorem 2.1.** *Let $g$ denote a continuous function. If $\mathbb{B}_n \xrightarrow{w} \mathbb{B}$, then $g(\mathbb{B}_n) \xrightarrow{d} g(\mathbb{B})$ as $n \to \infty$.*

Note that in the statement of Theorem 2.1 the term "function" is used in the general sense that it maps elements of the sample space of $\mathbb{B}_n$ to another metric space. This may have consequences for how continuity is defined. We refer to Shorack and Wellner (1986) for a careful study of weak convergences of the empirical process, continuous mapping theorems, and strong approximations. More recent accounts can be found in Van der Vaart and Wellner (2000) and Kosorok (2008).

### 2.2.3 Kac–Siegert Decomposition of Gausian Processes

Kac and Siegert (1947) suggested a very convenient decomposition of a Gaussian process which can be used, for instance, for simulating the process. Later in the book the decomposition is used to get a deeper understanding of the EDF-type tests such as, e.g., the Anderson–Darling test. We give here an intuitive introduction to the decomposition. More details can be found in Chapter 5 of Shorack and Wellner (1986).

Consider a zero-mean Gaussian process $\mathbb{P}$, defined over $[0, 1]$, and let $c(x, y)$ denote its covariance function. Suppose that the covariance function is a continuous and positive semidefinite symmetric kernel function; then Mercer's theorem applies. Mercer's theorem states that the kernel function $c(x, y)$ has the following expansion,

$$c(x, y) = \sum_{j=1}^{\infty} \lambda_j h_j(x) h_j(y), \tag{2.6}$$

for $0 \le x, y \le 1$, and where $\{\lambda_j\}$ and $\{h_j\}$ are the eigenvalues and the eigenfunctions of the kernel $c(x, y)$. The eigenvalues and the eigenfunctions are the solutions to the integral equation

$$\int_0^1 h(x)c(x,y)dx = \lambda h(y).$$

We refer to Kanwal (1971) for more detailed properties of kernel functions and on Mercer's theorem. Throughout this book, we assume that $\int_0^1 \int_0^1 c(x,y)dxdy < \infty$. This further implies that $\sum_{j=1}^{\infty} \lambda_j^2 < \infty$, and because proper covariance functions are positive semidefinite, we also know that all eigenvalues are nonnegative. Another important property is that the eigenfunctions form an orthonormal basis of the function space of all continuous square-integrateble functions on $[0, 1]$. This space, which is denoted as $L_2([0, 1])$ is a Hilbert space. See Section 2.5 for more details. In particular, for any eigenfunctions $h_j$ and $h_l$,

$$\int_0^1 h_j(x)h_l(x)dx = \delta_{jl},$$

where $\delta_{jl}$ is Knonecker's delta.

The main results of Kac and Siegert (1947) are summarised in the following theorem.

**Theorem 2.2.** *Consider the zero-mean Gaussian process* $\mathbb{P}$ *and its positive semidefinite covariance function* $c(x, y)$*, and suppose that Mercer's theorem applies to* $c(x, y)$*. Let* $Z_1, Z_2, \ldots$ *be i.i.d. standard normal random variables, and define*

$$\mathbb{K}_m(x) = \sum_{j=1}^{m} \sqrt{\lambda_j} h_j(x) Z_j. \tag{2.7}$$

*Then,*

$$\mathrm{E}\left\{(\mathbb{K}_m(x) - \mathbb{P}(x))^2\right\} \to 0 \text{ for each } x \text{ as } m \to \infty.$$

*Moreover the random variables*

$$\frac{1}{\sqrt{\lambda_j}} \int_0^1 \mathbb{P}(x)h_j(x)dx = Z_j \tag{2.8}$$

*(*$j = 1, \ldots, m$*) are standard normally distributed, and they are all mutually independent.*

This theorem importantly says that $\mathbb{K}_\infty = \sum_{j=1}^{\infty} \sqrt{\lambda_j} h_j(x) Z_j$ is an equivalent representation of the process $\mathbb{P}$, and the components $Z_j$ given by equation (2.8) are called the *principal components* of the process $\mathbb{P}$.

Although we do not give a formal proof of the theorem here, the core of the proof is easy to understand. First, we compute the covariance function of the process $\mathbb{K}_\infty$,

$$\begin{aligned}
\operatorname{Cov}\{\mathbb{K}_\infty(x), \mathbb{K}_\infty(y)\} &= \operatorname{Cov}\left\{\sum_{j=1}^{\infty}\sqrt{\lambda_j}h_j(x)Z_j, \sum_{l=1}^{\infty}\sqrt{\lambda_l}h_l(y)Z_l\right\} \\
&= \sum_{j=1}^{\infty}\sum_{l=1}^{\infty}\sqrt{\lambda_j}\sqrt{\lambda_l}h_j(x)h_l(y)\operatorname{Cov}\{Z_j, Z_l\} \\
&= \sum_{j=1}^{\infty}\lambda_j h_j(x)h_j(y)\operatorname{Var}\{Z_j\} \\
&= \sum_{j=1}^{\infty}\lambda_j h_j(x)h_j(y) \\
&= c(x,y).
\end{aligned}$$

The principal components arrise from Equation (2.8) as

$$\begin{aligned}
\frac{1}{\sqrt{\lambda_j}}\int_0^1 \mathbb{P}(x)h_j(x)dx &= \frac{1}{\sqrt{\lambda_j}}\int_0^1\left(\sum_{l=1}^{\infty}\sqrt{\lambda_l}h_l(x)Z_l\right)h_j(x)dx \\
&= \frac{1}{\sqrt{\lambda_j}}\sum_{l=1}^{\infty}\sqrt{\lambda_l}\left(\int_0^1 h_l(x)h_j(x)dx\right)Z_l \\
&= \frac{1}{\sqrt{\lambda_j}}\sqrt{\lambda_j}Z_j \\
&= Z_j.
\end{aligned}$$

Now that we have seen that $\mathbb{K}_\infty$ is an equivalent representation of the Gaussian process $\mathbb{P}$, we may understand, at least intuitively, the following important result, which may be useful in combination with the CMT.

**Theorem 2.3.** *Let $\mathbb{P}(x)$ denote a Gaussian process defined over $x \in [0,1]$ with continuous mean and covariance functions $m(x)$ and $c(x,y)$, respectively. Let $g(x)$ denote a continuous weight function so that $\int_0^1 |g(x)|dt < \infty$. Then,*

$$\int_0^1 \mathbb{P}(x)g(x)dx \sim P,$$

*where $P$ is a normally distributed random variable with mean and variance equal to*

$$\int_0^1 m(x)g(x)dx \quad \text{and} \quad \int_0^1\int_0^1 c(x,y)g(x)g(y)dxdy, \tag{2.9}$$

*respectively, provided the integrals involved exist.*

When $\mathbb{P} = \mathbb{B}$ is a standard Brownian bridge the mean and the variance used in Theorem 2.3 may be simplified. First, because the mean of $\mathbb{B}(x)$ is zero for all $x$, we have that the mean of $P$ is also zero. Let $\varphi(x) = \int_0^x g(t)dt$; i.e., $g(x)dx = d\varphi(x)$. The variance in (2.9) now becomes

$$\int_0^1 \int_0^1 (x \wedge y - xy) d\varphi(x) d\varphi(y) = \text{Var}\{\varphi(U)\}, \tag{2.10}$$

where $U \sim [0, 1]$. This last step makes use of an alternative formulation of the variance. See, e.g., pp. 116–117 in Shorack (2000) for more details.

## 2.3 The Quantile Function and the Quantile Process

### 2.3.1 The Quantile Function and Its Estimator

A distribution may also be characterised by its *quantile function*, which is usually defined as

$$Q(p) = F^{-1}(p) = \inf\{y \in \mathcal{S} : p \leq F(y)\}, \tag{2.11}$$

which is the inverse function of $F$, as $F$ is always a right-continuous function. For $p = 0.25$, $p = 0.50$, and $p = 0.75$, the quantile function gives the three *quartiles*. In particular, $F^{-1}(0.5)$ is the median, and $F^{-1}(0.75) - F^{-1}(0.25)$ is the *interquartile range* (IQR).

Just as the EDF $\hat{F}_n$ is a natural estimator of $F$, the *empirical quantile function* (EQF) is defined as the empirical version of $F^{-1}(p)$; i.e.,

$$\hat{Q}(p) = \hat{F}_n^{-1}(p) = X_{(i)} \text{ if } \frac{i-1}{n} \leq p < \frac{i}{n} \text{ for some } 1 \leq i \leq n, \tag{2.12}$$

where $X_{(i)}$ is the $i$th order statistic of the sample $X_1, \ldots, X_n$. For a given $p$, $\hat{F}_n^{-1}(p)$ is recognised as a *sample quantile*. Note that $\hat{F}_n^{-1}(p)$ always equals one of the $n$ sample observations. With this definition of the EQF we immediately also have estimators of the median, the other quartiles, and any other individual quantile. A pointwise asymptotic distribution theory for $\hat{F}_n^{-1}(p)$ could thus be useful for inference on quantiles. However, for later purposes, it is more important to consider $\hat{F}_n^{-1}(p)$ as a process over $p \in [0, 1]$.

A more general definition was proposed by Hyndman and Fan (1996). They defined a class of empirical quantile functions, indexed by two parameters $m \in \mathbb{R}$ and $\gamma$. The "parameter" $\gamma$ is actually a function of $j = \lfloor pn+m \rfloor$, where $p$ is the percentile corresponding to which the quantile is to be calculated, and $g = pn + m - j$. Let

$$\hat{Q}_{m,\gamma}(p) = (1-\gamma)X_{(j)} + \gamma X_{(j+1)} \text{ where } \frac{j-m}{n} \leq p < \frac{j-m+1}{n}.$$

The parameter $m$ may be interpreted as a kind of continuity correction. With $m = 0$ and $\gamma = 1$ when $g > 0$, and $\gamma = 0$ otherwise, this quantile estimator coincides with the traditional estimator using $\hat{F}_n^{-1}$. Other choices of $\gamma$ result in averaging of two subsequent order statistics. Hyndman and Fan (1996) discussed nine types of quantile estimators using their class. In R (R Development Core Team (2008)) the default is "type 7", which is given by $\gamma = 0.5$ and $m = 1 - p$, so that $j = \lfloor p(n-1) \rfloor = 1$. This gives for the three most important quantiles that are often used as summary statistics in a data exploration, (i.e., the first quartile $Q_1$, the median $Q_2$, and the third quartile $Q_3$)

$$Q_1 = \frac{1}{2} X_{(\lfloor 1/4n+3/4 \rfloor)} + \frac{1}{2} X_{(\lfloor 1/4n+3/4 \rfloor + 1)} \tag{2.13}$$

$$Q_2 = \frac{1}{2} X_{(\lfloor 1/2n+1/2 \rfloor)} + \frac{1}{2} X_{(\lfloor 1/2n+1/2 \rfloor + 1)} \tag{2.14}$$

$$Q_3 = \frac{1}{2} X_{(\lfloor 3/4n+1/4 \rfloor)} + \frac{1}{2} X_{(\lfloor 3/4n+1/4 \rfloor + 1)}. \tag{2.15}$$

## *2.3.2 The Quantile Process*

The *quantile process* is defined as

$$\mathbb{Q}_n(p) = \sqrt{n} \left( \hat{F}_n^{-1}(p) - F^{-1}(p) \right) \quad p \in [0,1].$$

Although a quantile process is defined in terms of the inverse of the CDF, it is asymptotically related to the empirical process $\mathbb{B}_n$. The relation is established by using the *Bahadur representation* of a sample quantile, which is basically a strong approximation of the sample quantile. In particular, for continuous $F$ and positive differentiable density function $f$, the *Bahadur–Kiefer theorem* (Kiefer (1970)) shows that

$$\hat{F}_n^{-1}(p) = F^{-1}(p) - \frac{\hat{F}_n(F^{-1}(p)) - p}{f(F^{-1}(p))} + R_n(p), \tag{2.16}$$

where $\sup_{0 \leq p < 1} |R_n(p)| \xrightarrow{p} 0$ (the actual result is a refinement of this statement). A more general, but weaker result may be found in Doss and Gill (1992). Equation (2.16) implies the asymptotic equivalence of the quantile process and a transformed empirical process. In particular, we may write

$$\sup_{0 \leq p < 1} \left| \sqrt{n}(\hat{F}_n^{-1}(p) - F^{-1}(p)) - \sqrt{n}(\hat{F}_n(F^{-1}(p)) - p) \left( - \frac{1}{f(F^{-1}(p))} \right) \right| \xrightarrow{p} 0,$$

as $n \to \infty$. The process within the $|.|$ is generally known as the *empirical difference process.* This result almost immediately gives the weak convergence of $\mathbb{Q}_n(p)$, which is stated here as a theorem (see, e.g., Section 21.2 of van der Vaart (1998)).

**Theorem 2.4.** *Suppose that $F$ is continuously differentiable on an interval $[a,b] = [F^{-1}(p_1) - \varepsilon, F^{-1}(p_2) + \varepsilon]$ for some $0 < p_1 < p_2 < 1$ and some $\varepsilon > 0$, or, if $F$ has a compact support, suppose that $F$ is continuously differentiable on its compact support, and suppose that the derivative $f$ is positive; then, as $n \to \infty$,*

$$\mathbb{Q}_n(p) \xrightarrow{w} -\frac{\mathbb{B}(p)}{f(F^{-1}(p))}.$$

## 2.4 Comparison Distribution

An important tool in goodness-of-fit testing is the probability integral transformation (PIT) of a random variable $X$ to $U = G(X)$. If $X$ has CDF $G$, then $U$ is uniformly distributed over $[0,1]$. In some literature the PIT is also known as the *grade transformation.* Thus, with this transformation at hand, there is actually only need for goodness-of-fit methods for the uniform distribution. This partly explains its popularity, but, as we later show, it is often more informative to apply methods particularly designed for a given $G$.

The CDF of the random variable $U = G(X)$ may be obtained as follows. Suppose $X$ has CDF $F$ instead of $G$. Then $U = G(X)$ has CDF

$$R(u) = \Pr_f\{U \leq u\} = \Pr_f\{G(X) \leq u\} = \Pr_f\left\{X \leq G^{-1}(u)\right\} = F(G^{-1}(u)). \tag{2.17}$$

Later, in Section 3.2.1, we show that this is exactly the population PP plot. The interpretation as a distribution function was first used by Parzen (1979), who called it the *comparison distribution.* In some literature it appears also as the ordinal dominance curve (e.g., Bamber (1975); Carolan and Tebbs (2005)) or the relative distribution function in, e.g., Handcock and Morris (1999). Parzen and Handcock and Morris consider also the density function of $U$,

$$r(U) = \frac{f(G^{-1}(u))}{g(G^{-1}(u))}, \tag{2.18}$$

which is named the *relative density function.* Handcock and Morris (1999) have developed a complete methodology based on a graphical representation of the relative density function. Their technique is particularly suited to get a detailed and easily interpretable graphical description of how distributions differ. As Equation (2.18) shows, the relative density needs the unknown density $f$ rather than the CDF $F$ as was the case in the methods discussed up to now. In contrast to the CDF, a nonparametric estimator of a density function requires more complex statistical methods, such as kernel smoothers. The advantage, on the other hand, is that such methods give smoother graphical representations than does the step function $\hat{F}_n$. More details on nonparametric density estimators are given in Section 2.8. In Section 3.3 the use of the comparison density as a graphical tool for assessing the goodness-of-fit is further illustrated.

## 2.5 Hilbert Spaces

Many constructions and characteristics of GOF methods make much sense when they are embedded in a Hilbert space. For instance, in Section 2.1.2 we have related the GOF problem to a distance function $d(F, G)$. Hilbert spaces are used to define an appropriate distance measure.

A Hilbert space may be interpreted as an extension of an Euclidean space in the sense that it has an infinite-dimensional basis. It is also an inner product space which has *vectors* as elements. For the applications that we are interested in, these vectors are functions (e.g., density functions, score functions, or relative density functions). In the next few paragraphs we give a very brief introduction to Hilbert spaces. We only mention definitions and properties that are useful in further chapters. Because we only use Hilbert spaces for functions, we restrict the discussion to Hilbert function spaces.

Let $G$ denote a CDF. The Hilbert space $L_2(\mathcal{S}, G)$ is a set of functions $u : \mathcal{S} \subseteq \mathbb{R} \mapsto \mathbb{R}$. The space is equipped with the inner product

$$< u, v >_g = \int_{\mathcal{S}} u(x)v(x)g(x)dx = \int_{\mathcal{S}} u(x)v(x)dG(x),$$

and the $L_2$ norm of an element $u$ is defined as

$$||u||_g = \sqrt{< u, u >_g} = \sqrt{\int_{\mathcal{S}} u^2(x)g(x)dx}.$$

The $L_2(\mathcal{S}, G)$ space is the set of all functions $u : \mathcal{S} \subseteq \mathbb{R} \mapsto \mathbb{R}$ for which $||u||_g$ is finite.

Hilbert spaces may be defined for vector-valued functions. For example, consider a Hilbert space $L_2(\mathcal{S}, G)$ of functions $\boldsymbol{v} : \mathcal{S} \subseteq \mathbb{R} \mapsto \mathbb{R}^q$. For $\boldsymbol{u}, \boldsymbol{v} \in L_2(\mathcal{S}, G)$, the inner product is then defined as

$$< \boldsymbol{u}, \boldsymbol{v} >_g = \int_{\mathcal{S}} \boldsymbol{u}(x)\boldsymbol{v}^t(x)dG(x),$$

which is a $q \times q$ matrix.

The Hilbert space that is described here is actually a Lebesgue space because of the use of the measure $G$ in the definition of the inner product. This links the function space with the measure space on which the random variables (random sample observations) are defined. We do not further elaborate on this because we try to avoid measurability issues throughout the book.

We next list some useful properties of the inner product: for all $u, v, w \in L_2(\mathcal{S}, G)$, and all $x \in \mathbb{R}$,

$$\begin{aligned} < u, v >_g &= < v, u >_g \\ < u + v, w >_g &= < u, w >_g + < v, w >_g \\ < xu, w >_g &= x < u, w >_g \\ ||u||_g = 0 &\Leftrightarrow u \equiv 0, \end{aligned}$$

where 0 is the zero element; i.e., $0 \in L_2(\mathcal{S}, G)$ satisfies $< 0, u >_g = < u, 0 >_g = 0$ for all $u \in L_2(\mathcal{S}, G)$.

In the Hilbert space $L_2(\mathcal{S}, G)$ there exists an infinite-dimensional orthonormal basis, say $\{h_j\}_{k \in \mathbb{N}}$. The orthonormality condition gives for every $i, j \in \mathbb{N}$,

$$< h_i, h_j >_g = \delta_{ij}, \tag{2.19}$$

where $\delta_{ij} = 0$ if $i \neq j$ and $\delta_{ij} = 1$ if $i = j$. For every element $u \in L_2(\mathcal{S}, G)$ there exists a set of constants $\{a_j\}_{j \in \mathbb{N}}$ ($a_j \in \mathbb{R}$) so that

$$u = \sum_{j=1}^{\infty} a_j h_j. \tag{2.20}$$

Sometimes, when confusion between functions and scalars may appear, we write $u(x) = \sum_{j=1}^{\infty} a_j h_j(x)$ instead of Equation (2.20). Using the orthonormality property of the basis functions, we immediately find

$$< u, h_j >_g = \left\langle \sum_{i=1}^{\infty} a_i h_i, h_j \right\rangle_g = \sum_{i=1}^{\infty} < a_i h_i, h_j >_g = a_j < h_j, h_j >_g = a_j.$$

Equation (2.20) thus becomes

$$u = \sum_{j=1}^{\infty} < u, h_j >_g h_j. \tag{2.21}$$

The right-hand side of Equation (2.20) or (2.21) is known as an *expansion* of the function $u$.

The inner product and the norm in Hilbert spaces have similar geometric interpretations as in the Euclidean space. For instance,

$$u_v = < u, v >_g \frac{v}{||v||_g^2} = \left\langle u, \frac{v}{||v||_g} \right\rangle_g \frac{v}{||v||_g}$$

is the orthogonal projection of $u$ onto $v$. The *orthogonal projection* onto $v \in L_2(\mathcal{S}, G)$ is a transformation which is sometimes denoted by the operator $P_v$. It satisfies the relations

$$P_v v = v$$

and

$$
\begin{aligned}
< u - P_v u, v >_g &= < u, v >_g - < P_v u, v >_g \qquad (2.22)\\
&= < u, v >_g - \left\langle \left\langle u, \frac{v}{||v||_g} \right\rangle_g \frac{v}{||v||_g}, v \right\rangle_g \\
&= < u, v >_g - < u, v >_g \frac{< v, v >_g}{||v||_g^2} \\
&= < u, v >_g - < u, v >_g = 0. \qquad (2.23)
\end{aligned}
$$

The function $u^T = u - P_v u$ is the residual of $u$ after orthogonal projection onto $v$. Equation (2.23) shows that $u_v^\top = u - P_v u$ is orthogonal to $v$. Moreover, any element $u$ can be decomposed as $u = P_v u + (u - P_v u) = u_v + u_v^\top$, where the two components are orthogonal in $L_2(\mathcal{S}, G)$.

Let $v_1, \ldots, v_k \in L_2(\mathcal{S}, G)$ $(k > 1)$. A subspace $\mathcal{P}$ of $L_2(\mathcal{S}, G)$ can be defined as the space spanned by the vectors $v_1, \ldots, v_k$, denoted as $\mathcal{P} = \text{span}(v_1, \ldots, v_k)$. The *orthogonal complement* of $\mathcal{P}$ is given by

$$
\mathcal{P}^\top = \{u \in L_2(\mathcal{S}, G) : P_v u = 0 \text{ for all } v \in \mathcal{P}\}.
$$

Note that all $u_{v_i}^T \in \mathcal{P}^T$ $(i = 1, \ldots, k)$. In later chapters we often consider expansions of an element $u \in L_2(\mathcal{S}, G)$ of the form (2.20), where $a_j = \theta_j$ is a parameter to be estimated from the sample observations. Often only a finite number of parameters can be estimated and therefore a series expansion is truncated at some finite order, say $k$. Define

$$
w = \sum_{j=1}^{k} \theta_j h_j.
$$

Interestingly, $\theta_j = < w, h_j >_g = < u, h_j >_g$ is also the solution to minimising

$$
\left\| u - \sum_{j=1}^{k} \theta_j h_j \right\|_g .
$$

The latter formulation of the problem has a simple geometric interpretation. First note that $w = \sum_{j=1}^{k} \theta_j h_j$ is a vector in the subspace $\mathcal{P}_k = \text{span}(h_1, \ldots, h_k)$. Hence, $w$ is the vector in $\mathcal{P}_k$ that is closest to $u$ in the Hilbert space $L_2(\mathcal{S}, G)$, or, $w$ is the orthogonal projection of $u$ onto the subspace $\mathcal{P}_k$, which may be denoted as $w = P_{\mathcal{P}_k} u$. A more general discussion on orthogonal projections on subspaces is given in the next paragraph.

More generally we may want to know the orthogonal projection of $u \in L_2(\mathcal{S}, G)$ onto a subspace $\mathcal{P}_k = \text{span}(v_1, \ldots, v_k)$, where now the $v_1, \ldots, v_k$ are not necessary orthogonal w.r.t. the inner product in $L_2(\mathcal{S}, G)$. We only assume that the $v_1, \ldots, v_k$ are linearly independent. Let $w = P_{\mathcal{P}_k} u$ denote the orthogonal projection for which we are looking. By definition an orthogonal projection satisfies

$$\langle u - P_{\mathcal{P}_k} u, w \rangle_g = 0 \text{ for all } w \in \mathcal{P}_k, \tag{2.24}$$

i.e. the orthogonal complement of $u$ is orthogonal to each element of $\mathcal{P}_k$. Note the analogy with (2.22). Some simple algebra shows that the condition (2.24) immediately implies

$$P_{\mathcal{P}_k} u = \langle u, \boldsymbol{v} \rangle_g \langle \boldsymbol{v}, \boldsymbol{v} \rangle_g^{-1} \boldsymbol{v},$$

where $\boldsymbol{v}^t = (v_1, \ldots, v_k)$ and in which $< \boldsymbol{v}, \boldsymbol{v} >_g = \mathrm{E}_g \{\boldsymbol{v}\boldsymbol{v}^t\}$ is invertible because of the assumptions on $v_1, \ldots, v_k$.

## 2.6 Orthonormal Functions

In the previous section we said that the Hilbert space $L_2(\mathcal{S}, G)$ may be provided with a basis $\{h_j\}$ of orthonormal functions $h_j$ that satisfy the orthonormality condition (2.19). In this section we give some important examples of such functions.

### *2.6.1 The Fourier Basis*

The first example is the well-known Fourier or sine basis. When $g$ is the uniform density, the functions

$$\begin{aligned} h_0(x) &= 1 \\ h_{2j-1}(x) &= \sqrt{2}\sin(2\pi j x) \\ h_{2j}(x) &= \sqrt{2}\cos(2\pi j x) \end{aligned}$$

$(j = 1, \ldots)$ form an orthonormal basis of the Hilbert space $L_2([0,1], 1)$.

### *2.6.2 Orthonormal Polynomials*

A very important class of orthonormal functions in goodness-of-fit testing is the class of orthonormal polynomials; i.e., each function $h_j(x)$ is a polynomial of degree $j$ in $x$. The simplest system of polynomials is $h_j(x) = x^j$, but these do usually not form an orthonormal basis. We denote this simple choice by $p_j(x) = x^j$. These functions can, however, be orthogonalised by means of the *Gram–Schmidt orthogonalisation scheme*. In particular, in the previous section we have seen that for any $u, v, \in L_2(\mathcal{S}, G)$, the elements $v$ and $u - P_v u$ are orthogonal. Thus, with $v = p_{j-1}$ and $u = p_j$, we find that $p_{j-1}$ and

$$h'_j = p_j - P_{p_{j-1}} p_j = p_j - < p_j, p_{j-1} >_g \frac{p_{j-1}}{||p_{j-1}||_g^{1/2}} \tag{2.25}$$

are orthogonal. This scheme is often initialised by the choice $h_0(x) = p_0(x) = x^0 = 1$, and applying the recurrence relation (2.25) for $j = 1, \ldots$. To get the orthonormal system $\{h_j\}$, the polynomials $h'_j$ must be normalised, i.e., take $h_j = h'_j / ||h'_j||_g$.

If in the Gram–Schmidt recurrence relation (2.25) the density function $g$ is explicitly filled in, then for each density $g$ a particular system of recurrence relations for the construction of the orthonormal polynomials is obtained. Appendix C of Rayner et al. (2009) gives polynomials for many important distributions. Many of them have been given specific names. For instance, for the uniform, the normal, and the exponential distributions, the polynomials are referred to as the Legendre, Hermite, and Laguerre polynomials, respectively. An efficient method for finding the orthonormal polynomials for a given density function $g$ is based on simple recurrence relations and is described in Rayner et al. (2008). Most of the popular orthonormal polynomials are available in the `cd` R-package through the function `orth.poly`.

For further purposes it is interesting to note that all integrals of the type $< p_j, p_{j-1} >_g$ occurring in Equation (2.25), are of the form $\int_{\mathcal{S}} x^{j+j-1} g(x) dx$, which is equal to the $(2j-1)$th noncentral moment of $g$. The coefficients of the orthonormal polynomials $h_j$ are thus characterised by the moments of the distribution $g$.

Finally, we give the first five Legendre polynomials:

$$\begin{aligned} h_0(x) &= 1 \quad h_1(x) = \sqrt{12}\left(x - \frac{1}{2}\right) \\ h_2(x) &= \sqrt{5}\left(6x^2 - 6x + 1\right) \\ h_3(x) &= \sqrt{7}\left(20x^3 - 30x^2 + 12x - 1\right) \\ h_4(x) &= 3\left(70x^4 - 140x^3 + 90x^2 - 20x + 1\right). \end{aligned}$$

## 2.7 Parameter Estimation

### *2.7.1 Locally Asymptotically Linear Estimators*

Let $\boldsymbol{\beta}$ denote a $p$-dimensional parameter. One of the important assumptions that is used frequently in this book is that the estimator $\hat{\boldsymbol{\beta}}_n$ is *locally asymptotically linear*, which means that the following expansion holds,

$$\hat{\boldsymbol{\beta}}_n - \boldsymbol{\beta} = \frac{1}{n}\sum_{i=1}^{n} \boldsymbol{\Psi}(X_i; \boldsymbol{\beta}) + o_P(n^{-1/2}), \tag{2.26}$$

where $\boldsymbol{\Psi}^t = (\Psi_1, \ldots, \Psi_p)$ is a continuously differentiable vector function $\mathbb{R}^p \to \mathbb{R}^p$ and has $\mathrm{E}\left\{\boldsymbol{\Psi}(X; \boldsymbol{\beta})\right\} = 0$ and $\mathrm{E}\left\{\boldsymbol{\Psi}(X; \boldsymbol{\beta})\boldsymbol{\Psi}^t(X; \boldsymbol{\beta})\right\}$ is finite and nonsingular. This property holds for many well-known estimators (maximum likelihood estimators, moment estimators, M- and Z-estimators). As an example, we show how the expansion in Equation (2.26) is obtained for Z-estimators. M- and Z-estimators are well studied, starting with Huber (1967). Good references are the books by Huber (1974) and Hampel et al. (1986), which mainly focus on the use of M-estimators in robust statistics. In robust statistics the function $\boldsymbol{\Psi}$ is known as the *influence function*. Its name refers to the interpretation of $\boldsymbol{\Psi}(X_i; \hat{\beta}_n)$ as a measure for the influence of the $i$ th observation on the estimation of $\boldsymbol{\beta}$. A very concise and modern treatment is given by van der Vaart (1998) (Chapter 5).

A Z-estimator $\hat{\boldsymbol{\beta}}_n$ is defined as the solution to estimation equations,

$$\sum_{i=1}^{n} \boldsymbol{b}(X_i; \hat{\boldsymbol{\beta}}_n) = \boldsymbol{0}, \tag{2.27}$$

where $\boldsymbol{b} = (b_1, \ldots, b_p)$ is a vector function satisfying the same conditions as $\boldsymbol{\Psi}$ and for which its first-order derivative w.r.t. $\boldsymbol{\beta}$, say $\dot{\boldsymbol{b}}$, exists and satisfies $\mathrm{E}\left\{\dot{\boldsymbol{b}}(X; \boldsymbol{\beta})\right\}$ is finite and $\mathrm{E}\left\{\dot{\boldsymbol{b}}(X; \boldsymbol{\beta})\dot{\boldsymbol{b}}^t(X; \boldsymbol{\beta})\right\}$ is finite and nonsingular. For this class of estimators the asymptotic linear representation is obtained with

$$\boldsymbol{\Psi}(X) = \mathrm{E}\left\{-\dot{\boldsymbol{b}}(X)\right\}^{-1} \boldsymbol{b}(X).$$

Under the conditions given above, the estimator $\hat{\boldsymbol{\beta}}$ is asymptotically normal, i.e., as $n \to \infty$,

$$\sqrt{n}(\hat{\boldsymbol{\beta}}_n - \boldsymbol{\beta}) \xrightarrow{d} N(0, \boldsymbol{\Sigma}_{\beta}),$$

where

$$\boldsymbol{\Sigma}_{\beta} = \mathrm{E}\left\{-\dot{\boldsymbol{b}}(X)\right\}^{-1} \mathrm{E}\left\{\boldsymbol{b}(X)\boldsymbol{b}^t(X)\right\} \mathrm{E}\left\{-\dot{\boldsymbol{b}}(X)\right\}^{-t}.$$

Note that MLE belongs to the class of M-estimators with $\boldsymbol{b}(X) = (\partial \log f(X))/\partial\boldsymbol{\beta}$, which is the score function of $\boldsymbol{\beta}$. In the next section we briefly introduce method of moments estimators.

### 2.7.2 Method of Moments Estimators

In the context of goodness-of-fit testing the *method of moments estimators* (MME) of a $p$-dimensional parameter $\boldsymbol{\beta}$, say $\tilde{\boldsymbol{\beta}}_n$, is basically the estimator that makes $p$ moments of the fitted density coincide with the corresponding sample moments. Suppose the objective is the estimation of $\boldsymbol{\beta}$ of the density function $g(.;\boldsymbol{\beta})$. Let $\mu = \mu_1 = \mu_1(\boldsymbol{\beta})$ denote the mean of $g(.;\boldsymbol{\beta})$, and $\mu_m = \mu_m(\boldsymbol{\beta})$ $(m = 2,\ldots)$ is the $m$th central moment; i.e., $\mu_m(\boldsymbol{\beta}) = \int_S (x-\mu)^m g(x;\boldsymbol{\beta})dx$. The corresponding sample moments are denoted as $M_1 = \bar{X}$ for the sample mean, and $M_m = (1/n)\sum_{i=1}^n (X_i - \bar{X})^m$ for the central moments. The MME of $\boldsymbol{\beta}$ is described in the following definition.

**Definition 2.1.** The MME of $\boldsymbol{\beta}$ is given by $\tilde{\boldsymbol{\beta}}_n$ that is the solution of the estimation equations

$$\mu_m(\tilde{\boldsymbol{\beta}}_n) = M_m \tag{2.28}$$

for $m = 1,\ldots p$.

For $m = 2,\ldots,p$ Equation (2.28) may be expressed as

$$\sum_{i=1}^n \left\{ \left(X_i - \bar{X}\right)^m - \mu_m(\tilde{\boldsymbol{\beta}}_n) \right\} = 0,$$

from which we immediately read the estimating function $\boldsymbol{b}$ of (2.27).

### 2.7.3 Efficiency and Semiparametric Inference

In parameter estimation the concept of *efficiency* is very important. Because this book is about hypothesis testing, we are not much concerned with efficiency.

An efficient parameter estimator is basically an estimator that among a wider class of estimators and within a specified class of distributions has the smallest variance (sometimes defined in an asymptotic sense, related to rates of convergence). It is, for example, well known that under quite mild regularity conditions the MLE is efficient. The MME, on the other hand, is sometimes not efficient. Why would MME be used then? Later in the book two reasons are made clear. One important reason is that MME will often improve the interpretability of the goodness-of-fit test. This is often stressed later. The other reason is that MLE can only be defined when the density function of the observations is specified. The MME requires only the knowledge of $p$ moments, and may thus be used in a less parametric setting, i.e., in a semiparametric model. In this sense the MME is actually a semiparametric estimator. For such estimators the efficiency concept is extended to the *semiparametric efficiency bound*, which is basically the smallest variance an estimator can obtain within a class of semiparametric models. This semiparametric class is typically larger than the full parametric class that contains at most a family of density functions indexed by a nuisance parameter. We refer to Tsiatis (2006) and Kosorok (2008) for recent accounts on semiparametric inference. The relation between locally asymptotically linear estimators and efficiency is also well explained in Hall and Mathiason (1990).

In Section 1.3.1, while discussing Pearson's $\chi^2$ tests, some results were given that required the nuisance parameter estimator to be *best asymptotically normal* (BAN). This estimator is also locally asymptotically linear, as well as asymptotically efficient. The BAN estimator allows a particular expansion, which is used, e.g., in a proof presented in Appendix A.2.

## 2.8 Nonparametric Density Estimation

### *2.8.1 Introduction*

Nonparametric density estimation is, just as goodness-of-fit, a very old and important field of statistical research with many applications. The objective of density estimation is the estimation of a density function based on a sample of $n$ observations (here we consider only the i.i.d. case). When no restrictive distributional assumptions are made, the problem is referred to as *nonparametric density estimation* (NDE). In some sense NDE and GOF are two approaches to the same set of statistical inference problems. The latter is the hypothesis testing approach, whereas the former the estimation. Strangely enough, the literature about both techniques is almost completely separated. In this section we only give a very brief introduction to NDE, and we limit the overview to the NDE methods that are relevant for the GOF tests treated in the book.

First we introduce some notation. When $f$ is the density function of the $n$ i.i.d. sample observations $X_1, \ldots, X_n$, then we use $\hat{f}_n$ to denote a NDE of $f$. Many papers present NDE methods and study their properties. We mention here some of the properties that are desirable for the estimators.

1. The NDE $\hat{f}_n$ should be a *bona fide* density function; i.e., with probability one, for all $n$

$$\hat{f}_n(x) \geq 0 \text{ for all } x \in \mathcal{S} \text{ and } \int_{\mathcal{S}} \hat{f}_n(x)dx = 1.$$

2. The NDE $\hat{f}_n$ should be unbiased; i.e.,

$$\mathrm{E}_f\left\{\hat{f}_n(x)\right\} = f(x) \text{ for all } x \in \mathcal{S} \text{ and for all } n.$$

   It is, however, more realistic to look for an NDE that has this property asymptotically; i.e.,

$$\lim_{n\to\infty} \mathrm{E}_f\left\{\hat{f}_n(x)\right\} = f(x) \text{ for all } x \in \mathcal{S}.$$

3. Weak pointwise consistency is expressed as

$$\hat{f}_n(x) \xrightarrow{p} f(x) \text{ for all } x \in \mathcal{S},$$

as $n \to \infty$, and strong pointwise consistency as

$$\hat{f}_n(x) \xrightarrow{\text{a.s.}} f(x) \text{ for all } x \in \mathcal{S},$$

as $n \to \infty$.

These properties are all pointwise, whereas one typically wants estimators that behave well over the whole support $\mathcal{S}$. The study of such properties is usually done based on an *error criterion*. As there are many error criteria described in the literature, we here only describe a couple of criteria that may be useful in other parts of this book too.

1. The *integrated squared error* (ISE) is defined as

$$\text{ISE}_n = \int_{\mathcal{S}} \left(\hat{f}_n(x) - f(x)\right)^2 dx,$$

which is considered to be a good criterion when one want to measure how good an estimate $\hat{f}_n$ is for a given dataset.
2. When an estimator has to be theoretically evaluated, it is better to use the *mean integrated squared error* (MISE), which is defined as

$$\text{MISE}_n = \text{E}_f \{\text{ISE}_n\} = \int_{\mathcal{S}} \text{E}_f \left\{ \left(\hat{f}_n(x) - f(x)\right)^2 \right\} dx, \tag{2.29}$$

and which is sometimes also known under the name *integrated mean squared error*. It measures the average performance of an estimator.
3. The integrand of the right-hand side of (2.29) may also be written as $\text{MISE}_n = \int_{\mathcal{S}} \text{MSE}_n(x)dx$, with $\text{MSE}_n$ the (pointwise) *mean squared error* given by

$$\text{MSE}_n(x) = \text{E}_f \left\{ \left(\hat{f}_n(x) - f(x)\right)^2 \right\} = \text{Var}_f \left\{ \hat{f}_n(x) \right\} + \left\{ \text{bias}(\hat{f}_n(x)) \right\}^2. \tag{2.30}$$

All these criteria are based on $L_2$ norms. All the criteria listed here may be extended to weighted versions. Without introducing extra notation we define the weighted ISE as $\text{ISE}_n = \int_{\mathcal{S}} w(x) \left(\hat{f}_n(x) - f(x)\right)^2 dx$, the weighted MISE as $\text{MISE}_n = \int_{\mathcal{S}} \text{E}_f \left\{ w(x) \left(\hat{f}_n(x) - f(x)\right)^2 \right\} dx$ and the weighted MSE as $\text{MSE}_n(x) = \text{E}_f \left\{ w(x) \left(\hat{f}_n(x) - f(x)\right)^2 \right\}$, where $w(x)$ is a weight function that is positive for all $x \in \mathcal{S}$ and integrates to 1 over the domain of $f$. As a last error criterion we mention the *expected Kullback–Leibler loss*,

$$\mathrm{E}_f\left\{\int_{\mathcal{S}} f(x)\log\frac{f(x)}{\hat{f}_n(x)}dx\right\},$$

which was studied by, among others, Hall (1987).

Many papers in the NDE literature study the rate of convergence of an estimator $\hat{f}_n$ in terms of the convergence rate of one of these error criteria to zero; the MISE is particularly popular. For example, Farrell (1972) showed that the fastest convergence rate of the MISE for a bona fide NDE is $O(n^{-4/5})$.

## 2.8.2 Orthogonal Series Estimators

Orthogonal series estimators are most closely related to the GOF methods described in this book. They were introduced by Cencov (1962), and studied since by many others. In their simplest form they are based on an expansion of $f$ using orthogonal series expansions. First suppose that $f$ has finite support, say $[0,1]$ without loss of generallity. When $\{h_j\}$ denotes a complete system of orthonormal functions on the uniform $[0,1]$ distribution, then $f(x)$ has expansion

$$f(x) = \sum_{j=0}^{\infty} \theta_j h_j(x),$$

which is basically (2.20). An unbiased estimator of $\theta_j$ is given by

$$\hat{\theta}_j = \frac{1}{n}\sum_{i=1}^{n} h_j(X_i), \tag{2.31}$$

because

$$\mathrm{E}_f\left\{\hat{\theta}_j\right\} = \int_0^1 h_j(x)\left(\sum_{m=0}^{\infty}\theta_m h_m(x)\right)dx = \theta_j.$$

A natural estimator of $f$ is thus

$$\hat{f}_{n\infty} = \sum_{j=0}^{\infty}\hat{\theta}_j h_j(x),$$

but unfortunately with $n$ observations the estimator based on an infinite number of $\hat{\theta}_j$ $(j = 1, \ldots)$ is useless, because it has infinite variance. Before we give typical solutions for this problem, we first introduce some other types of orthogonal series expansions.

Let $\{h_j\}$ now be a set of orthogonal functions on some distribution function $G$ (i.e., $h_j \in L_2(\mathcal{S}, G)$), and consider the expansion

$$f(x) = g(x)\left\{1 + \sum_{j=1}^{\infty} \theta_j h_j(x)\right\}; \tag{2.32}$$

then $f$ can be estimated by replacing the $\theta_j$ in the expansion by $\hat{\theta}_j$ of (2.31). This estimator is again unbiased, as may be seen from

$$\begin{aligned}
\mathrm{E}_f\left\{\hat{\theta}_j\right\} &= \int_{\mathcal{S}} h_j(x)g(x)\left(1 + \sum_{m=1}^{\infty} \theta_m h_m(x)\right)dx \\
&= \int_{\mathcal{S}} \theta_j h_j(x) h_j(x) g(x) dx = \theta_j.
\end{aligned}$$

The density function $g$ in (2.32) has received several names in the statistical literature. For example, Hjort and Glad (1995) referred to it as a *parametric start*, Buckland (1992) called it a *parametric key*, and Efron and Tibshirani (1996) gave it the name *carrier density*.

Yet another method is to consider $\{h_j\}$ to be a set of orthonormal functions on the uniform distribution; consider the expansion

$$f(x) = g(x)\left\{1 + \sum_{j=1}^{\infty} \theta_j h_j(G(x))\right\},$$

and estimate $\theta_j$ by $\hat{\theta}_j = (1/n)\sum_{i=1}^{n} h_j(G(X_i))$. Unbiasedness follows from

$$\begin{aligned}
\mathrm{E}_f\left\{\hat{\theta}_j\right\} &= \int_{\mathcal{S}} h_j(G(x))g(x)\left(1 + \sum_{m=1}^{\infty} \theta_m h_j(G(x))\right)dx \\
&= \int_{\mathcal{S}} \theta_j h_j(G(x)) h_m(G(x)) dG(x) \\
&= \theta_j \int_0^1 h_j(p) h_j(p) dp = \theta_j.
\end{aligned}$$

The problem mentioned earlier about the infinite variance of $\hat{f}_{n\infty}$ originates from the use of an infinite number of parameter estimators, all based on $n$ observations. A general solution exists in *tapering* or *modulating* the estimator, proposed by Watson (1969). Consider the orthogonal series estimator

$$\hat{f}_n(x) = \sum_{j=0}^{\infty} b_j \hat{\theta}_j h_j(x)$$

(or using one of the other types of expansions), where $\{b_j\}$ is a set of *tapering coefficients* that basically shrink the tapered estimators $b_j\hat{\theta}_j$ towards zero. We list a few tapering systems:

1. A partial sum orthogonal series estimator results from $b_j = 1$ for $j \leq k$, and $b_j = 0$ for $j > k$, with $k$ some constant. The constant $k$ may be chosen by the user prior to observing the data, or it can be selected in an adaptive fashion, in which case we denote it by $K_n$ to stress its dependence on the sample size and its randomness.
2. A parameterised weighting system was suggested by Wahba (1958),
$$b_j = \frac{1}{1 + \lambda(2\pi j)^{2m}},$$
with $\lambda$ and $m$ some tuning parameters.
3. Among others, Hall (1983) and Hall (1986) proposed weighting schemes that depend on the variance of $\hat{\theta}_j$ so that the terms with larger variance get shrunken more than those with small variance. This is related to the variance–bias trade-off, that could, for example, be seen from (2.30).
4. The modulators $b_j$ can also be chosen from a large set of modulators so that a certain risk function is minimised. For example, the ISE or a estimator of the MISE.

We conclude this section by remarking that an orthogonal series estimator is not necessarily bona fide. There is particularly no guarantee that the NDE is positive over $\mathcal{S}$. This can be repaired by using the correction methods of Gajek (1986) or Glad et al. (2003). First note that the perhaps most intuitive method, that consists in truncating the NDE at zero and renormalisation of the truncated NDE, is not a good method. See, for example, Hall and Murison (1993) and Kaluszka (1998).

We now present the correction method of Gajek (1986), but we limit the exposition to orthogonal series estimators based on the expansion (2.32) that includes only terms of order $j \in S$, with $S$ a finite index set. This will be sufficiently general for later purposes. Let $\hat{f}_n$ denote the uncorrected NDE. The Gajek-corrected NDE then becomes

$$\hat{f}_n^c(x) = g(x) \max\left\{0, 1 + \sum_{j\in S} \hat{\theta}_j h_j(x) - a\right\},$$

where $a$ is chosen so that the corrected NDE integrates to one. Gajek (1986) showed that in terms of the weighted MISE, defined as

$$\text{MISE}_n(p) = \text{E}_f\left\{\int_{\mathcal{S}} \frac{(p(x) - f(x))^2}{g(x)} dx\right\},$$

the corrected NDE possesses the property

$$\text{MISE}_n(\hat{f}_n^c) \leq \text{MISE}_n(\hat{f}_n).$$

### 2.8.3 Kernel Density Estimation

A very popular type of NDE is kernel density estimation. They were first studied by Rosenblatt (1956) and Parzen (1962). Let $K$ denote a symmetric *kernel* function which satisfies

$$K(x) \geq x \text{ for all } x \in \mathcal{S} \text{ and } \int_{\mathcal{S}} K(x)dx = 1.$$

All bona fide density functions, for example, are proper kernel functions. The kernel density estimator is then given by

$$\hat{f}_{nh}(x) = \frac{1}{n}\sum_{i=1}^{n}\frac{1}{h}K\left(\frac{x - X_i}{h}\right),$$

where $h$ is the *bandwidth*, which determines the roughness of the estimator or the variance–bias balance. Convergence rates of the MISE for this estimator have been studied in detail. These studies often allow the bandwidth to decrease with increasing sample size. In this case the bandwidth is denoted by $h_n$. Optimal bandwidths have been found, depending on the type of kernel function and on the true distribution $F$ of the data. For example, when $K$ is the *Gaussian kernel*, and when $F$ is the normal distribution with variance $\sigma^2$, the optimal bandwidth is given by $h_n = 1.06\sigma^{-1/5}$. The best achievable convergence rate of the MISE is shown to be $O(n^{-4/5})$.

We refer to the books of Scott (1992) and Silverman (1986) for good introductions to the field of kernel estimation.

### 2.8.4 Regression-Based Density Estimation

Fan and Gijbels (1996) described a NDE method that makes use of regression methods that are available in many software packages. Because this method is particularly based on the histogram as a NDE, and because the histogram is a very commonly used graphical exploratory tool, we have chosen to postpone its description to Chapter 3. Thus this regression-based density estimator is also discussed there.

## 2.9 Hypothesis Testing

Because this book is merely about hypothesis testing, we cannot avoid having an introductory section on this topic. Despite the importance of the basic theory of hypothesis testing we cannot, however, go into deep detail, as this would require another lengthy book. In this section we only present some very basic minimal theory that is needed to understand some of the topics treated later in this book. We restrict the discussion to one-sided tests, but the generalisation of the concepts presented here is mostly straightforward. For a more detailed treatment of hypothesis testing, we refer to the textbook of Lehmann and Romano (2005).

### *2.9.1 General Construction of a Hypothesis Test*

Let $X_1, \ldots, X_n$ denote the $n$ sample observations, which are i.i.d. with density function $f$. Suppose the null and alternative hypotheses are formulated as

$$H_0 : f \in \mathcal{F}_0 \quad \text{and} \quad H_1 : f \in \mathcal{F}_1,$$

where the disjoint sets $\mathcal{F}_0$ and $\mathcal{F}_1$ can contain one or more densities. In the former case the hypothesis is called *simple*, otherwise it is *composite*. In general a statistical hypothesis test is defined through a test statistic which is a function of the $n$ sample observations, say $T_n = T_n(X_1, \ldots, X_n)$. We further assume that the function $T_n$ is invariant to permutations of the entries $X_1, \ldots, X_n$ under the null hypothesis. Let $\boldsymbol{X}^t = (X_1, \ldots, X_n)$. A second ingredient of a statistical test is the *test function* $\phi$, which is in general defined as

$$\phi(\boldsymbol{X}) = \begin{cases} 0 & \text{if } T_n(\boldsymbol{X}) < c_\alpha \\ \gamma & \text{if } T_n(\boldsymbol{X}) = c_\alpha \, , \\ 1 & \text{if } T_n(\boldsymbol{X}) > c_\alpha \end{cases}$$

where $\gamma$ and $c_\alpha$ are chosen so that the test has size $\alpha$, i.e., so that

$$\sup_{f \in \mathcal{F}_0} \mathrm{E}_f \left\{ \phi(\boldsymbol{X}) \right\} = \sup_{f \in \mathcal{F}_0} \Pr{}_f \left\{ \text{reject } H_0 | H_0 \right\} = \alpha. \tag{2.33}$$

Let $M = M(\alpha, \mathcal{F}_0)$ denote the set of test functions for which (2.33) holds true. Note that if $H_0$ is a simple null hypothesis, say $H_0 : f = g$, Equation (2.33) reduces to $\mathrm{E}_g \left\{ \phi(\boldsymbol{X}) \right\} = \alpha$. Sometimes the equality in (2.33) only holds asymptotically. The *power function* for a fixed density $f$ of a level $\alpha$ test $\phi$ is defined as

$$\beta_n(\alpha, \phi, f) = \mathrm{E}_f \left\{ \phi(\boldsymbol{X}) \right\};$$

it is the probability to reject the null hypothesis at level-$\alpha$, when the $n$ sample observations are i.i.d. $f$. A level-$\alpha$ test $\phi_0$ is called *consistent* for testing $H_0$ versus $H_1$ if

$$\lim_{n \to \infty} \beta_n(\alpha, \phi_0, f) = 1 \text{ for all } f \in \mathcal{F}_1.$$

An *unbiased test* is a test for which (2.33) holds, as well as

$$\inf_{f \in \mathcal{F}_1} \mathrm{E}_f \left\{ \phi(\boldsymbol{X}) \right\} = \inf_{f \in \mathcal{F}_1} \beta_n(\alpha, \phi_0, f) = \inf_{f \in \mathcal{F}_1} \Pr{}_f \left\{ \text{reject } H_0 | H_1 \right\} \geq \alpha,$$

which says that under the alternative hypothesis, the power may never be smaller than the level of the test. Again, sometimes this property is only obtained asymptotically.

It is often the aim to find the level-$\alpha$ test that has maximal power to test $H_0$ versus $H_1$, therefore we need to know what the maximal achievable power is. This is measured by the *envelope power function*

$$\beta_n^\star(\alpha, M, \mathcal{F}_1) = \sup_{\phi \in M(\alpha, \mathcal{F}_0)} \inf_{f \in \mathcal{F}_1} \mathrm{E}_f \{\phi(\boldsymbol{X})\}.$$

The infimum in this expression is needed for composite alternative hypotheses; for a given level-$\alpha$ test $\phi$ the power is defined as the minimal power of that test over all densities covered by the alternative hypothesis ($\mathcal{F}_1$). The power envelope measures thus the power of the best level-$\alpha$ test under the worst detectable alternative.

## 2.9.2 Optimality Criteria

### 2.9.2.1 Finite Sample Criteria

A test $\phi_0$ for testing $H_0$ versus a simple alternative, say $\mathcal{F}_1 = \{f_1\}$, is said to be the *most powerful test* (MPT) at level $\alpha$ when

$$\beta_n(\alpha, \phi_0, f_1) = \beta_n^\star(\alpha, M, f_1). \tag{2.34}$$

When testing $H_0$ versus a composite alternative, a level-$\alpha$ test $\phi_0$ is called a *uniformly most powerful test* (UMPT) if (2.34) holds for all $f_1 \in \mathcal{F}_1$. A level-$\alpha$ test $\phi_0$ is a *maximin most powerful test* if

$$\inf_{f \in \mathcal{F}_1} \beta_n(\alpha, \phi_0, f) = \beta_n^\star(\alpha, M, \mathcal{F}_1).$$

The optimality criteria described in the previous paragraph are strong in the sense that they hold for all sample sizes, and for alternatives within a large class $\mathcal{F}_1$. Later it becomes clear that it is often very hard to prove these optimalities because (1) no small sample theory is available, or (2) power evaluation under a fixed alternative $f_1 \in \mathcal{F}_1$ is very hard. There are basically two solutions to get around this problem. A first solution exists in studying the power only against *local* alternatives, i.e., alternatives $f_1 \in \mathcal{F}_1$ that are very close to densities in $\mathcal{F}_0$. More details are provided in the next paragraph. The second way around is to study the behaviour of tests in an asymptotic sense, i.e., for large sample sizes. Also here only local alternatives are of importance, because usually a (consistent) test has asymptotically a trivial power equal to one under an alternative that is far away from $H_0$.

A *locally most powerful test* (LMPT) is defined as follows. Let $f(.;\theta)$ denote a family of densities indexed by a parameter $\theta \geq 0$, and assume that $f(.;\theta) \in \mathcal{F}_0$ if and only if $\theta = 0$. Otherwise $f(.;\theta) \in \mathcal{F}_1$. Consider now a small subset of $\mathcal{F}_1$ given by $\mathcal{F}_{1\varepsilon} = \{f(.;\theta) \in \mathcal{F}_1 : 0 < \theta < \epsilon\}$. In this setting the null hypothesis can be rephrased as $H_0 : \theta = 0$. A level-$\alpha$ test $\phi_0$ is said to be *locally most powerful* for testing $H_0$ against $H_1 : f \in \mathcal{F}_{1\varepsilon}$ if there exists an $\varepsilon > 0$ so that $\phi_0$ is uniformly most powerful for this testing problem.

#### 2.9.2.2 Asymptotic Criteria

To further stress the dependence of the testing procedure on the sample size, we write $\phi_n$ for the test function. All properties discussed in this section are defined for test sequences $\phi_n$ for which

$$\limsup_{n\to\infty} \sup_{f\in\mathcal{F}_0} \mathrm{E}_f\left\{\phi_n(\boldsymbol{X})\right\} \leq \alpha.$$

Tests that satisfy this condition are referred to as asymptotic level-$\alpha$ tests. We still use the notation $M = M(\alpha, \mathcal{F}_0)$ to denote the set of such tests.

Because many tests are consistent, and they therefore have asymptotic power one against fixed alternatives, we need to study here sequences of alternatives that approach $\mathcal{F}_0$ as the sample size increases. To keep the exposition general here, we use the notation $f_n$ for such a sequence of alternatives. In particular, $f_n \in \mathcal{F}_1$ so that for some $f_0 \in \mathcal{F}_0$, $f_n \to f_0$ as $n \to \infty$. At this point we do not give details on the mode of convergence. To introduce the asymptotic version of the envelope power function we also need a sequence of sets $\mathcal{F}_{1n} \subseteq \mathcal{F}_1$ so that, as $n \to \infty$, there is an $f_0 \in \mathcal{F}_0$ so that

$$\lim_{n\to\infty} \sup_{f\in\mathcal{F}_{1n}} ||f - f_0|| = 0.$$

The envelope power function now becomes

$$\beta^\star(\alpha, M, \mathcal{F}_{1n}) = \lim_{n\to\infty} \beta_n^\star(\alpha, M, \mathcal{F}_{1n}).$$

Note that the index $n$ in $\mathcal{F}_{1n}$ in $\beta^\star(\alpha, M, \mathcal{F}_{1n})$ is only used to stress that the envelope power function depends on the sequence of alternatives chosen.

An asymptotic level-$\alpha$ test $\phi_n$ is said to be *asymptotically most powerful* (AMP) for testing $H_0$ against $H_1 : f = f_n$ if

$$\lim_{n\to\infty} \beta_n(\alpha, \phi_n, f_n) = \lim_{n\to\infty} \beta_n^\star(\alpha, M, \{f_n\}).$$

The notion of an asymptotically uniformly most powerful test (AUMP) also exists. It is a test which is AMP for any sequence $f_n \in \mathcal{F}_1$ that approaches some $f \in \mathcal{F}_0$. A special case of an AMP test is a *locally asymptotically most powerful test* (LAMPT), which is an AMP test against alternatives $f_n$ that approach $f \in \mathcal{F}_0$ at the rate $n^{-1/2}$.

An asymptotically maximin test $\phi_n$ of asymptotic level $\alpha$ for testing $H_0$ versus $\mathcal{F}_{1n}$ satisfies

$$\lim_{n \to \infty} \beta_n^\star(\alpha, \phi_n, \mathcal{F}_{1n}) = \beta^\star(\alpha, M, \mathcal{F}_{1n}).$$

The (asymptotic) power performance of a test is also often quantified by its *asymptotic relative efficiency* (ARE) or *asymptotic efficiency.* The ARE is also known as the *Pitman efficiency.* The central idea here is to compute the limit of the ratio of the minimal sample sizes of two level-$\alpha$ tests so that asymptotically they have equal power. Most of the interesting tests are consistent, thus they have asymptotic power equal to one under a fixed alternative $f_1 \in \mathcal{F}_1$. Therefore, asymptotic test performance is measured under sequences of alternatives, $f_n$, that converge to $f_0$ at a suitable rate so that the asymptotic power is kept away from 0 and 1. For many regular testing problems for testing $H_0 : \theta = 0$ versus $H_1 : \theta > 0$, the rate equals $1/\sqrt{n}$; i.e., the sequence of alternatives in terms of the parameter $\theta$ may be represented as $\theta_n = \delta/\sqrt{n}$ $(\delta > 0)$.

Consider now two level-$\alpha$ tests, say $\phi_{1n}$ and $\phi_{2n}$; then let $\nu$ denote a "time" parameter $(\nu > 1)$, and let $n_{1\nu}$ and $n_{2\nu}$ denote the minimal sample sizes so that, for some $\alpha < \gamma < 1$ and for all $\nu > 1$,

$$\beta_n(\alpha, \phi_{1n_{1\nu}}, f_\nu) = \beta_n(\alpha, \phi_{2n_{2\nu}}, f_\nu) = \gamma.$$

When the sample sizes $n_{1\nu}$ and $n_{2\nu}$ increase with increasing time $\nu$, the ARE of test 1 relative to test 2 is then defined as

$$\mathrm{ARE}_{1,2} = \lim_{\nu \to \infty} \frac{n_{2\nu}}{n_{1\nu}}.$$

An ARE larger than one means that test 2 requires more observations than test 1 for obtaining the same power $\gamma$; test 1 is thus better than test 2. Similarly, an ARE smaller than one indicates that test 2 is asymptotically more powerful than test 1. Although the definition of ARE depends on $\alpha$, $\gamma$, and $\delta$, it has been shown that for a large class of important tests the ARE is independent of these parameters. For example, this happens for many test statistics that have an asymptotic normal distribution under both the null hypothesis and under the sequence of alternatives. Le Cam's third lemma may be very useful in establishing this asymptotic normality under sequences of alternatives. See, e.g., Chapter 7 in van der Vaart (1998) and Chapter 7 in Hájek et al. (1999) for good introductions to local asymptotic normality and Le Cam's third lemma. In Chapter 4 we give theory that uses Le Cam's third lemma.

The *asymptotic efficiency* of a test is the ARE of that test, relative to the AMP test.

## 2.9.3 The Neyman–Pearson Lemma

The Neyman–Pearson lemma, which is also known as the *fundamental lemma of Neyman and Pearson*, shows the existence and the construction of a MPT for testing a simple null versus a simple alternative hypothesis. Although this setting is the most simple setting, which hardly ever occurs in real situations, it is considered as THE basis of statistical hypothesis testing. Many of the extensions of this lemma to more complicated settings (composite hypotheses) still possess a flavour of this original lemma. For this reason, we state the lemma here. We use here the notation $f_{0n}(\boldsymbol{x})$ and $f_{1n}(\boldsymbol{x})$ for the joint densities of $\boldsymbol{X}^t = (X_1, \ldots, X_n)$ under the null and the alternative hypothesis, respectively.

**Lemma 2.1.** *The most powerful level-$\alpha$ test for testing $H_0 : f = f_0$ versus $H_1 : f = f_1$ is given by*

$$\phi(\boldsymbol{X}) = \begin{cases} 0 & \text{if } \frac{f_{1n}(\boldsymbol{X})}{f_{0n}(\boldsymbol{X})} < c_\alpha \\ \gamma & \text{if } \frac{f_{1n}(\boldsymbol{X})}{f_{0n}(\boldsymbol{X})} = c_\alpha \\ 1 & \text{if } \frac{f_{1n}(\boldsymbol{X})}{f_{0n}(\boldsymbol{X})} > c_\alpha \end{cases},$$

*where $c_\alpha$ and $\gamma$ can always be chosen so that* $\mathrm{E}_{f_0}\{\phi(\boldsymbol{X})\} = \alpha$.

The Neyman–Pearson lemma shows thus that the likelihood ratio test statistic

$$T_n = \frac{f_{1n}(\boldsymbol{X})}{f_{0n}(\boldsymbol{X})} = \prod_{i=1}^{n} \frac{f_1(X_i)}{f_0(X_i)}$$

is the MPT for this simple testing problem.

# Chapter 3
# Graphical Tools

A graphical presentation of the data is typically one of the first steps in an exploratory data analysis. This is not different in the goodness-of-fit context. Although many of the graphs presented in the chapter are well known by most statisticians, we think it is still important to give some further details on those methods, particularly because some of the goodness-of-fit tests are very closely related to some of the graphs presented here. We start in Section 3.1 with the description of the histogram and the boxplot, of which the former is basically a nonparametric density estimator. Probability plots (PP and QQ) and comparison distributions are the topics of Sections 3.2 and 3.3, respectively. Both types of plots are related to very important goodness-of-fit tests, and we therefore spend quite some space on these methods.

## 3.1 Histograms and Box Plots

Among the simplest graphical techniques we find the histogram and the box plot. Although they are well known we give a brief description.

### *3.1.1 The Histogram*

#### 3.1.1.1 The Construction

The histogram is basically a nonparametric density estimator, and could thus just as well have been described in Section 2.8. It can be considered as an estimator of the categorised distribution of $X$. For simplicity in this section we further assume that $X$ has a bounded support, denoted by $\mathcal{S} = [l, u]$. It becomes clear that this does not affect the practical implementation of

O. Thas, *Comparing Distributions,* Springer Series in Statistics,
DOI 10.1007/978-0-387-92710-7_3, 

the histogram. The construction of the histogram may proceed along the following steps.

1. Construct a partition of the support:

$$\mathcal{S} = [l, d_{n1}) \cup [d_{n1}, d_{n2}) \cup \cdots \cup [d_{nc-1}, u],$$

   where $d_{n0} = l, d_{n1}, \ldots, d_{nc-1}, d_{nc} = u$ are the *bin edges.* The $c$ intervals are referred to as the *bins.* The index $n$ stresses that the edges may depend on the sample size.
2. Count the number of sample observations within each of the $c$ bins. In particular, let $(j = 1, \ldots, c)$

$$N_j = \#\{X_i, i = 1, \ldots, n : X_i \in [d_{nj-1}, d_{nj})\}. \tag{3.1}$$

3. Let $h_{nj} = d_{nj} - d_{nj-1}$ $(j = 1, \ldots, c)$ denote the bin widths.

With this notation the histogram is given by the following nonparametric density estimator,

$$\hat{f}_n(x) = \frac{1}{n} \sum_{j=1}^{c} \frac{N_j}{h_{nj}} \mathrm{I}\left(x \in [d_{nj-1}, d_{nj})\right). \tag{3.2}$$

The partition is frequently chosen so that $h_{n1} = \cdots = h_{nc} = h_n$, i.e., equal bin widths. Equation (3.2) then simplifies to

$$\hat{f}_n(x) = \frac{1}{nh_n} \sum_{j=1}^{c} N_j \mathrm{I}\left(x \in [d_{nj-1}, d_{nj})\right). \tag{3.3}$$

Sometimes $nh_n\hat{f}_n(x)$ is plotted versus $x$, so that the observation counts can be read directly from the vertical axis.

#### 3.1.1.2 Some Properties

When the sample sizes increase it is natural to let the bin width decrease. This is similar to the bandwidth of the kernel density estimators of Section 2.8. The rate of convergence of the MISE can again be optimised by choosing $h_n$ appropriately. As for many nonparametric density estimators, the optimal choice of $h_n$ depends not only on the sample size, but also on the true unknown distribution of $X$. For the histogram density estimator it has been shown that the convergence rate of MISE can never be faster than $O(n^{-2/3})$, which is slower than many other types of estimators. This demonstrates that the histogram is not the best choice for density estimation. On the other hand it is still a very popular graphical tool for exploring the shape of the distribution of $X$.

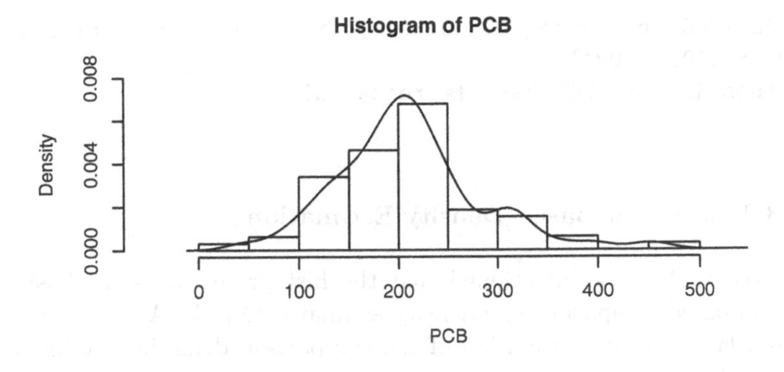


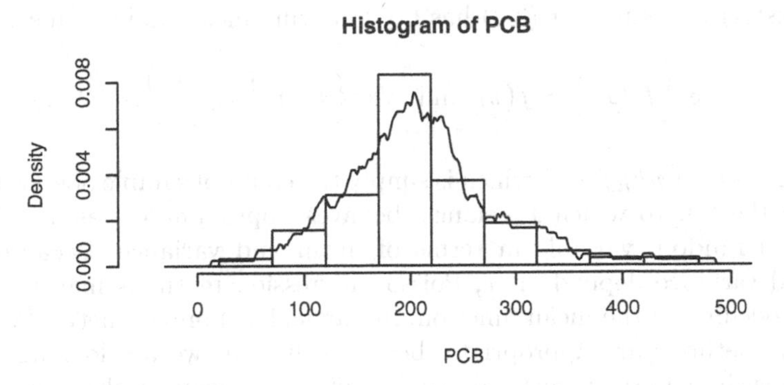


**Fig. 3.1** Two histograms of the PCB concentration data, with different bin locations and equal bin widths. Kernel density estimates with a Gaussian kernel (upper pannel) and rectangular kernel (lower pannel) are added

The simplicity of the histogram is definitely an advantage, but it also suffers from a few shortcomings. We name a few. The first is the slow convergence rate that was mentioned in the previous paragraph. A second undesirable characteristic is that the histogram strongly depends on the choice of the bin edges, even for a fixed bin width. This is illustrated in Figure 3.1 which shows two histograms of the same data (PCB concentration data), but the locations of the bins have been shifted 30 units. The lower panel suggests that the distribution is quite peaked, but this feature is not observed in the upper panel. Such problems are avoided when, for example, kernel density estimators are used. For illustration purposes we have added two different kernel density estimates to the histograms. For the Gaussian kernel used in the upper panel the Gaussian density is used as kernel function $K$. The rectangular kernel can be considered as a moving window version of the histogram, so that at least the bin location choice problem is resolved. The R code follows.

```
> par(mfrow=c(2,1))
> hist(PCB,breaks=seq(0,550,50),xlim=c(-50,550),prob=T,
+ ylim=c(0,0.008))
> lines(density(PCB,kernel="gaussian"))
```

```
> hist(PCB,breaks=seq(0,550,50)-30,xlim=c(-50,550),prob=T,
+ ylim=c(0,0.008))
> lines(density(PCB,kernel="rectangular"))
```

#### 3.1.1.3 Regression-Based Density Estimation

In Section 2.8.4 it was mentioned that the histogram forms the basis of a regression-based nonparametric density estimator (NDE). As this method is also used later for the estimation of the comparison densities, we introduce the method here.

The histogram estimator (3.3) has the following mean and variance,

$$\mathrm{E}\left\{\hat{f}_n(x)\right\} \approx f(x) \text{ and } \mathrm{Var}\left\{\hat{f}_n(x)\right\} \approx \frac{f(x)}{nh_n}.$$

This suggests that $nh_n\hat{f}_n(x)$, which is simply the count of sample observations falling in the bin to which $x$ belongs, behaves approximately as a Poisson distributed random variable in terms of mean and variance. Because the mean and variance depend on $x$, Poisson regression methods may be used for the modelling of the mean function. In particular, nonparametric Poisson regression methods are appropriate, because after all we are looking for a NDE. In such regression methods the conditional mean of the counts $N_j$ (3.1) is modelled as a function of $r_j$, where $r_j$ is the center of the $j$th bin ($j = 1, \ldots, c$). We could write

$$\mathrm{E}\{N_j\} = m(r_j),$$

where the mean function $m$ is estimated by means of smoothing splines or local polynomial regression. We refer to Fan and Gijbels (1996) and Simonoff (1996) for more details on these regression methods. The estimator resulting from the nonparametric Poisson regression, say $\hat{m}$, is, however, not normalised and should thus be normalised before it can be used as a density estimator.

### *3.1.2 The Box Plot*

The box plot, or the box and whisker plot, was originally suggested by Tukey (1977). It is typically used in a data exploration phase of the statistical analysis, and is used to get a rough idea of the shape of the distribution sample observations. It is particularly useful for assessing the asymmetry of the distribution and for detecting outliers.

Although many versions of the box plot have been described in the literature, we focus here on the implementation of boxplot in the R statistical software (R Development Core Team (2008)).

The box plot is essentially a one-dimensional plot of a few sample quantiles and statistics derived from the sample quantiles. The median and the first and third quartiles, denoted by $Q_2$, $Q_1$, and $Q_3$, respectively, are computed as in (2.14), (2.13), and (2.15), respectively. This corresponds to "type 7" of the class of quartile estimators of Hyndman and Fan (1996). The box plot also makes use of the *interquartile range* (IQR), defined as $IQR = Q_3 - Q_1$. The IQR serves sometimes as the basis for the calculation of robust estimators of the variance. In the box plot the IQR is used for the calculation of the *whiskers*. First define the statistics

$$P_l = Q_1 - k \times IQR \text{ and } P_u = Q_3 + k \times IQR,$$

where $k = 1.5$ in the R implementation, as well as in most other box plot constructions. Let $Mn$ and $Mx$ denote the smallest and largest sample observation, respectively. The lower and upper whiskers are then defined as

$$W_l = \max\{P_l, Mn\} \text{ and } W_u = \min\{P_u, Mx\},$$

respectively. A further interpretation and the definition of outliers are provided by means of two artificial examples.

*Example 3.1 (Two toy examples).* The left panel in Figure 3.2 illustrates how these statistics are depicted in the box plot. The box plot presented here is based on a sample of 100 observations from a standard normal distribution. The summary statistics used in the box plot are provided by the summary function in R. The output is shown below

```
> summary(x)
   Min. 1st Qu.  Median    Mean 3rd Qu.    Max.
-2.0650 -0.4909  0.1121  0.2208  0.9200  2.6970
```

Because $Q_1 - 1.5 \times IQR = -2.607 < -2.065 = Mn$, the lower whisker corresponds to the smallest sample observation. Similarly, because $Q_3 + 1.5 \times IQR = 3.036 > 2.697 = Mx$ the upper whisker is plotted at $Mx$. The box plot looks quite symmetric (median in the middle of the two other quartiles, and box in the middle of the two whiskers), suggesting that the sample comes from a symmetric distribution, which is indeed the case here (normal distribution). It is important to stress that this plot does not give any information about the number of observations on which it is based. Obviously, the more observations it is based on, the more confidence may be placed on the graph.

For a second toy example we have sampled 100 observations from a standard lognormal distribution. The summary statistics are shown below.

```
> summary(x)
   Min. 1st Qu.  Median    Mean 3rd Qu.    Max.
0.04848 0.47690 1.07800 1.44500 1.76900 5.78000
```

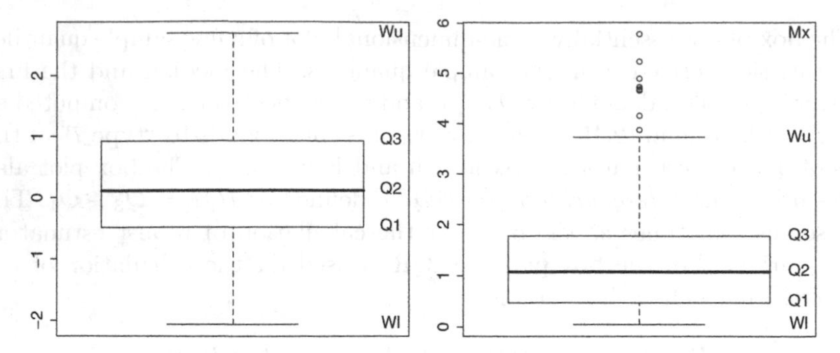


**Fig. 3.2** Box plots of a sample of $n = 100$ observations from a standard normal distribution (left panel) and from a standard lognormal distribution (right panel). Quartiles and whiskers are indicated

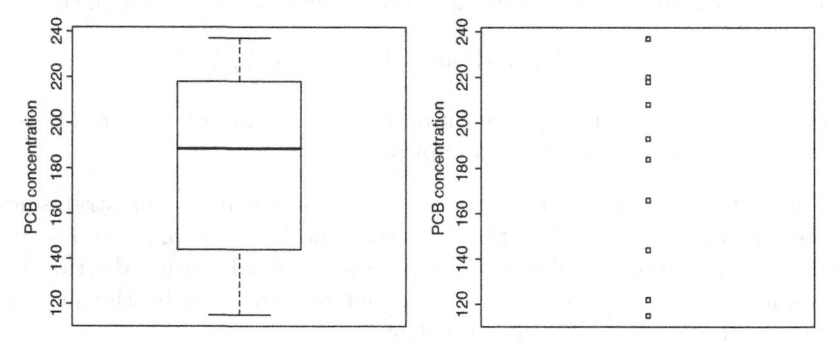


**Fig. 3.3** Box plots of a subsample of $n = 10$ of the PCB dataset (left panel) and a strip chart of the same sample (right panel)

The box plot is shown in the right panel of Figure 3.2. The lower whisker coincides again with the smallest sample observation, but now the upper whisker is located at $W_u = Q_3 + 1.5 \times IQR = 3.707$, because this is smaller than the largest observation $Mx = 5.780$. The box plot now suggests that the sample data come from an asymmetric distribution with a long right tail. This is suggested by the asymmetric placement of the box relative to the whiskers. The plot further shows a few dots located above the upper whisker $W_u$. These correspond to observations identified as outliers according to the definition given further below.

Box plots may also give information on the tails of the distribution. For example, in the left panel of Figure 3.2 we notice that the box, of which the width is given by the IQR, is well separated from the two whiskers. This indicates that the tails are not very short. The left panel of Figure 3.3, on the other hand, shows a box of which the lower and upper borders are very close to the whiskers. This is an indication of short tails.

**Definition 3.1 (outliers).** All observations smaller than the lower whisker $W_l$ are referred to as *outliers*, and, similarly, all observations larger than the upper whisker $W_u$ are also referred to as outliers. This happens when $Mn < W_l$ or $W_u < M_x$.

Finally we stress once more that the traditional visualisation of the box plot does not provide any information on the sample size, so that one should always be careful not to overinterpret the graph. When the sample size is really small, it may be better to make a *strip chart*, using the stripchart function in R. This is illustrated in the next example.

*Example 3.2 (PCB concentration data).* The box plot of the PCB concentration data has been shown already in Figure 1.2 (right panel). To illustrate the danger of overinterpreting the box plot with small samples, we have sampled at random ten observations from the PCB dataset, and used these ten observations for constructing the box plot in the left panel of Figure 3.3. Based on this plot, one could perhaps conclude that the tails of the distribution are short (see earlier in this section). This conclusion is definitely not confirmed by the plot based on the complete dataset (Figure 1.2). Thus the shape of the plot may be misleading, but this is of course a simple consequence of the large variance on the quartile estimators (and maximum and minimum observations) used for the construction. For small samples it may be safer to plot a *strip chart.* The strip chart of the subsample of ten observations is presented in the right panel of Figure 3.3. This is basically a plot of the individual observations, with no adding of the sample quartiles so that overinterpretation is harder. The plot just shows ten points, quite evenly distributed over the range, but by observing that there are only ten observations, one should be warranted that there is not much information in the sample.

*Example 3.3 (Combined plots).* To conclude we demonstrate the use of the Boxplot function in the R-package accompanying the book. The first graph, which is presented in the left panel of Figure 3.4 shows a combined graphical

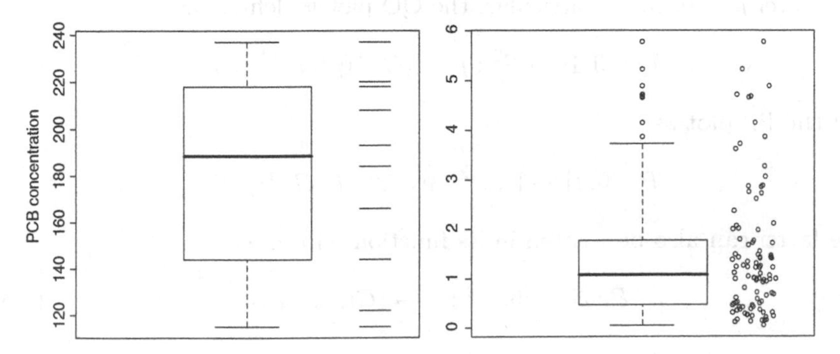


**Fig. 3.4** Box and bar code plots of a subsample of $n = 10$ of the PCB dataset (left panel) and a box and jitter plot of 100 standard lognormal observations (right panel)

display of the box plot and a bar code plot (similar to strip chart) of the subsample of ten PCB concentration observations. In the right panel of Figure 3.4 the box plot of the 100 lognormal observations of Example 3.1 is shown. The right margin of the plot shows the individual observations, but their horizontal position is randomly jittered so as to give a better view on the individual observations. This visualisation thus gives a better impression of the size of large samples. Variants similar to these graphs were also presented by Lee and Tu (1997). The R-code follows.

```
> Boxplot(s,ylab="PCB concentration",side="bar")
> Boxplot(x,side="jitter")
```

## 3.2 Probability Plots and Comparison Distribution

### *3.2.1 Population Probability Plots*

Together with the boxplot and the histogram, the probability plots are among the most widely used graphical methods for exploring the distribution of the data. There are two types of probability plots: probability–probability (PP) and quantile–quantile (QQ) plots. Whereas boxplots and histograms are purely exploratory in the sense that they only visualise the data at hand without trying to answer a particular question, QQ and PP plots are used in a more directional manner. QQ and PP plots are graphs used to compare the EDF with the hypothesised distribution function $G$, and so they are extremely useful in the setting of the one-sample problem.

Before giving the sample versions of the probability plots we give their definitions in the more general setting of comparing two distribution functions $F$ and $G$, both defined on the same support $\mathcal{S}$. To stress the distinction from the sample versions, we call them *population probability plots*. Probability plots are curves in a two-dimensional plane indexed by a probability parameter $p \in [0,1]$. In particular, the QQ plot is defined as

$$Q : [0,1] \mapsto \mathcal{S}^2 : p \to (G^{-1}(p), F^{-1}(p)), \tag{3.4}$$

and the PP plot as

$$P : [0,1] \mapsto [0,1]^2 : p \to (p, F(G^{-1}(p))).$$

The latter can also be written in its functional form as

$$P : \mathcal{S} \mapsto [0,1]^2 : x \to (G(x), F(x)). \tag{3.5}$$

Both plots have the property that they show a straight 45 degree line through the origin if and only if $F = G$. Any deviation from this straight line indicates that $F \neq G$, and the shape of the curve tells something about how the two distributions differ. This property makes the probability plots suited for one-sample goodness-of-fit purposes where $G$ is the hypothesised distribution and $F$ is replaced by its sample version (see next section), and for two-sample goodness-of-fit problems where both $F$ and $G$ are replaced by their respective sample versions (see Section 8.1). Also note that the PP plot is related to the comparison distribution (Section 2.4). In particular (3.5) is a plot of the CDF of the comparison distribution (2.17).

### *3.2.2 PP and QQ plots*

The sample versions of the probability plots are obtained by replacing the true, but unknown distribution $F$ by the EDF $\hat{F}_n$. The convention is to draw the probability plot as a scatterplot with exactly $n$ points. This makes sense because the EDF and its inverse are piecewise constant functions. Moreover, an advantage of a scatterplot representation is that the number of points gives visually an appreciation of the amount of information present in the sample. The reduction of a line plot to a scatterplot leaves the question open for which values of $p$ to construct the plot. These values are referred to as the *plotting positions*. A general form for the plotting positions is proposed by Blom (1958),

$$p_i = \frac{i - c_i}{n + 1 - 2c_i}, \tag{3.6}$$

where $0 \leq c_i < 1$ ($i = 1, \ldots, n$). Equation (3.6) is often simplified by setting $c_1 = \cdots = c_n = c$. Note that for all $c$, $(i-1)/n < p_i \leq i/n$. This has the following consequences for the QQ plot: $\hat{F}_n^{-1}(p_i) = X_{(i)}$ for all $0 \leq c < 1$ and $i = 1, \ldots, n$. Every observation is thus plotted exactly once on the vertical axis. A similar property does not hold for the PP plot.

The choice of $c$ turns out to be much more important for the QQ plot than for the PP plot. Kimball (1960), Mage (1982), and Thode (2002) give overviews of popular choices for $c$, which are summarised in Table 3.1. Kimball (1960) stresses that the choice of $c$ depends on the goal of the data analysis. He recognises three purposes: (1) goodness-of-fit, (2) parameter estimation, and (3) extrapolation to the extremes. Because our primary aim is goodness-of-fit, we only briefly describe (2) and (3) in the next paragraphs.

QQ plots have also been popular for parameter estimation in location-scale families (e.g., the normal, logistic, and extreme-value distributions). Suppose both $F$ and $G$ belong to the same location-scale family. A location-scale family is characterised by the property $F(x; \mu, \sigma) = F((x - \mu)/\sigma; 0, 1)$, where $\mu$ and $\sigma$ are the location and scale parameters, respectively.

**Table 3.1** Plotting positions proposed in the statistical literature

| Value of $c$ | Plotting position $p_i$ | References |
|---|---|---|
| 0 | $\frac{i}{n+1}$ | Kimball (1960); Filliben (1975) |
| 0.3 | $\frac{i-0.3}{n+0.4}$ | Bernard and Bos-Levenbach (1953) |
| 0.3175 | $\frac{i-0.3175}{n+0.365}$ | Filliben (1975) |
| 0.375 (=3/8) | $\frac{i-0.375}{n+0.25}$ | Filliben (1975) |
| 0.4 | $\frac{i-0.4}{n+0.2}$ | Cunnane (1978) |
| 0.44 | $\frac{i-0.44}{n+0.12}$ | Gringorten (1963); Mage (1982) |
| 0.5 | $\frac{i-0.5}{n}$ | Blom (1958); Hazen (1930) |
| 0.567 | $\frac{i-0.567}{n-0.124}$ | Larsen et al. (1980); Mage (1982) |
| 1 | $\frac{i-1}{n-1}$ | Filliben (1975) |

Let $G$ be the standard distribution (i.e., location parameter $\mu = 0$ and scale parameter $\sigma = 1$), and let $F$ have arbitrary parameters $\mu$ and $\sigma$. Then,

$$F^{-1}(p) = \mu + \sigma G^{-1}(p),$$

and the population QQ plot now still shows a straight line, but with intercept $\mu$ and slope $\sigma$. Thus, fitting a linear regression model to the points in the QQ plot results immediately in estimates of the location and scale parameters. Plotting positions can be determined by adopting an optimality criterion to which the estimators should apply, for instance, unbiased and minimum variance. The optimal plotting positions, however, depend on the family to which $F$ and $G$ belong.

The third kind of purpose that can be served by QQ plots is extrapolation. In particular in the analysis of extreme events the focus is often on quantiles corresponding to very small or very large probabilities. For instance, based on a data set of the yearly maximum water levels at a fixed location, a QQ plot with respect to a standard extreme value distribution may be constructed, and the location and scale parameters may be estimated by fitting a regression line to the QQ plot. This fitted regression line is subsequently used to predict the quantile (water level), say $\hat{q}$, that corresponds to a return period of 10,000 years, which is equivalent to a probability of $p = 1/10000 = 0.0001$ that an extreme high water level of $q$ occurs at most once every 10,000 years. Because it is very likely that an extreme water level as high as $\hat{q}$ is not observed in the period that the data were collected, this prediction is clearly an extrapolation. Plotting positions may now be found to give the most accurate predictions at small or large $p$.

Back to goodness-of-fit. The following values of $c$ are often used:

- $c = 0.375$ or $c = 0.5$, resulting in

$$p_i = \frac{i - 0.375}{n + 0.25} \quad \text{or} \quad p_i = \frac{i - 0.5}{n},$$

respectively. The rationale is found in trying to give the QQ plot the following interpretation: observed quantiles ($X_{(i)}$) are plotted against their expectations, so one expects the points to lie on the diagonal. Thus on the horizontal axis we need the expectation of the order statistics under the null hypothesis $F = G$; i.e., we need to find $\mathrm{E}_g\left\{X_{(i)}\right\}$. Because the horizontal axis of a QQ plot is generally determined by $G^{-1}(p_i)$, plotting positions $p_i$ can be found by solving $\mathrm{E}_g\left\{X_{(i)}\right\} = G^{-1}(p_i)$. Exact solutions exist for the normal distribution, but they require the $c_i$s to be nonconstant. Fortunately, the $c_i$s show only small variation, so that usually an approximate solution is used. Both $c = 0.375$ and $c = 0.5$ give good approximations. When the standard error is to be estimated from the fitted regression line in the QQ plot, these solutions also give a nearly unbiased estimator, and a biased estimator with minimum variance, respectively.

- The choice of $c = 0.5$, resulting in $p_i = (i-0.5)/n$, has also another origin. This plotting position is also found as the middle point between $(i-1)/n$ and $i/n$ (note that $\hat{F}_n^{-1}$ remains constant within this interval).
- $c = 0$, resulting in $p_i = i/(n+1)$. This corresponds to the exact solution to the equation $\mathrm{E}_g\left\{X_{(i)}\right\} = G^{-1}(p_i)$, when $G$ is the uniform distribution over $[0,1]$. Thus, when $X$ is uniformly distributed, the order statistics $X_{(i)}$ are plotted against their expected values $i/(n+1)$. Because any random variable can be transformed to be uniformly distributed by applying the probability integral transformation (i.e., when $X \sim G$, then $G(X) \sim U(0,1)$), this plotting position system is quite generally applicable. It is also the most common choice when constructing PP plots.

*Example 3.4 (Pseudo-random generator data).* The R function qqplot only works with the normal distribution as the reference distribution $G$, but in the cd package there is a more generic function QQplot that can be used with any distribution $G$. The function PPplot is used to plot the PP plot. The R code is shown below, and the QQ and PP plots are presented in Figure 3.5. From these plots it is again concluded that the runif function generates uniformly distributed numbers.

```
> QQplot(PRG,distr=qunif,pars=c(0,1),blom=0)
> PPplot(PRG,distr=qunif,pars=c(0,1),blom=0)
```

*Example 3.5 (PCB concentration data).* A QQ plot is plotted for the PCB concentration to assess whether the data are normally distributed. The mean and the standard deviation are estimated from the sample. We have now used $c = 0.375$ because the reference distribution is the normal distribution. See Figure 3.6

```
> QQplot(PCB,distr=qnorm,pars=c(mean(PCB),sd(PCB)),
  blom=0.375)
> PPplot(PCB,distr=qnorm,par=c(mean(PCB),sd(PCB)),blom=0.375)
```

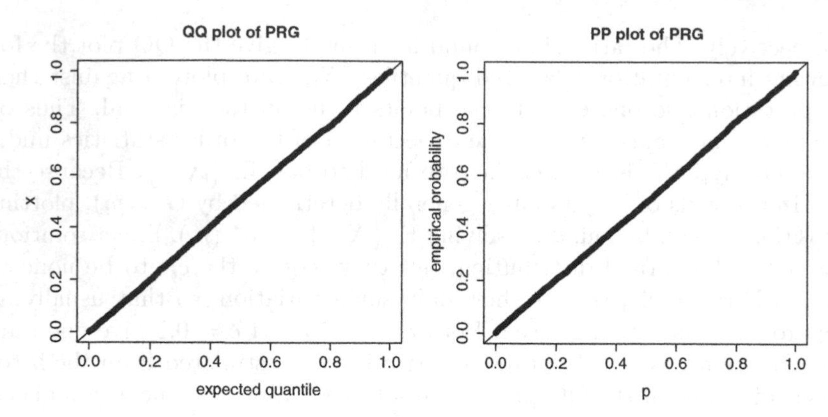


**Fig. 3.5** QQ (left panel) and PP (right panel) plots of the PRG dataset

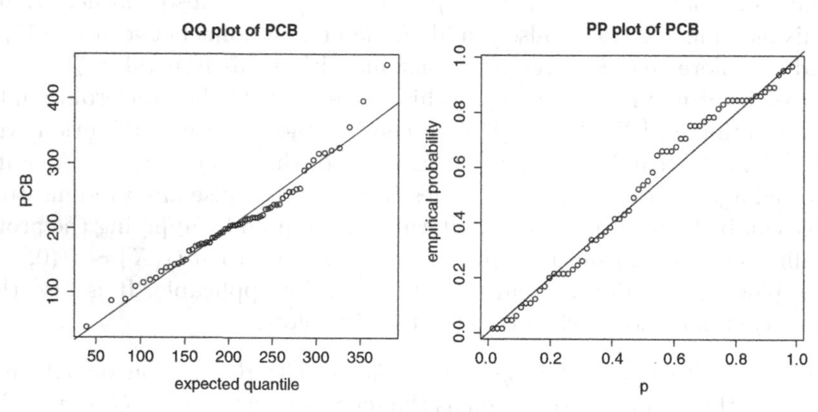


**Fig. 3.6** QQ (left panel) and PP (right panel) plots of the PCB dataset

One of the comments often given to QQ plots, is that the variability of the order statistics $X_{(i)}$ is not constant over the range of plotting positions. Theoretically, it is easy to show that the variance of the $i$th order statistic with distribution $G$ can be approximated by

$$\text{Var}\left\{X_{(i)}\right\} = \frac{p_i(1-p_i)}{ng(G^{-1}(p_i))},$$

where $g$ is the density function of $X$. In Figure 3.7 we show QQ plots of 100 independent samples of 20 observations from a normal distribution with mean 10 and standard deviation 2. The graph clearly illustrates that the variances of the more extreme order statistics are larger than those of the order statistics close to the median. This phenomenon is present in individual QQ plots. Figure 3.8 shows a few of the individual QQ plots. In the two upper panels and the lower left panel there seems to be a deviation from the 45 degree line in the tails of the distribution, from which one may be inclined

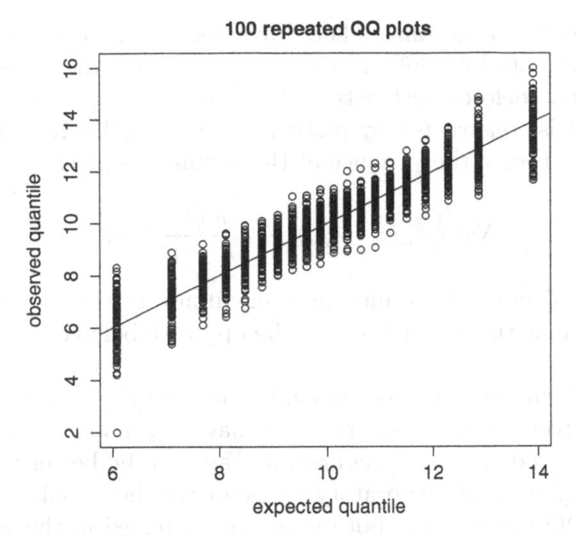


**Fig. 3.7** QQ plots of 100 independent samples of 20 observations from a normal distribution

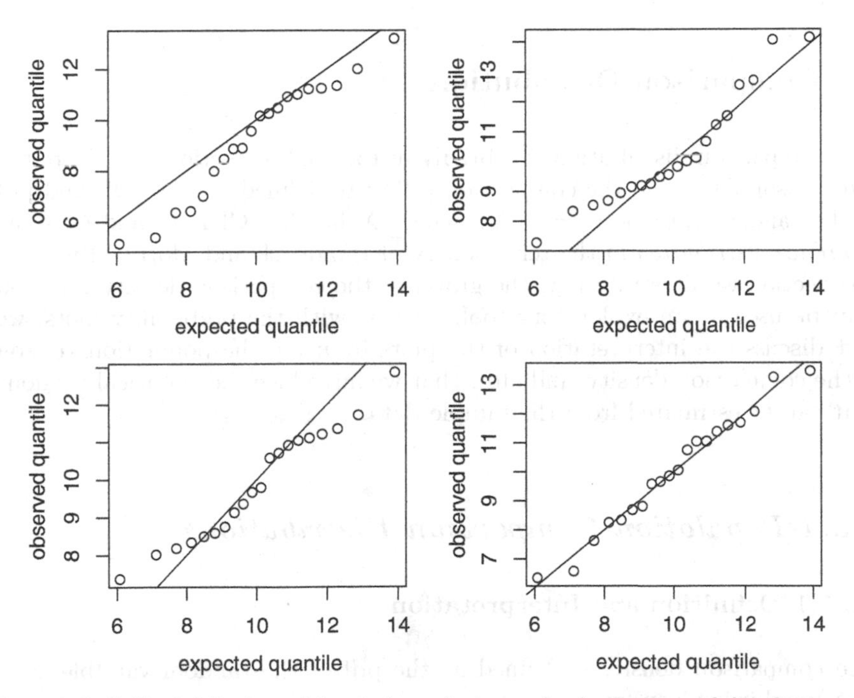


**Fig. 3.8** Individual QQ plots of four independent samples from a normal distribution

to conclude that the sample did not arise from the hypothesised normal distribution. This is definitely not the case here for we used simulated data. The larger variability in the tails has led practitioners to focus less on what the QQ plot shows in the tails. There have been some attempts to stabilise the probability plots by applying transformations on the observations and plotting positions before constructing the plots (see, e.g., Michael (1983)).

The PP plot is constructed by plotting $\hat{F}_n(G^{-1}(p_i))$ versus $p_i$. Under the simple null hypothesis, the variance of the ordinate is given by

$$\text{Var}\left\{\hat{F}_n(G^{-1}(p_i))\right\} = \frac{p_i(1-p_i)}{n},$$

which is minimal in the tails and has a maximum at the median. Thus, for many distributions the PP plot shows the opposite behavior in terms of the variability.

From this discussion we may conclude that the choice between QQ and PP plots is driven by the importance of having a good fit in the tails or rather near the median. But, of course, it may even be better to plot both a QQ and a PP plot in an exploratory phase of the data analysis. Finally, an advantage of QQ plots is that both axes are expressed in the same units as the observations.

## 3.3 Comparison Distribution

The comparison distribution was briefly introduced in Section 2.4. There the comparison CDF and the comparison pdf were defined as the CDF and pdf of the random variable $U = G(X)$, where $X$ has has CDF $F$, and $G$ is the *reference distribution* in the terminology of Handcock and Morris (1999). In this section we illustrate how the graph of the comparison density function may be used as an exploratory tool. Just as with the probability plots, we first discuss the interpretation of the plots by using the population version of the comparison density, and after that we introduce the empirical versions that can be estimated from the sample data.

### *3.3.1 Population Comparison Distributions*

#### 3.3.1.1 Definition and Interpretation

The comparison density is defined as the pdf of the random variable $U = G(X)$ and it is given by

$$r(u) = \frac{f(G^{-1}(u)}{g(G^{-1}(u)}. \tag{3.7}$$

The population comparison density is then defined as

$$C : [0,1] \mapsto [0,1] \times \mathbb{R}^+ : u \to (u, r(u)).$$

This plot has the property that it shows a horizontal line at $r(u) = 1$ if and only if $F = G$. Any deviation from this straight line implies that the null hypothesis is not true, and the type of the deviation is informative about how the two distributions differ. Because $r(u)$ is basically a ratio of two densities evaluated at the percentile $0 \leq u \leq 1$, it may be interpreted as follows.

- $r(u) > 1$: We expect a larger frequency of observations around the $u$th percentile of the reference distribution $G$, than if the observations had distribution function $G$.
- $r(u) < 1$: We expect a smaller frequency of observations around the $u$th percentile of the reference distribution $G$, than if the observations had distribution function $G$.

Although, for instance, Handcock and Morris (1999) prefer to plot $r(u)$ versus the percentile $u$, the interpretation may sometimes be easier when $x = G^{-1}(u)$ is used instead, leading to an alternative definition of the population comparison density,

$$C' : \mathcal{S} \mapsto \mathcal{S} \times \mathbb{R}^+ : x \to (x, r(G(x))).$$

We further illustrate the interpretation through some hypothetical examples. Figure 3.9 presents the densities and the population comparison densities of two normal distributions with equal variance (1) and means 0 and 0.5. To make the two plots comparable, both horizontal axes are on the same scale (according to the alternative definition $C'$). The plot shows the population comparison density which is here an increasing function. It can be proven that this always holds for location-shift models, irrespective of the parent distribution. A detailed interpretation of the plot says that for values smaller than 0.25 (the vertical reference line) the true density $f$ is smaller than the reference density $g$, but for observations larger than 0.25 the opposite is true. The second example is presented in Figure 3.10. Here two normal distributions with equal mean but with different scales are shown. The comparison density has now a typical $U$-shape. It further demonstrates that within the interval $[-1.2, +1.2]$ the true density $f$ is smaller than the reference density $g$. Outside of the interval, we may expect a larger frequency of observations than expected under the hypothesised distribution. Thus observations under $f$ show a larger variability.

#### 3.3.1.2 Decomposition of the Comparison Density

Although the comparison densities can always be interpreted in terms of the ratio of two densities, as in Equation (3.7), it is not always easy to recognise shifts in means or variances, particularly when there is no pure shift in mean or variance. This is illustrated in Figure 3.11, where again two normal distributions are considered, but now they differ both in mean and variance. The comparison density looks like an asymmetric $U$-shape, and it is hard, if not

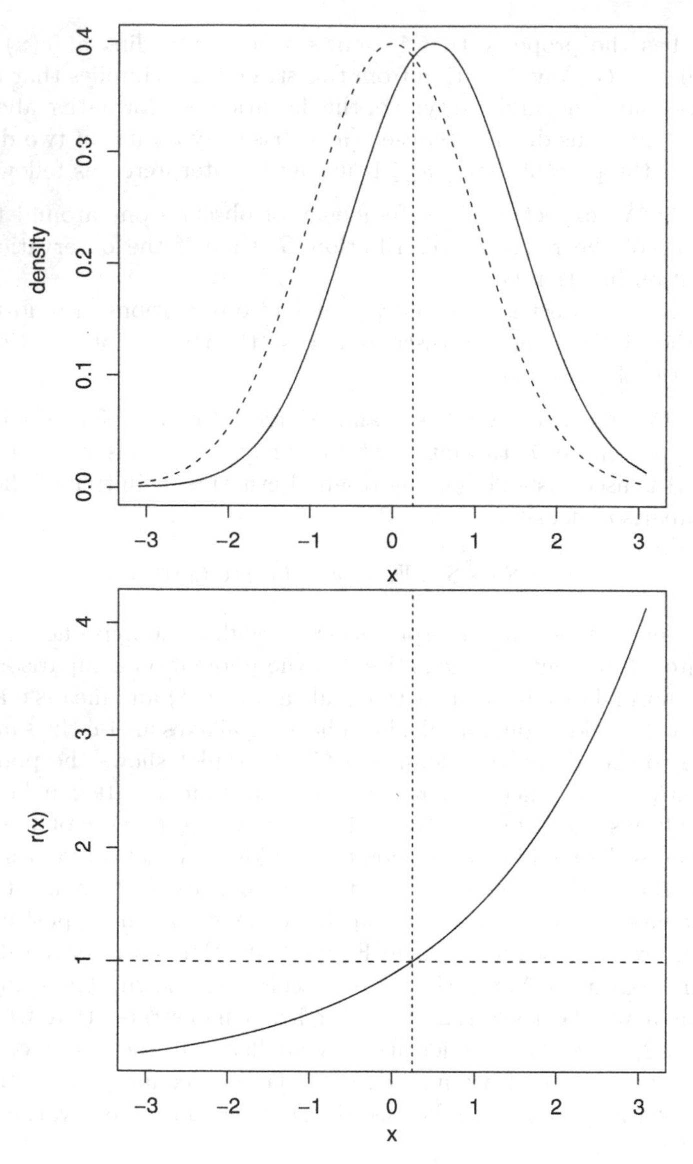


**Fig. 3.9** The upper panel shows the densities $f$ (full line) and $g$ (dotted line) of two normal distributions with mean shifted over 0.5, and the lower pannel shows the population comparison density. The vertical reference line indicates the position where the comparison density equals 1

impossible, to uniquely conclude that this is caused by a difference in mean and variance. Therefore, Handcock and Morris (1999) proposed to decompose the comparison density in factors that can be attributed to mean, scale, and more general shape differences. Consider the identity (with $x = G^{-1}(u)$)

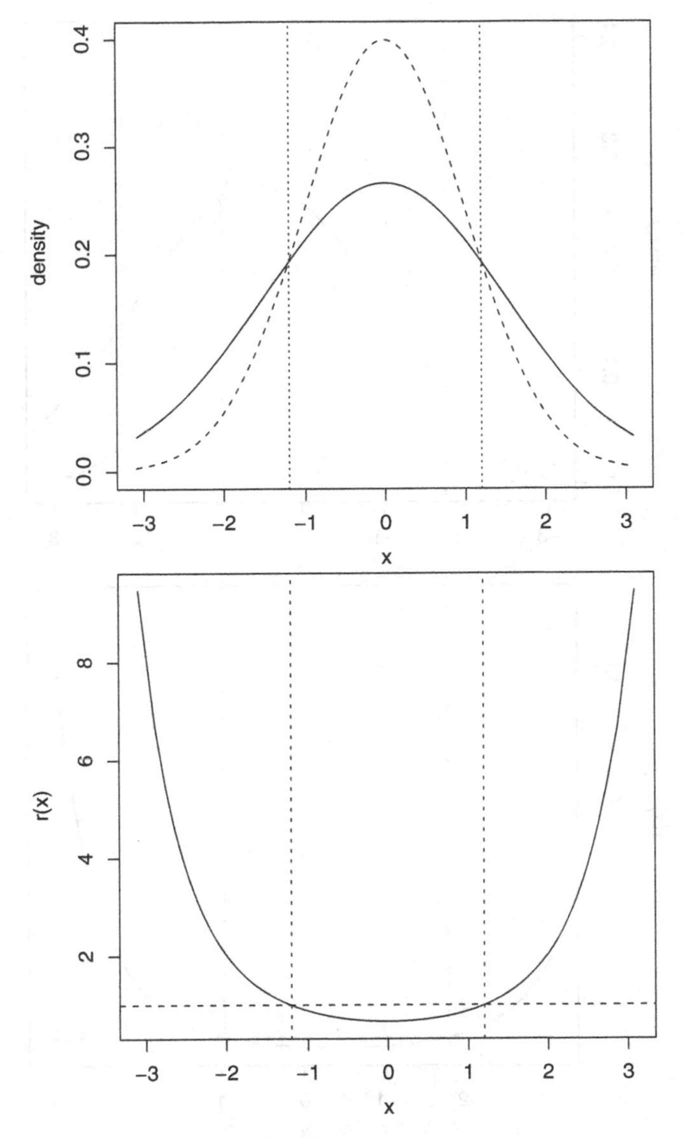


**Fig. 3.10** The upper panel shows the densities $f$ (full line) and $g$ (dotted line) of two normal distributions with variances 1.5 and 1, respectively, and the lower panel shows the population comparison density. The vertical reference line indicates the position where the comparison density equals 1

$$r(x) = \frac{f(x)}{g(x)} = \frac{g_L(x)}{g(x)} \times \frac{g_{LS}(x)}{g_L(x)} \times \frac{f(x)}{g_{LS}(x)},$$

where $g_L$ is the density of a random variable $X+\delta$, where $X$ has density $g$ and $\delta$ is such that the mean of $X+\delta$ equals the mean of distribution $f$. Similarly,

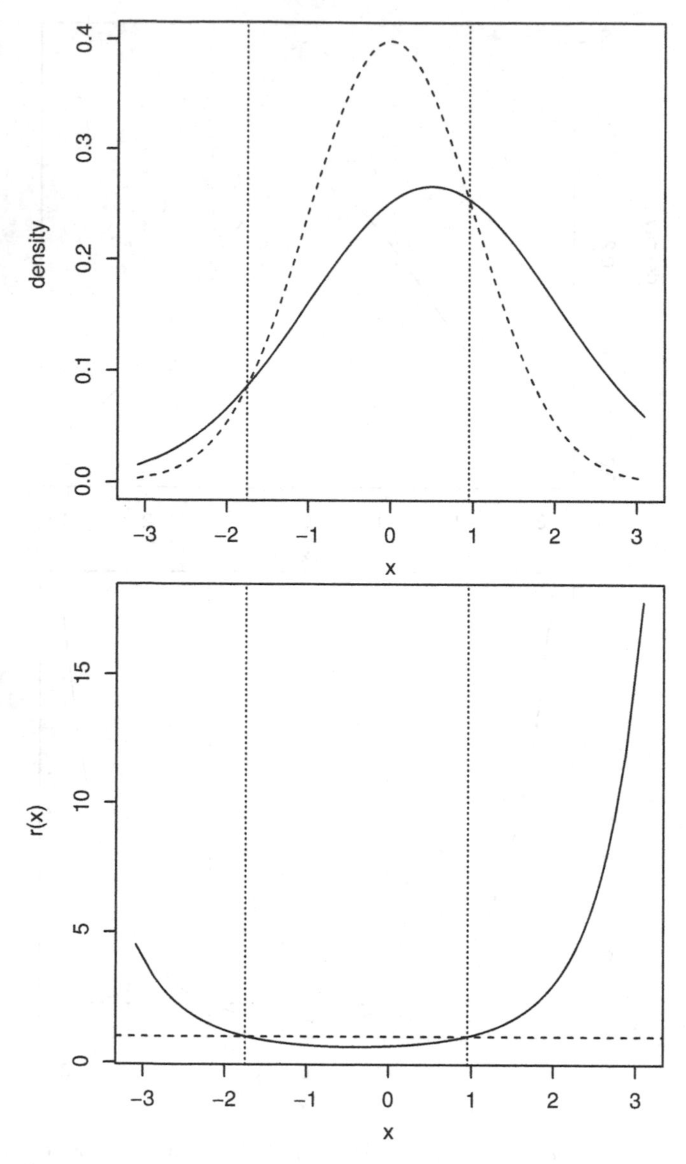


**Fig. 3.11** The upper panel shows the densities $f$ (full line) and $g$ (dotted line) of two normal distributions with mean (variances) equal to 0.5 (1.5) and 0 (1), respectively, and the lower panel shows the population comparison density. The vertical reference line indicates the position where the comparison density equals 1

$g_{LS}$ is the density of the random variable $\gamma(X+\delta)$, where $X$ has again density $g$, and $\gamma$ and $\delta$ are so that the mean and variance of $\gamma(X+\delta)$ equal those of the distribution $f$. With these definitions, the ratio $r_L(x) = g_L(x)/g(x)$ contains only information about a mean difference, and $r_{LS}(x) = g_{LS}(x)/g_L(x)$ only tells something about a difference in scale. The "residual" ratio $r_R(x) = f(x)/g_{LS}(x)$ contains all the information on shape differences not caused by a mean shift or a difference in scale. Note that when both $f$ and $g$ belong to the same location-scale family, $r_R(x) = 1$ for all $x \in \mathcal{S}$.

As an illustration we consider the comparison of a normal distribution with mean and variance equal to 2.15 and 1, respectively, with a standard log-normal distribution, which is a right-skewed distribution. The normal distribution acts as the reference distribution $g$. Figure 3.12 shows the densities and the population comparison density. From the latter it is hard to determine if there is a shift in mean and/or a difference in scale. Nevertheless, the comparison density is interpretable as the ratio of densities. More explanatory plots are given in Figure 3.13 which shows the graphs representing the components $r_L$, $r_{LS}$, and $r_R$. From $r_L$ and $r_{LS}$ we learn that there is both a difference in mean and scale between $f$ and $g$. The graph of $r_R$ shows the pure shape differences that are not due to mean and scale differences.

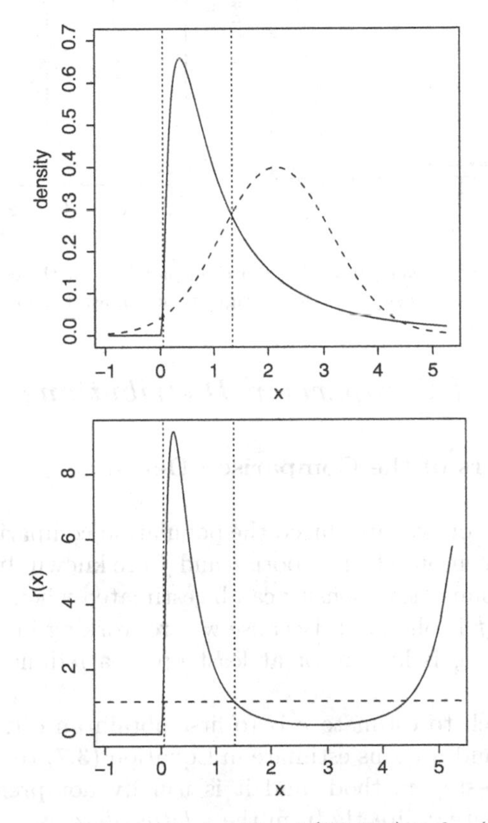


**Fig. 3.12** The upper panel shows the densities $f$ (full line) and $g$ (dotted line) of a log-normal and a normal distribution, respectively, and the lower panel shows the population comparison density. The vertical reference line indicates the position where the comparison density equals 1

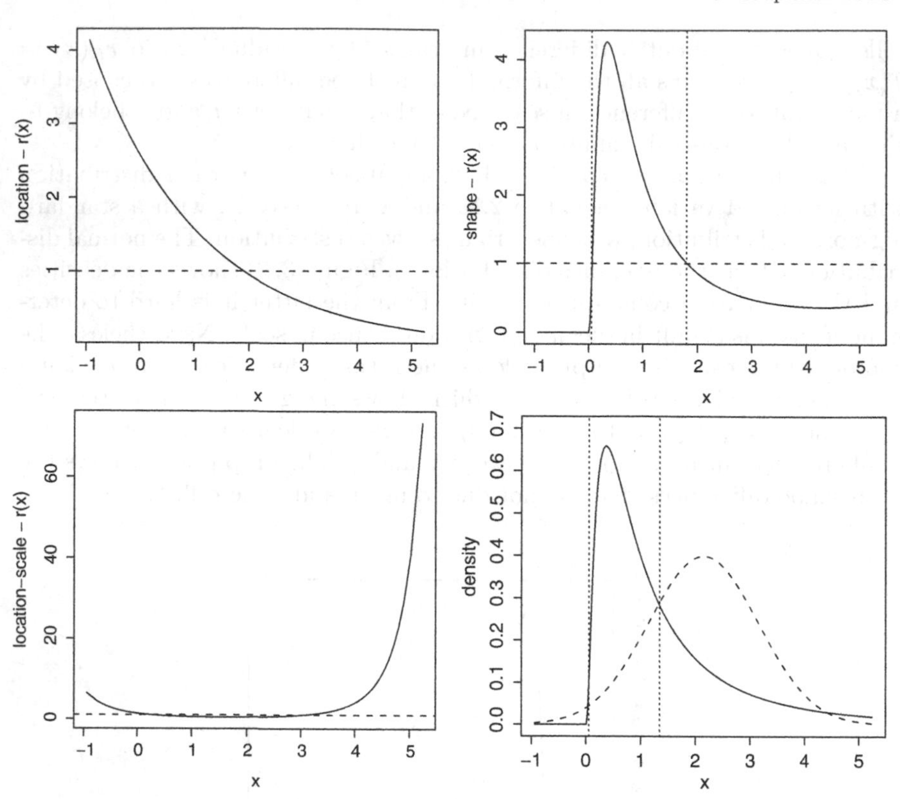


**Fig. 3.13** The upper left, lower left, and upper right panels show the components $r_L$, $r_{LS}$, and $r_R$ of the comparison density. The lower right panel shows the densities $f$ and $g_{LS}$

## 3.3.2 Empirical Comparison Distributions

### 3.3.2.1 Estimators of the Comparison Density

In the previous section we introduced the population comparison distribution which can only be computed when both $g$ and $f$ are known. In this section we explain how this comparison density can bc cstimated when only a sample of observations from $f$ is observed. Because we are working in a goodness-of-fit setting, the density $g$ is known, or at least up to a $p$-dimensional nuisance parameter $\boldsymbol{\beta}$.

A naive approach to estimate $r$ is to first obtain an estimate of the unknown density $f$, and use this estimate in Equation (3.7) to compute $r$. This is, however, a two-step method, and it is usually not preferred. A better method is to estimate $r$ directly from the *relative data*,

$$U_i = G(X_i; \boldsymbol{\beta}),$$

where $\beta$ is replaced by its estimator $\hat{\beta}$ if it is not specified under the null hypothesis. At this point we do not impose any assumptions on this estimator, as we are using the comparison density only as an exploratory tool. As soon as inference based on $r$ is needed, some restrictions on $\beta$ are required. The rationale behind the use of the relative data is given in Section 2.4 where it was shown that the comparison density $r$ is exactly the density of $U = G(X)$. The problem is thus reduced to finding a good nonparametric density estimator of $r$ based on the relative data. All methods described in Section 2.8 may be applied here.

Handcock and Morris (1999) gave several techniques for this estimation problem. In particular they discussed a histogram estimator, kernel density estimators, regression based density estimators, and exponential series density estimators. We, however, do not go into the details of the properties of these estimators in the present context. We only mention a few general comments.

- The kernel density estimators suffer from edge or boundary effect; i.e., near the boundaries $u \approx 0$ and $u \approx 1$ the estimators have a downward bias.
- The regression-based estimators are based on a Poisson approximation of the counts in the histogram estimator. The Poisson mean is modelled by means of smoothing splines or local polynomial estimators. These estimators do not suffer from the boundary bias, but they have a larger variance near the boundaries (cfr. bias–variance trade-off). See Section 3.1.1.3 for more details.
- Both the nonparametric density estimator and the local polynomial regression estimator require the specification of a bandwidth. For the former the Sheater–Jones bandwidth selector is recommended, and for the latter, a bandwidth minimising the generalised cross-validation criterion or a corrected Akaike's information criterion (AIC) are recommended by Handcock and Morris (1999).
- The exponential series density estimator is closely related to the smooth tests presented in Chapter 4. We therefore postpone the discussion of this estimation technique to that chapter.

In the remainder of this section we use the regression-based estimator in combination with local quadratic polynomials.

#### 3.3.2.2 Confidence Intervals of the Comparison Density

Just as with the PP and QQ plots, care should be taken with the interpretation because these diagnostic graphs only show point estimates. There is no information in these graphs about the sampling variability. Here we first illustrate some aspects of this variability and we mention briefly something about confidence intervals that can be added to the plots. The importance of confidence intervals is greater with comparison densities than with PP plots because with the latter the scales on both axes are always fixed at $[0,1]$, whereas the range of the comparison density is in itself informative.

As an illustration of the variance of the comparison density estimator, we have simulated ten random samples of 100 observations from a standard normal distribution. For each sample, we have computed the comparison density estimator based on local quadratic polynomials for which the bandwidth was selected by minimising the generalised cross-validation criterion. The results are shown in Figure 3.14. The plot shows ten different lines. Most of them look more or less straight. Some suggest a positive shift in mean, others a negative shift. This is because the bandwidth selector often results in large bandwidths, particularly when the distribution of the sample is close to $g$. Still there are a few comparison densities that suggest a scale difference or an even more complicated difference in shape. The graph further illustrates that the estimator has a larger variance near the boundaries $u \approx 0$ and $u \approx 1$. An important observation is the following. Although none of the comparison densities is a horizontal line at $r(u) = 1$, as is the population version, they all are bounded between approximately 2/3 and 3/2. Thus, despite the shapes are often quite distinct from a horizontal line at $r(u) = 1$, the

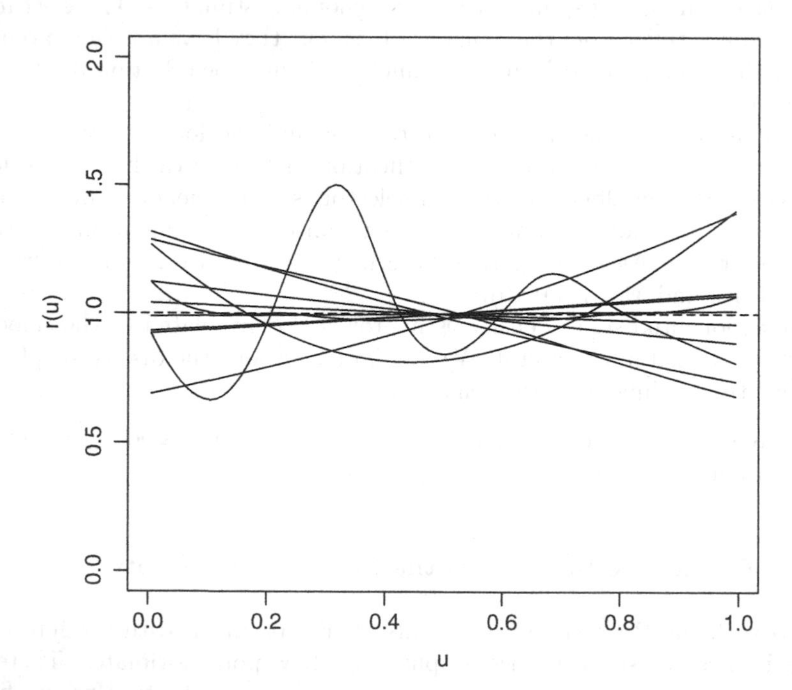


**Fig. 3.14** Comparison densities of ten random samples of 100 observations from a standard normal distribution. The horizontal axis is on the scale of the relative data; i.e., $u = G(x)$

range of $r(u)$ is rather small. On the other hand, for instance, in Figure 3.12, where $f$ and $g$ are very different (normal and lognormal), the range $r(u)$ is much larger. Therefore, we suggest that the user should look very carefully at this range before interpreting the graph. This remark becomes very important when software is used which automatically determines the "optimal" range. For many applications a user-defined range of $[0, 2]$ up to $[0, 10]$ is very informative.

To find out if an observed pattern of the comparison density is not just a consequence of sampling variability, confidence bands can be computed and added to the graph of the comparison density. We should make a distinction between two types of confidence bands: pointwise and simultaneous confidence bands. The former have a *pointwise* interpretation in the sense that their coverage only holds pointwise for a given value of $u$. The coverage of simultaneous confidence bands holds simultaneously for all $u \in [0, 1]$. It is actually the latter type that is most convenient in the present setting, but this falls outside the scope of this book. Handcock and Morris (1999) give some theory about pointwise confidence intervals, but only for the case where $g$ is completely determined. Nevertheless, we think that it is still better to plot these pointwise intervals than to plot nothing. Pointwise bands are typically more narrow than simultaneous bands. Thus if the line $r(u) = 1$ is contained in the pointwise bands, then it is very likely that it is also contained in the simultaneous bands. The opposite is, however, not true.

We further illustrate the use of the comparison density by applying it to some example datasets.

*Example 3.6 (PCB).* We want to graphically assess normality of the PCB data, but the mean $\mu$ and the variance $\sigma^2$ are not specified. Thus before we can compute the relative data $U_i = G(X_i; \mu, \sigma^2)$, we need to replace the unknown nuisance parameters by their estimates. Here we take the sample mean $\hat{\mu} = 210$ and sample variance $\hat{\sigma}^2 = 5303.656$. The upper panel in Figure 3.15 shows the histogram of the PCB data, a nonparametric density smoother and the density $g(.; \hat{\mu}, \hat{\sigma}^2)$.

With the next R-code the comparison density in the middle panel of Figure 3.15 is constructed. We have used the reldist function which is available in the reldist package. Here we have used the default bandwidth selector, which is the minimiser of the generalised cross-validation criterion. The reldist function actually plots the percentiles $u$ on the horizontal axis, but here we have used the original scale of the data.

```
> PCB<-sort(PCB)
> sd.PCB<-sd(PCB)
> m.PCB<-mean(PCB)
> n<-length(PCB)
> rd<-reldist(y=PCB,yo=qnorm((1:n+0.5)/(n+1),sd=sd.PCB,
+     mean=m.PCB),ci=T,graph=F)
Smoothing the maximum amount
```

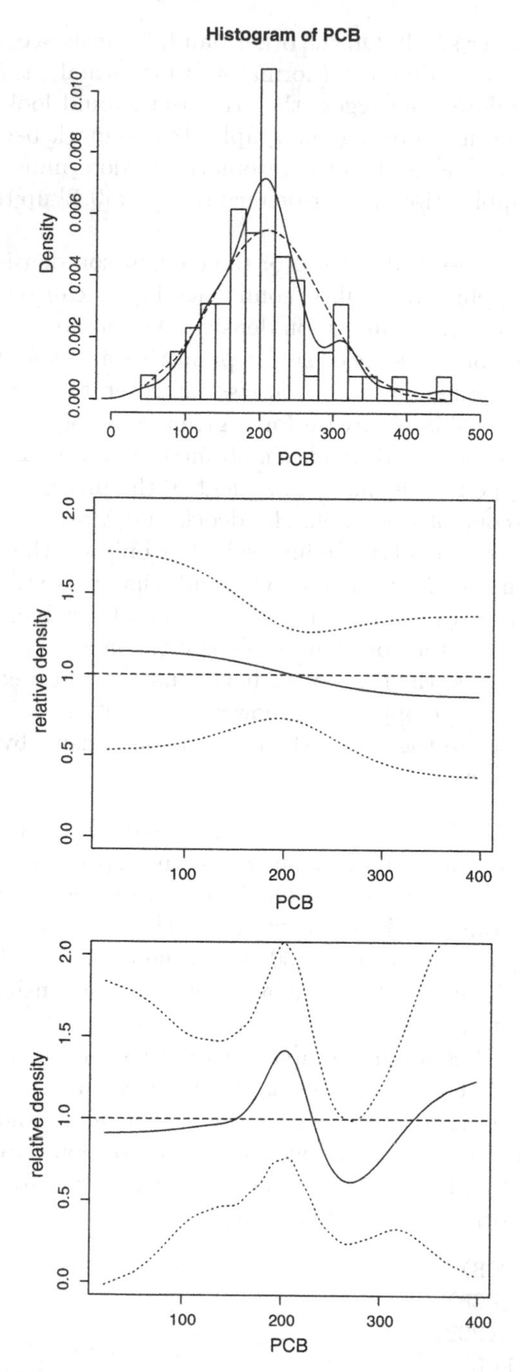


**Fig. 3.15** The histogram, nonparametric density estimator, and fitted normal density (upper panel) of the PCB data. The middle and lower panels show the comparison densities of the PCB data based on the generalised cross-validation and AIC bandwidth selector, respectively. The horizontal axis is on the scale of the relative data; i.e., $u = G(x; \hat{\mu}, \hat{\sigma}^2)$

```
> x<-qnorm(rd$x,sd=sd.PCB,mean=m.PCB)
> plot(x,rd$y,type="l",xlab="PCB",ylab="relative density",
+     ylim=c(0,2))
> lines(x,rd$ci$l,lty=3)
> lines(x,rd$ci$u,lty=3)
> abline(h=1,lty=2)
```

The graph in Figure 3.15 shows no substantial deviation from the hypothesised normal distribution. This conclusion is supported by the observation that the line $r(x) = 1$ lies well in between the (pointwise) confidence bands.

The next R-code provides the comparison density based on the minimum AIC bandwidth selector (option smooth=-1). The result is shown in the lower panel of Figure 3.15. Again the null hypothesis seems to be supported, but now the comparison density is less smooth (this seems to be a general characteristic of the AIC-based bandwidth selection). The upper limit of the confidence band goes slightly below $r(x) = 1$ for values of $x$ around 270. For these values, the plot suggests weakly that the frequency of PCB concentrations is smaller than expected under the hypothesis of normality. This can also be seen from the (smoothed) histogram in the upper panel in Figure 3.15.

```
> rd<-reldist(y=PCB,yo=qnorm((1:n+0.5)/(n+1),sd=sd.PCB,
+     mean=m.PCB),ci=T,graph=F,smooth=-1)
Smoothing using 5
> x<-qnorm(rd$x,sd=sd.PCB,mean=m.PCB)
> plot(x,rd$y,type="l",xlab="PCB",ylab="relative density",
+     ylim=c(0,2))
> lines(x,rd$ci$l,lty=3)
> lines(x,rd$ci$u,lty=3)
> abline(h=1,lty=2)
```

Inasmuch the mean and the variance of the hypothesised normal distribution are estimated from the data, it makes no sense to perform the decomposition of $r(u)$. Finally note that this example illustrates that the conclusions depend largely on the choice of the bandwidth.

### *3.3.3 Comparison Distribution for Discrete Data*

Suppose the random variable $X$ is discrete and it takes values in the ordered set $\{x_1, \ldots, x_m\}$ ($x_1 < \cdots < x_m$), where $m > 1$ may be infinite (e.g., for a Poisson distribution). We assume that the distributions $f$ and $g$ have the same outcome set. We use the notation

$$f_i = f(x_i) = \Pr_f\{X = x_i\} \text{ and } g_i = g(x_i) = \Pr_g\{X = x_i\}.$$

The CDF of $g$ is given by

$$G(x) = \sum_{i:x \leq x_i} g_i,$$

which is a step function with jumps of size $g_i$ at position $x_i$. Therefore, the PIT transformation $U = G(X)$ is not appropriate here. To make $G$ a continuous transformation, we define $G(x_0) = 0$ for any $x_0 < x_1$ and

$$G_d(x) = U\left[G(x_{i-1}), G(x_i)\right] \text{ for } x_{i-1} < x \leq x_i, \; i = 1, \ldots, m,$$

where $U[a, b]$ is a uniform random variable over the interval $[a, b]$. With this definition, $G_d$ is a continuous transformation, and $U = G_d(X)$ has a continuous distribution. Moreover, $X$ has distribution $g$ if and only if $U = G_d(X)$ is uniformly distributed with density function $r(u) = 1$ for $0 \leq u \leq 1$. Based on this argument, it again makes sense to define the *discrete comparison density* for a discrete random variable as the density of $U = G_d(X)$. It can be shown that this density function is a step function given by

$$r(u) = \frac{f_i}{g_i} \text{ for } G(x_{i-1}) < u \leq G(x_i) \text{ for } 0 \leq u \leq 1.$$

A natural estimator of $r(u)$ is obtained by replacing $f_i$ by its empirical probability estimator,

$$\hat{f}_i = \hat{f}(x_i) = \frac{1}{n} \sum_{j=1}^{n} \mathrm{I}\left(X_j = x_i\right).$$

The estimator of the discrete comparison density then becomes

$$\hat{r}(u) = \frac{\hat{f}_i}{g_i} \text{ for } G(x_{i-1}) < u \leq G(x_i) \text{ for } 0 \leq u \leq 1.$$

When the hypothesised distribution $G$ (and hence $g$) depends on a nuisance parameter $\boldsymbol{\beta}$, then $\boldsymbol{\beta}$ is replaced by an estimator.

*Example 3.7 (Pulse rate).* The pulse rate is measured as the number of pulses per minute. This is thus a discrete variable, and because counts are often distributed as a Poisson distribution, we assess the fit of the data to a Poisson distribution with mean equal to the sample mean ($\bar{x} = 82.3$). The upper panel in Figure 3.16 shows the histogram of the data, the fitted Poisson distribution, and a nonparametric kernel density estimator, and the comparison density is shown in the lower panel. This graph suggests that as compared to a Poisson distribution with mean 82.3, there are far too many counts around

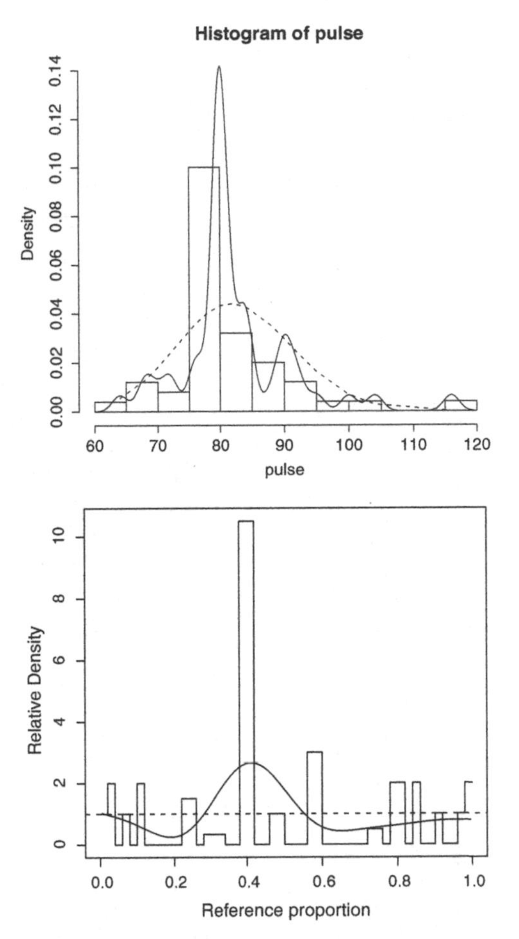


**Fig. 3.16** The histogram, nonparametric density estimator, and fitted Poisson density (upper panel) of the pulse rate data. The lower panel shows the comparison density of the pulse rate data. The horizontal axis is the percentile $u = G(x)$

the 40th percentile of the hypothesized Poisson distribution, i.e., counts close to $G^{-1}(0.4; \hat{\lambda} = 82.3) = 80$.

The R-code for the comparison density is presented below.

```
> pulse<-pulse$Pulse
> pulse<-sort(pulse)
> n<-length(pulse)
> wgt<-rep(1,length(pulse))
> rd<-reldist(y=pulse,yo=qpois((1:n+0.5)/(n+1),lambda=mean
+ (pulse)),discrete=T,ywgt=wgt,yogt=wgt)
```

# Chapter 4
# Smooth Tests

In this chapter we discuss *smooth tests.* This class of tests actually dates back from Neyman (1937), who developed a smooth test as a score test for which he proved some optimality properties. Although smooth tests are considered as nonparametric tests, they are actually constructed by first considering a $k$-dimensional *smooth family of alternatives* in which the hypothesised distribution is embedded. These smooth alternatives are the subject of Section 4.1. The tests are given in Section 4.2. The power of the test depends on how well the true distribution is approximated by the $k$-dimensional smooth alternative. In particular, for each data-generating distribution there exists an optimal order $k$. In Section 4.3 we discuss adaptive smooth tests of which the order is estimated from the data so that often the power is improved. Sections 4.1 up to 4.3 are limited to continuous distributions; smooth tests for discrete distributions are the topic of Section 4.4, and in Section 4.5 we show how smooth tests may be viewed from within a semiparametric framework. Finally, in Section 4.7 we give a brief summary from a practical viewpoint.

## 4.1 Smooth Models

### 4.1.1 Construction of the Smooth Model

The general idea behind smooth tests is to first embed the hypothesised distribution $g$ into a larger family of distributions, say $g_k$, which is indexed by a $k$-dimensional parameter vector $\boldsymbol{\theta}^t = (\theta_1, \dots, \theta_k)$ in such a way that $\boldsymbol{\theta} = \mathbf{0}$ corresponds to the hypothesised distribution. Such a family is called a *smooth model*, or a *smooth order k alternative* to the null. The adjective "smooth" was invented by Neyman (1937) to indicate that these alternatives depart smoothly from the hypothesised density. The exact construction of $g_k$ is important. It should be large enough so that it is likely to include the true

O. Thas, *Comparing Distributions*, Springer Series in Statistics,
DOI 10.1007/978-0-387-92710-7_4, © Springer Science+Business Media, LLC 2010

density $f$. By doing so, the original null hypothesis $H_0 : f(x) = g(x)$ (for all $x$) reduces to $K_0 : \boldsymbol{\theta} = \mathbf{0}$. In some sense, the former null hypothesis requires a nonparametric test because the unknown density $f$ may involve an infinite number of nuisance parameters, whereas the latter is clearly parametric for which actually any parametric test may be considered. The alternative hypothesis is usually formulated as $H_1 : f(x) \neq g(x)$ (for at least one $x$). As it is soon shown, the alternative is implied by $K_1 : \boldsymbol{\theta} \neq \mathbf{0}$; i.e., $K_1 \Rightarrow H_1$, but the implication does not necessarily hold true the other way around. This argument is important later on to understand the power and the limitations of the smooth tests. In particular, when $H_0$ is tested by means of a parametric test for $K_0$, then the rejection of $K_0$ will immediately imply the rejection of the original $H_0$, but when $K_0$ cannot be rejected, it is still possible that $f$ does not equal $g$, even with very large sample sizes. In Section 4.5, where we discuss a semiparametric framework, we come back to the interpretation of tests for the original hypotheses $H_0$ and $H_1$ versus tests for hypotheses $K_0$ and $K_1$.

To keep the exposition general, we allow $g$ to depend on a $p$-dimensional vector $\boldsymbol{\beta}$ of nuisance parameters. At this point only the models are defined, therefore it is not important to know whether $\boldsymbol{\beta}$ is specified (simple null) or has to be estimated (composite null).

We define a first type of an order $k$ family of smooth alternatives given by the density function

$$g_k(x; \boldsymbol{\theta}, \boldsymbol{\beta}) = C(\boldsymbol{\theta}, \boldsymbol{\beta}) \exp\left(\sum_{j=1}^{k} \theta_j h_j(x; \boldsymbol{\beta})\right) g(x; \boldsymbol{\beta}), \tag{4.1}$$

where $C(\boldsymbol{\theta}, \boldsymbol{\beta})$ is a normalisation constant, and $\{h_j\}$ represents a set of orthonormal functions in the Hilbert space $L_2(\mathcal{S}, G)$. See Sections 2.5 and 2.6 for more details on Hilbert spaces and orthonormal functions. The orthogonality condition has two important consequences: (1) in Section 4.2 we show that the smooth test statistic is often decomposable in asymptotically independent components that say something about how the true distribution may depart from the hypothesised distribution; (2) it allows the construction of a flexible $k$-order family of distributions with a minimal number of uniquely identifiable $\theta$-parameters.

Note that model (4.1) is only well defined if the functions $h_j$ $(j = 1, \ldots, k)$ are bounded in $L_2(\mathcal{S}; G)$. In his original paper, Neyman (1937) actually only considered testing for uniformity, i.e., $g(x) = 1$ for all $x \in [0, 1]$, and so the $h_j$ are the Legendre polynomials over $[0, 1]$, which are bounded functions. Although any distribution may be transformed to a uniform, we show in the next section that taking $g$ explicitly into account, as in (4.1), results in a more informative assessment. Thus, methods based on model (4.1) are actually extensions of the work of Neyman. They were first introduced by Rayner and Best (1986). The book of Rayner et al. (2009) is completely devoted to smooth tests of this class.

A similar model has been given by Barton (1953). He defines the order $k$ smooth alternative as

$$g_k(x;\boldsymbol{\theta},\boldsymbol{\beta}) = \left(1 + \sum_{j=1}^{k} \theta_j h_j(x;\boldsymbol{\beta})\right) g(x;\boldsymbol{\beta}), \tag{4.2}$$

where $\{h_j\}$ is as before. This model has the advantage that no normalisation constant is needed. On the other hand, the model does not guarantee that $g_k(x)$ is positive for all $x \in \mathcal{S}$. A possible solution consists in restricting the parameter space of $\boldsymbol{\theta}$, but this usually brings many complications along. Other solutions have been discussed by, e.g., Gajek (1986), Glad et al. (2003), Hall and Murison (1993), and Kaluszka (1998). See Section 2.8.2 for details on the method of Gajek (1986).

From Section 2.8 we know that in the nonparametric density estimation literature models (4.2) and (4.1) are known as an *orthonormal series expansion* and a *log-linear orthogonal series expansion*. However, in the present context we prefer to call them the Barton and the Neyman model, respectively.

Both the Neyman and the Barton alternatives may be motivated from a function approximation theory point of view within the framework of Hilbert spaces. As before, $f$ denotes the true density function, and suppose $f, g \in L_2(\mathcal{S};G)$, where $L_2(\mathcal{S};G)$ is a Hilbert space spanned by a system of orthonormal functions $h_0, h_1, \ldots$, where for later convenience $h_0 \equiv 1$.

Rather than by applying the expansion directly to the density function $f$, the Barton model appears when it is applied to the comparison density $f(x)/g(x)$,

$$\frac{f(x)}{g(x)} = \sum_{j=0}^{\infty} \theta_j h_j(x) = \theta_0 + \sum_{j=1}^{\infty} \theta_j h_j(x). \tag{4.3}$$

Because $\theta_0 = 1$ guarantees that $\int_{\mathcal{S}} f(x)dx = 1$, we further restrict the discussion to

$$\frac{f(x)}{g(x)} = 1 + \sum_{j=1}^{\infty} \theta_j h_j(x). \tag{4.4}$$

With this choice, we find $f(x) = (1+\sum_{j=1}^{\infty} \theta_j h_j(x))g(x)$, which is the Barton model (4.2) with order $k \to \infty$. The Barton density $g_k$ is thus to be interpreted as an expansion truncated at some finite order $k$. Note that the $\theta$s in (4.4) are still found as $\theta_j = < h_j, (f/g) >_g$, i.e. they are the orthogonal projections of the comparison density $f/g$ onto the basis functions of the Hilbert space $L_2(\mathcal{S};G)$. The $\theta$ parameters in the Barton model have also an interpretation in terms of the *Pearson $\phi^2$ measure* (Lancaster (1969), pp. 86–87) which is defined as

$$\phi^2 = \int_{\mathcal{S}} \frac{(f(x) - g(x))^2}{g(x)} dx.$$

Note that $\phi^2$ measures in some sense how far $f$ and $g$ are apart, but because it is asymmetric it is not a real distance measure. In information theory this is called an *information divergence.* Also note that if $X$ were a discrete random variable, then $\phi^2$ would be similar to the Pearson $\chi^2$ statistic. Straightforward algebra shows

$$\begin{aligned}
\phi^2 &= \int_S \left(\frac{f(x)-g(x)}{g(x)}\right)^2 g(x)dx \\
&= \int_S \left(\sum_{j=1}^{\infty} \theta_j h_j(x)\right)^2 g(x)dx \\
&= \sum_{i=1}^{\infty}\sum_{j=1}^{\infty} \theta_i\theta_j \int_S h_i(x)h_j(x)g(x)dx \\
&= \sum_{j=1}^{\infty} \theta_j^2. \qquad (4.5)
\end{aligned}$$

This relationship is known as *Parseval's relation.* In the present context of goodness-of-fit testing, this was also recognised by Eubank et al. (1987).

A further argument which shows the usefulness of (4.5) is obtained by having a closer look at the Argmin definition of the $\theta$s in the truncated series representation, which, applied to $f/g$, gives the following. The solution $\theta_j = < h_j, f/g >_g$ minimises

$$\begin{aligned}
&\left\|\frac{f}{g} - \left(1+\sum_{j=1}^{k}\theta_j h_j\right)\right\|_g^2 \\
&= \int_S \frac{\left(f(x) - (1+\sum_{j=1}^{k}\theta_j h_j(x))g(x)\right)^2}{g(x)}dx \\
&= \int_S \frac{(f(x) - g_k(x))^2}{g(x)}dx. \qquad (4.6)
\end{aligned}$$

If we now replace $f(x)$ by the $k=\infty$ Barton model (4.4), we find

$$\int_S \frac{\left(\sum_{j=k+1}^{\infty}\theta_j h_j(x)g(x)\right)^2}{g(x)}dx = \sum_{j=k+1}^{\infty}\theta_j^2. \qquad (4.7)$$

Hence, the $\theta$ parameters are the solutions to minimising the Pearson $\phi^2$ measure, and the minimised $\phi^2$ is the sum of the squared $\theta$ parameters not included in the order $k$ Barton model. Similarly, $r_k(x) = 1+\sum_{j=1}^{k}\theta_j h_j(x)$ is the orthogonal projection of $f/g$ onto the subspace $\mathcal{P}_k$ spanned by $1, h_1, \ldots, h_k$. It is the comparison density in $\mathcal{P}_k$ that is closest to the true $f/g$. The orthogonal complement of $\mathcal{P}_k$ in $L_2(\mathcal{S};G)$, denoted by $\mathcal{P}_k^{\perp}$, is spanned by $h_{k+1},\ldots$. The squared length of this residual $r(x)-r_k(x)$, which lies in $\mathcal{P}_k^{\perp}$, is exactly (4.7).

Also the Neyman model is an example of a truncated expansion. Consider the log comparison distribution

$$\log\frac{f(x)}{g(x)}=\sum_{j=1}^{\infty}\theta_j h_j(x), \tag{4.8}$$

where again $\{h_j\}$ is a set of orthonormal function w.r.t. $g$. Truncating expansion (4.8) at order $k$ results in an approximation of $f$ given by

$$\exp\left(\sum_{j=1}^{\infty}\theta_j h_j(x)\right)g(x), \tag{4.9}$$

which is, however, not necessarily normalised. This explains the presence of the normalisation constant $C$ in the Neyman model. The $\theta$s are now given by $\theta_j=<h_j,\log(f/g)>_g$ which minimise

$$\left\|\log\frac{f}{g}-\sum_{j=1}^{k}\theta_j h_j\right\|_g^2=\int_{\mathcal{S}}\left(\log f(x)-\log\tilde{g}_k(x)\right)^2 g(x)dx, \tag{4.10}$$

where $\tilde{g}_k(x)=g(x)\exp(\sum_{j=1}^{k}\theta_j h_j(x))$, which is the nonnormalised Neyman model as in Equation (4.9). This formulation is as expected when function approximation is the only intention, but here we want to approximate $f$ by a true density function $g_k$ which is related to $\tilde{g}_k$ by $g_k(x)=C(\boldsymbol{\theta})\tilde{g}_k(x)$. Equation (4.10) may be written as

$$\begin{aligned}\left\|\log\frac{f}{g}-\sum_{j=1}^{k}\theta_j h_j\right\|_g^2&=\int_{\mathcal{S}}\left(\log f(x)-\log\frac{g_k(x)}{C(\boldsymbol{\theta}}\right)\right)^2 g(x)dx\\&=\int_{\mathcal{S}}\left(\log f(x)-\log g_k(x)\right)^2 g(x)dx\\&\quad+\left(\log C(\boldsymbol{\theta})\right)^2.\end{aligned} \tag{4.11}$$

Replacing $f(x)$ in (4.11) with the $k=\infty$ Neyman models we find

$$\left\|\log\frac{f}{g}-\sum_{j=1}^{k}\theta_j h_j\right\|_g^2=\sum_{j=k+1}^{\infty}\theta_j^2+\left(\log C(\boldsymbol{\theta})\right)^2,$$

which is, up to the $(\log C)^2$ term, equal to (4.7). The second term is here to be interpreted as a penalty term to make $g_k$ a proper normalised density function. Both terms are functions of $\theta_1,\ldots,\theta_k$.

It is also worth noting that the Barton model is basically a first-order approximation of the Neyman model obtained by a series expansion of the exponential function.

## 4.2 Smooth Tests

### 4.2.1 Simple Null Hypotheses

#### 4.2.1.1 Test Statistics and Null Distributions

In this section we restrict the problem to testing a simple null hypothesis; i.e., we assume that the parameter $\boldsymbol{\beta}$ indexing the hypothesised distribution $g(x;\boldsymbol{\beta})$ is completely known. Examples include testing for a standard normal distribution ($\boldsymbol{\beta}^t = (\mu,\sigma) = (0,1)$) and testing for the uniform distribution over $[0,1]$). In the next section, we extend the methodology to testing composite null hypotheses where $\boldsymbol{\beta}$ has to be estimated from the sample data.

As explained in Section 4.1, Neyman (1937) started with embedding the hypothesised density function in an order $k$ smooth alternative. In doing so, $k$ embedding parameters $\theta_j$ ($j = 1,\ldots,k$) are introduced, and the original null hypothesis corresponds to $\boldsymbol{\theta} = \mathbf{0}$. Neyman argued that if the true density is a member of this order $k$ alternative, an optimal testing procedure is based on the score test for testing $\boldsymbol{\theta} = \mathbf{0}$. Thus he reduced a genuine nonparametric testing problem to a finite-dimensional parametric testing problem. The next theorem is about this score test. Because the score test based on the Barton model is exactly the same, we combine both into one theorem. This particular score test is called the smooth test. The proof of the theorem is given in Section A.3.

**Theorem 4.1.** *Let $X_1,\ldots,X_n$ denote a sample of i.i.d. observations which have under the null hypothesis density function $g$, and let $\{h_j\}$ denote a set of orthonormal functions w.r.t. $g$.*

*(1) The score test statistic for testing $H_0 : \boldsymbol{\theta} = \mathbf{0}$ in the order $k$ Neyman smooth model is given by*

$$T_k = \sum_{j=1}^{k} U_j^2,$$

*where $U_j = (1/\sqrt{n})\sum_{i=1}^{n} h_j(X_i)$. Let $\boldsymbol{U}^t = (U_1,\ldots,U_k)$. Under the null hypothesis, as $n \to \infty$,*

$$\mathrm{E}_0\left\{\boldsymbol{U}\right\} = \mathbf{0} \text{ and } \mathrm{Var}_0\left\{\boldsymbol{U}\right\} = \boldsymbol{I}$$

*and*

$$\boldsymbol{U} \xrightarrow{d} MVN(0,\boldsymbol{I}) \text{ and } T_k \xrightarrow{d} \chi_k^2.$$

*(2) The score test statistic for testing $H_0 : \boldsymbol{\theta} = \mathbf{0}$ in the order $k$ Barton smooth model is exactly the same as presented in part (1) of this theorem.*

We often write $T_k = \boldsymbol{U}^t\boldsymbol{U}$, where $\boldsymbol{U}^t = (U_1, \ldots, U_k)$. The individual $U_i$ (or $U_i^2$) are referred to as the *components* of the test statistic. A test based on a single component is called a *component test.*

#### 4.2.1.2 Interpretation of Components

In the previous section we have seen that the test statistic has the representation $T_k = \sum_{j=1}^k U_j^2$, and the components $U_j$ are asymptotically i.i.d. under the null hypothesis. The decomposition is particularly interesting when the behaviour of $T_k$ is studied under alternatives. The null hypothesis is rejected for large values of $T_k$, or, equivalently, for large values of the squared components $U_j^2$ ($j = 1, \ldots, k$), thus we may try to identify the distributions $F \neq G$ for which one or more components are expected to be very distinct from zero. The expected value of the $j$th order component is given by

$$\mathrm{E}_f\left\{U_j\right\} = \sqrt{n}\int_{\mathcal{S}} h_j(x)f(x)dx = \sqrt{n}\left\langle \frac{f}{g}, h_j \right\rangle_g, \tag{4.12}$$

which is up to a factor $\sqrt{n}$ equal to the length of the orthogonal projection of the comparison density onto the $j$th order function $h_j$ in $L_2(\mathcal{S}, G)$. When the Barton model is adopted we saw in Section 4.1.1 that this expectation equals $\theta_j$. A deeper interpretation depends on the choice of the orthonormal system $\{h_j\}$. In the next section a detailed discussion is given for the cases where the $h_j$ are polynomials. Polynomial-based smooth tests form the largest class of smooth tests. There is, however, one very important nonpolynomial-based smooth test, which appears often in the important situation of testing for uniformity. Because all random variables can be transformed to have a uniform distribution by the PIT (Section 2.4) the uniform distribution plays a very important role. Therefore we spend the next paragraph on the interpretation of the nonpolynomial Fourier basis.

For the uniform density $g(x) = 1$ for $x \in [0, 1]$, we find, for instance, the cosine basis, $h_j(x) = \sqrt{2}\cos(j\pi x)$. Suppose the data are not uniformly distributed, but they have CDF $F \neq G$ instead; then the $j$th order component has expectation

$$\mathrm{E}_f\left\{U_j\right\} = \sqrt{2n}\int_0^1 \cos(j\pi x)f(x)dx = \sqrt{2n}\left\langle \frac{f(x)}{g(x)}, \cos(j\pi x) \right\rangle_g .$$

It is proportional to the length of the orthogonal projection of the comparison density onto the basis function $\cos(j\pi x)$. A large expectation thus appears when $f(x)/g(x)$ shows some kind of oscillation with period equal to $2/j$. Then we may expect that *slowly oscillating* alternatives with period larger than $2/k$

can be detected with the order $k$ smooth test, but *fast oscillating* alternatives with period smaller than $2/k$ may be missed. A similar interpretation is also common in lack-of-fit tests for regression functions. We specifically refer to Hart (1997) because he often uses the same basis functions, and he also uses the term *oscillating* for characterising alternatives.

#### 4.2.1.3 Interpretation of Components when Orthonormal Polynomials Are Used

One of the reasons that explain the popularity of smooth tests is the interpretability of the components $U_j$ when orthonormal polynomials are used for their construction. As we illustrate later these components are often considered as diagnostic in terms of moment deviations. In particular, when the component test based on $U_j$ rejects the null hypothesis $\theta_j = 0$, it is often concluded that the true distribution deviates from the hypothesised distribution in the $j$th moment. Later, however, we show that this reasoning is slightly oversimplified, and that some more care is needed, but at this point we first give some arguments supporting the moment interpretation of component tests.

**Lemma 4.1.** *Let $\mu_m$ denote the mth central moment of the hypothesised distribution, and let $\mu = \mathrm{E}_0\left\{X\right\}$. Suppose the order k Barton model holds. Then, with $m \leq k$,*

$$\theta_m = 0 \Leftrightarrow \mathrm{E}_k\left\{(X-\mu)^m\right\} = \mu_m.$$

The proof of this lemma is given in Appendix A.4. This result only shows that $\theta_j$ is related to the $j$th moment, but it still does not relate the component $U_j$ to the $j$th moment. The next lemma sheds some light on $U_j$ through the polynomial $h_j$. The proof can be found in Appendix A.5.

**Lemma 4.2.** *Let $\{h_j\}$ denote a set of orthonormal polynomials w.r.t. the hypothesised density g, and let $h_0(x) = 1$. Then, for all j, there exists a set of constants $\{c_{ij}\}$ with $c_{jj} \neq 0$ so that*

$$h_j(x) = \sum_{i=1}^{j} c_{ij}\left[(x-\mu)^i - \mu_i\right]. \tag{4.13}$$

To get a deeper insight into the use of a component test based on $U_j$ for arriving at conclusions in terms of moment deviations, we need to say something in more detail about the null hypothesis. Neyman started with approximating the true density $f$ with a smooth order $k$ alternative $g_k$. For simplicity, we even assume that $f$ belongs to this order $k$ alternative. In doing so, the null hypothesis $H_0 : f = g$ is reduced to $H_0 : \theta_1 = \cdots = \theta_k = 0$. In Theorem 4.1 it was shown that under this null hypothesis, all $U_j$ have asymptotically standard normal null distributions. Let us first focus on the

the mean of $U_j$. Of course, under $H_0$, we have $\mathrm{E}_0\{U_j\} = 0$. Now suppose that exactly one moment does not agree with $g$, say the $m$-th moment. Then Lemma 4.2 shows that $\mathrm{E}\{U_m\} \neq 0$, but also other $U_j$ $(j > m)$ do not have necessarily mean zero. Thus, having a significant $U_m$ component test only proves that at least one of the first $m$ moments differs from what is expected under $H_0$. Later we refine this reasoning by taking also the variance of $U_m$ into account.

From this argument we may deduce a practical rule: first test the overall null hypothesis that all $k$ $\theta$s are zero with the order $k$ smooth test statistic $T_k$. When this test rejects the null hypothesis at the $\alpha$ level of significance, it is informative to look at the components $U_1, \ldots, U_k$ and the corresponding $p$-values, say $p_1, \ldots, p_k$. Suppose $p_1, \ldots, p_{m-1} \geq \alpha$, and $p_m < \alpha$ (the other $p$-values do not matter at this point); then it may be concluded that the distribution $f$ follows $g$ in its first $m-1$ moments, but not in the $m$th moment. Suppose now that also higher-order component tests $(U_j, j > m)$ are significant; then it is not necessarily true that the corresponding higher-order moments deviate from what is hypothesised. However, if for a given distribution $g$, one finds how the polynomials $h_j$ are turned into the form presented in Lemma 4.2, and if many of the constants $c$ in Equation (4.13) are zero, then a clearer interpretation is possible. Table 4.1 shows these $c$ coefficients for the normal distribution.

*Example 4.1 (Testing for a standard normal distribution).* Suppose we have to test the null hypothesis that $f$ is a standard normal distribution. Table 4.1 gives for the normal distribution the $c$ coefficients of the Hermite polynomials. We conclude:
(1) For the second-order component: because $c_{12} = 0$, this component is insensitive to the wrong specification of the mean.
(2) For the third-order component: because $c_{23} = 0$, this component is insensitive to the wrong specification of the variance, but because $c_{13} \neq 0$, it is sensitive to the specification of the mean.
(3) And so on.
For the normal distribution it turns out that components of odd (even) degree are only sensitive to deviations in odd (even) moments.

The discussion of the previous paragraphs was based on the assumption that $f$ belongs to a finite-order $k$ family of smooth alternatives to $g$. We

**Table 4.1** The coefficients $c_{ij}$ as in Equation (4.13) for the normal distribution

| Degree of Polynomial | | | | |
|---|---|---|---|---|
| 1 | 2 | 3 | 4 | 5 |
| $c_{11} = 1$ | $c_{12} = 0$ | $c_{13} = -3$ | $c_{14} = 0$ | $c_{15} = 15$ |
| | $c_{22} = 1$ | $c_{23} = 0$ | $c_{24} = -6$ | $c_{25} = 0$ |
| | | $c_{33} = 1$ | $c_{34} = 0$ | $c_{35} = -10$ |
| | | | $c_{44} = 1$ | $c_{45} = 0$ |
| | | | | $c_{55} = 1$ |

now relax this assumption. Thus, now $g_k$ is only an approximation to the true $f$. As in Klar (2000), we say that the original null hypothesis $H_0 : f = g$ is the *full parametric null hypothesis* in the sense that it implies that all moments of $f$ and $g$ are equal. When $f$ is approximated by $g_k$ with a finite-order $k$, the null hypothesis $H_0 : \theta_1 = \cdots = \theta_k = 0$ is called the *semiparametric* null hypothesis, because the smooth test $T_k$ is only consistent against alternatives having moments of order $\leq k$ in disagreement with the density $g$ (this is a direct consequence of Lemma 4.2). In this case we no longer call $g$ the hypothesised distribution. It only serves as a moment-generating density and only the first $k$ of these moments are part of the semiparametric null hypothesis.

In a series of papers (Henze (1997), Henze and Klar (1996), Klar (2000)), Henze and Klar went one step further in examining the *diagnostic properties* of the component tests. Instead of only looking at the mean of $U_j$, as we have done above, they also studied the relation between the variance of $U_j$ and moment deviations.

Under the parametric null hypothesis, Theorem 4.1 states $\text{Var}_0\left\{U_j\right\} = 1$. To explain the arguments of Henze and Klar, we first take a closer look at the variance. Some results are summarised in the following lemma, the proof of which is given in Appendix A.6.

**Lemma 4.3.** *(1) If $f$ agrees with $g$ in all first $2j$ moments, then* $\text{Var}\left\{U_j\right\}=1$*; (2) if $f$ disagrees with $g$ in at least one moment of degree $\leq 2j$, and let $m$ denote the smallest order of such moments, then*

$$\text{Var}\left\{U_j\right\} = \begin{cases} 1 - \left(\text{E}_f\left\{h_j(X)\right\}\right)^2 + \sum_{l=m}^{2j} c_l \, \text{E}_f\left\{h_l(X)\right\} & \text{if } m \leq j \\ 1 + \sum_{l=m}^{2j} c_l \, \text{E}_f\left\{h_l(X)\right\} & \text{if } j < m \leq 2j \\ 1 & \text{if } 2j < m \;, \end{cases} \tag{4.14}$$

*where $c_m, \ldots, c_{2j}$ are constants which are not necessarily zero.*

Henze and Klar went further by showing that for a wide class of alternatives, say $\mathcal{A}$,

$$\sup_{f \in \mathcal{A}} \text{Var}_f\left\{U_j\right\} = +\infty \quad \text{and} \quad \inf_{f \in \mathcal{A}} \text{Var}_f\left\{U_j\right\} = 0.$$

With this information we come to a quite drastic conclusion: within the class $\mathcal{A}$, the infimum of the power function is zero; i.e., $\inf_{f \in \mathcal{A}} \Pr_f\left\{|U_j| > z_{1-\alpha/2}\right\} = 0$. Based on this extreme result, Henze and Klar concluded that $U_j$ is never guaranteed to have *diagnostic properties* w.r.t. moment differences. Their solution to the problem consists in replacing the asymptotic variance of $U_j$, which is, as for all score tests, determined under the full parametric null hypothesis, by the empirical variance estimator,

$$S_j^2 = \frac{1}{n} \sum_{i=1}^{n} h_j^2(X_i).$$

The intuitive explanation is easy: $S_j^2$ is a consistent estimator of the variance, both under the null and the alternative hypothesis, and its correctness does not depend on any other moment restriction.

In doing so, they showed that the asymptotic null distribution of $U_j$ is not changed (it is basically the replacement of a variance by a consistent estimator).

The use of the empirical variance estimator is particularly important when the individual components are used in a diagnostic manner, but the idea may also be applied to the order $k$ smooth test,

$$T_k = \boldsymbol{U}^t \hat{\boldsymbol{\Sigma}}^{-1} \boldsymbol{U},$$

where $\hat{\boldsymbol{\Sigma}} = (1/n)\sum_{i=1}^n \boldsymbol{h}(X_i)\boldsymbol{h}^t(X_i)$, and $\boldsymbol{h}^t(x) = (h_1(x), \ldots, h_k(x))$. Another, but similar solution is proposed by Chervoneva and Iglewicz (2005). They suggest to estimate $\boldsymbol{\Sigma}$ with a U-statistic based on a symmetric kernel of degree two. This estimator is slightly more computationally intensive, but they prove that under mild assumptions their estimator is optimal in the sense that it is minimum variance unbiased and $n^{1/2}$-consistent.

Replacing $\boldsymbol{\Sigma}$ with its empirical estimator $\hat{\boldsymbol{\Sigma}}$ results in a slowing down of the convergence of $T_k$ to its asymptotic null distribution. The parametric bootstrap is not an option here, because it again implies that the full parametric null hypothesis is true. Bickel and Ren (2001) suggested a modification of the nonparametric bootstrap that forces the simulated null distribution of the test statistic to be centred as it would be expected if the semiparametric null hypothesis holds. The method is explained in Appendix B.3. Finally, we mention that, despite the correct criticism of Henze and Klar, we have the experience that in most situations the traditionally standardised component tests are quite good in detecting the right moment deviations, particularly when the true distribution is not too distinct from the hypothesised.

In this section on the simple null hypothesis, we present only one example. More examples are given later for the more interesting situation in which a nuisance parameter is to be estimated. We also see that for composite hypotheses the problem of the diagnostic property becomes more complicated.

*Example 4.2 (Pseudo-random generator data).* In the cd package, the R function smooth.test may be used to perform smooth tests. Here we consider $k = 4$. Because the null hypothesis of uniformity is to be tested, the smooth test is based on the Legendre polynomials. For the PRG data the output is given below.

```
> smooth.test(PRG,order=4,distr="unif",B=NULL)
Smooth goodness-of-fit test
Null hypothesis:  unif  against  4 th order alternative
Nuisance parameter estimation:  NONE
Parameter estimates:  no parameter estimation necessary
```

```
Smooth test statistic S_k =  4.402007  p-value =  0.3543
      1 th component V_k = -1.026668  p-value =  0.3046
      2 th component V_k = -0.324184  p-value =  0.7458
      3 th component V_k = -1.442003  p-value =  0.1493
      4 th component V_k = -1.078652  p-value =  0.2807

All p-values are obtained by the asymptotic approximation
```

Clearly, the null hypothesis is accepted. Thus, it is not informative to examine the individual components.

## 4.2.2 Composite Null Hypotheses

### 4.2.2.1 Maximum Likelihood and Method of Moments Estimators

In most realistic situations the null hypothesis only specifies a family of distributions, indexed by a $p$-dimensional parameter vector $\boldsymbol{\beta}^t = (\beta_1, \ldots, \beta_p)$ which is referred to as the nuisance parameter. The null hypothesis now becomes $H_0 : F \in \{G(.;\boldsymbol{\beta}) : \boldsymbol{\beta} \in B\}$, where $B$ denotes the parameter space which is an open subset of $\mathbb{R}^p$. The null hypothesis is sometimes written as $H_0 : F(x) = G(x;\boldsymbol{\beta})$ for all $x \in \mathcal{S}$, and it is referred to as the *composite* null hypothesis. Typical examples are the normal distribution indexed by the mean $\mu$ and the variance $\sigma^2$ ($\boldsymbol{\beta}^t = (\mu, \sigma^2)$), the exponential distribution indexed by the rate $\lambda$ ($\beta = \lambda$), etc. An intuitively appealing solution exists in adopting a two-step approach: (1) estimate the nuisance parameters. Let $\hat{\boldsymbol{\beta}}$ denote the estimator. And, (2), proceed as in the simple null hypothesis case, with $\boldsymbol{\beta}$ replaced by $\hat{\boldsymbol{\beta}}$. Up to a certain extent, this is indeed a correct solution. In particular, using $\hat{\boldsymbol{\beta}}$ in the specification of the orthonormal polynomials results in the correct orthonormality criterion of Equation (2.19). Also the statistics $\hat{U}_j = U_j(\hat{\boldsymbol{\beta}}) = (1/\sqrt{n}) \sum_{i=1}^n h_j(X_i;\hat{\boldsymbol{\beta}})$ are meaningful in the sense that $\mathrm{E}\left\{U_j(\hat{\boldsymbol{\beta}})\right\} = \mathrm{E}\left\{U_j(\boldsymbol{\beta})\right\}$ under both the null and the alternative hypotheses. However, an important consequence of imputing nuisance parameter estimators is that the variance–covariance matrix of $\boldsymbol{U}(\hat{\boldsymbol{\beta}})$ is no longer the identity matrix. More specifically, we now very often have $\mathrm{Var}\left\{\boldsymbol{U}(\hat{\beta})\right\} \neq \mathrm{Var}\left\{\boldsymbol{U}(\beta)\right\}$. The exact expression of the variance–covariance matrix depends on the method of estimation. We restrict the discussion to asymptotically linear estimators.

Although much of the theory that we present here is valid for all asymptotically linear estimators, we focus on two particular types of Z-estimators: maximum likelihood estimators (MLE) and method of moments estimators (MME). Both are solutions of estimation equations of the form $\sum_{i=1}^n \boldsymbol{b}(X_i;\hat{\boldsymbol{\beta}}) = \boldsymbol{0}$, where the estimation function $\boldsymbol{b}$ satisfies some regularity conditions (see Section 2.7).

For MLE, $\boldsymbol{b}$ is the score function, and for MME the estimation functions express the equality of the moments of $g$ to the sample moments; i.e., the $j$th estimation function is of the form $b_j(x;\boldsymbol{\beta}) = (x-\mu)^j - \mathrm{E}_0\left\{(X-\mu)^j\right\}$, where $\mathrm{E}_0\{.\}$ denotes the expectation w.r.t. the hypothesised distribution $g(.;\boldsymbol{\beta})$. The next lemma shows an important consequence of the use of MME in smooth tests.

**Lemma 4.4.** *Without loss of generality we set $\mu = 0$. If the p-dimensional nuisance parameter $\boldsymbol{\beta}$ is estimated by means of MME, i.e., $\hat{\boldsymbol{\beta}}$ is the solution to the estimation equations*

$$\sum_{i=1}^{n} b_j(X_i;\hat{\boldsymbol{\beta}}) = \sum_{i=1}^{n} X_i^j - n\,\mathrm{E}_{g(.;\hat{\beta})}\left\{X^j\right\} = 0 \quad j = 1,\ldots,p, \tag{4.15}$$

*then $U_j(\hat{\beta}) \equiv 0$ $(j = 1,\ldots,p)$ with probability one.*

This lemma is almost a direct consequence of Lemma 4.2 which shows that $U_j(\hat{\beta}) = \sum_{i=1}^{n} h_j(x;\hat{\boldsymbol{\beta}})$ is a linear combination of contrasts between the first $j$ sample moments and the matching moments of $g$. All these contrasts are exactly zero according to (4.15).

This lemma suggests that it makes no sense to include the first $p$ components $\hat{U}_j$ in the construction of a goodness-of-fit test statistic. Or, put in another way, the $p$ first $\theta$ parameters in the smooth alternatives of (4.1) or (4.2) can be omitted because their role is replaced by the nuisance parameters $\boldsymbol{\beta}$. It also has consequences for the interpretation of a MME-based smooth test. First, no deviations in the first $p$ moments can be detected because the density $g(.;\hat{\boldsymbol{\beta}})$ fits exactly in terms of these moments. When one of the higher-order component tests turns out significant, a higher-order moment interpretation can be given, but always conditional on an exact fit of the first $p$ moments. Within a semiparametric framework for smooth tests, Klar (2000) gives some further arguments that bring him to the conclusion that MME is the only meaningful estimation method in goodness-of-fit testing. More details follow in Section 4.5.

As illustrated in the examples to come, sometimes MME and MLE coincide. Klar (2000) showed that this always occurs when $g$ belongs to a subclass of the exponential family for which the sufficient statistics are polynomial in the observations (e.g., the normal and the exponential distributions, but not the gamma distribution).

*Example 4.3 (MLE and MME in the normal distribution).* For the normal distribution with nuisance parameters $\boldsymbol{\beta}^t = (\mu, \sigma^2)$, the score functions are

$$\boldsymbol{u}_\beta(x;\boldsymbol{\beta}) = \frac{\partial \log g(x;\boldsymbol{\beta})}{\partial \boldsymbol{\beta}} = \begin{pmatrix} \frac{x-\mu}{\sigma^2} \\ \frac{(x-\mu)^2}{\sigma^4} - \frac{1}{\sigma^2} \end{pmatrix} = \begin{pmatrix} u_\mu(x) \\ u_\sigma(x) \end{pmatrix}.$$

So we have $b_\mu = u_\mu$ and $b_\sigma = u_\sigma$, but usually the estimation equations are simplified to $b_\mu(x) = x - \mu = 0$ and $b_\sigma(x) = (x-\mu)^2 - \sigma^2 = 0$. Although strictly speaking these are no longer true *score functions*, we often do not make a distinction in terminology when they are used in the context of estimation equations.

Because $\mathrm{E}\{X\} = \mu$ and $\mathrm{Var}\{X\} = \sigma^2$, the MME estimation equations are directly given by

$$b_\mu(x) = x - \mu = 0 \text{ and } b_\sigma(x) = (x-\mu)^2 - \sigma^2 = 0.$$

*Example 4.4 (MLE and MME in the logistic distribution).* A logistic distribution is a symmetric distribution with density function

$$g(x;\boldsymbol{\beta}) = \frac{\exp(-(x-\mu)/\sigma)}{\sigma\left(1+\exp(-(x-\mu)/\sigma)\right)^2} \text{ for } -\infty < x < +\infty,$$

where $\boldsymbol{\beta}^t = (\mu, \sigma)$ contains a location parameter $\mu$ and a scale parameter $\sigma$. The MLE estimation functions are given by

$$b_\mu(x) = \frac{1}{2} - \frac{\exp\left(-\frac{x-\mu}{\sigma}\right)}{\left(1+\exp\left(-\frac{x-\mu}{\sigma}\right)\right)^2}$$

$$b_\sigma(x) = \sigma - (x-\mu)\frac{1-\exp\left(-\frac{x-\mu}{\sigma}\right)}{1+\exp\left(-\frac{x-\mu}{\sigma}\right)}.$$

The MLE estimation equations $b_\mu = b_\sigma = 0$ need to be solved iteratively.

Because $\mathrm{E}\{X\} = \mu$ and $\mathrm{Var}\{X\} = (\pi^2\sigma^2)/3$, we find the MME estimation functions

$$b_\mu(x) = x - \mu \text{ and } b_\sigma(x) = (x-\mu)^2 - \frac{\pi^2}{3}\sigma^2.$$

Now MME and MLE are clearly distinct. The MME have explicit solutions,

$$\tilde{\mu} = \bar{X} \text{ and } \tilde{\sigma} = \frac{\sqrt{3}}{\pi}\sqrt{\frac{1}{n}\sum_{i=1}^n (X_i - \tilde{\mu})^2}.$$

#### 4.2.2.2 The Efficient Score Test

When nuisance parameters are estimated by their MLE (i.e., $\boldsymbol{b} = \boldsymbol{u}_\beta$), score tests are usually based on the efficient score function

$$\boldsymbol{v}(x) = \boldsymbol{h}(x;\boldsymbol{\beta}) - \boldsymbol{\Sigma}_{h\beta}\boldsymbol{\Sigma}_{\beta\beta}^{-1}\boldsymbol{u}_\beta(x;\boldsymbol{\beta}), \tag{4.16}$$

which in a Hilbert space is interpreted as the orthogonal projection of $\boldsymbol{h}$ on the orthogonal complement space of $\boldsymbol{u}_\beta$; i.e.,

$$\boldsymbol{v} = \boldsymbol{h} - \frac{<\boldsymbol{h}, \boldsymbol{u}_\beta>_g}{<\boldsymbol{u}_\beta, \boldsymbol{u}_\beta>_g}\boldsymbol{u}_\beta = \boldsymbol{h} - \left\langle \boldsymbol{h}, \frac{\boldsymbol{u}_\beta}{||\boldsymbol{u}_\beta||_g}\right\rangle_g \frac{\boldsymbol{u}_\beta}{||\boldsymbol{u}_\beta||_g}.$$

In the Hilbert space $\boldsymbol{h}$ represents the direction of the alternative for which the smooth test has power. This was explained in Section 4.1.1, where it was shown that the comparison density has the representation

$$\frac{f(x;\boldsymbol{\beta})}{g(x;\boldsymbol{\beta})} = 1 + \sum_{j=1}^{k} \theta_j h_j(x;\boldsymbol{\beta}) \tag{4.17}$$

in the $k$-dimensional subspace spanned by $\{h_1,\ldots,h_k\}$. When $\boldsymbol{\beta}$ is held constant at its true value, say $\boldsymbol{\beta}_0$, the density $g(x;\boldsymbol{\beta}_0)$ corresponds to one point in the Hilbert space, but when we let $\boldsymbol{\beta}$ vary in $\mathbb{R}^p$, the set

$$\mathcal{G}_\beta = \left\{ \frac{f(x;\boldsymbol{\beta})}{g(x;\boldsymbol{\beta})} : \boldsymbol{\beta} \in \mathbb{R}^p \right\}$$

represents a line or a (hyper)plane in $L_2(\mathcal{S}, G(.;\boldsymbol{\beta}_0))$, indexed by $\boldsymbol{\beta}$. The space $\mathcal{G}_\beta$ is typically not linear, but a Taylor series expansion may be used for obtaining a local (i.e., for $\boldsymbol{\beta}$ close to $\boldsymbol{\beta}_0$) approximation. In the next paragraph we first illustrate this idea on the comparison density $g(x;\boldsymbol{\beta})/g(x;\boldsymbol{\beta}_0)$, which compares the members within the composite null hypothesis to the true density $g(x;\boldsymbol{\beta}_0)$ under the null hypothesis.

A Taylor expansion gives

$$\begin{aligned} g(x;\boldsymbol{\beta}) &= g(x;\boldsymbol{\beta}_0) + (\boldsymbol{\beta}-\boldsymbol{\beta}_0)^t \frac{\partial g}{\partial \boldsymbol{\beta}}(x;\boldsymbol{\beta}_0) + O((\boldsymbol{\beta}-\boldsymbol{\beta}_0)^t(\boldsymbol{\beta}-\boldsymbol{\beta}_0)) \\ &= g(x;\boldsymbol{\beta}_0) + (\boldsymbol{\beta}-\boldsymbol{\beta}_0)^t \frac{\partial \log g}{\partial \boldsymbol{\beta}}(x;\boldsymbol{\beta}_0) g(x;\boldsymbol{\beta}_0) + O((\boldsymbol{\beta}-\boldsymbol{\beta}_0)^t(\boldsymbol{\beta}-\boldsymbol{\beta}_0)). \end{aligned}$$

Thus, locally (i.e., for $\boldsymbol{\beta}$ close to $\boldsymbol{\beta}_0$), we find approximately

$$\frac{g(x;\boldsymbol{\beta})}{g(x;\boldsymbol{\beta}_0)} = 1 + (\boldsymbol{\beta}-\boldsymbol{\beta}_0)^t \boldsymbol{u}_\beta(x),$$

which is linear in $\boldsymbol{\beta}$, and the score function $\boldsymbol{u}_\beta$ is now interpretable as the vector that spans the subspace that is (locally) consistent with the composite null hypothesis.

We now apply the Taylor expansion to (4.17), but now with $\boldsymbol{\beta} = \boldsymbol{\beta}_0$ and $\boldsymbol{\theta} = \mathbf{0}$. We eventually arrive at

$$\frac{f(x;\boldsymbol{\beta})}{g(x;\boldsymbol{\beta})} = \frac{f(x;\boldsymbol{\beta}_0)}{g(x;\boldsymbol{\beta}_0)} + (\boldsymbol{\beta}-\boldsymbol{\beta}_0)^t \left( \boldsymbol{u}_\beta^f(x) - \boldsymbol{u}_\beta^g(x) \right) + \boldsymbol{\theta}^t \boldsymbol{h}(x;\boldsymbol{\beta}_0),$$

where $\boldsymbol{u}_\beta^f(x)$ and $\boldsymbol{u}_\beta^g(x)$ denote the score functions of $\boldsymbol{\beta}$ w.r.t. $f$ and $g$, respectively. This approximation demonstrates that the comparison density lives in a subspace which is spanned by the $k$-dimensional $\boldsymbol{h}$, but also by the score functions $\boldsymbol{u}_\beta = \boldsymbol{u}_\beta^g$ of the nuisance parameter $\boldsymbol{\beta}$, and the latter actually spans the $p$-dimensional subspace of comparison densities that are consistent with the null hypothesis.

Suppose that a $k$-dimensional $\boldsymbol{h}$ is not orthogonal to $\boldsymbol{u}_\beta$; then not all of the spanned $k$-dimensional subspace is relevant for the alternative. It is therefore more efficient to transform $\boldsymbol{h}$ so that it spans a $k$-dimensional subspace that is exclusively relevant for the alternative, i.e., a subspace that has an empty intersection with the linearised $\mathcal{G}_\beta$ (note: the intersection actually only contains the 1-element). To guarantee this, $\boldsymbol{h}$ is transformed to the orthogonal complement after its orthogonal projection onto $\boldsymbol{u}_\beta$.

Because for MLE $\boldsymbol{b} = \boldsymbol{u}_\beta$ this means that the efficient score function is orthogonal to the estimation equation, which translates to independence in the world of statistics. Despite the independence, Theorem 4.2 (see later) shows that the variance–covariance matrix of $\hat{\boldsymbol{V}}$ is generally not diagonal,

$$\boldsymbol{\Sigma}_{\hat{v}} = \boldsymbol{I}_k - \boldsymbol{\Sigma}_{h\beta}\boldsymbol{\Sigma}_{\beta\beta}^{-1}\boldsymbol{\Sigma}_{\beta h}. \tag{4.18}$$

To obtain a decomposition of $T_k$ into asymptotically independent components the last term in Equation (4.18) must be zero. This happens in the important case where the score function $\boldsymbol{u}_\beta$ and the polynomials $\boldsymbol{h}$ are orthogonal ($<\boldsymbol{u}_\beta, \boldsymbol{h}>_g = \boldsymbol{\Sigma}_{\beta h} = \boldsymbol{0}$). The polynomials form by definition a set of orthogonal functions, therefore a diagonal variance–covariance matrix is obtained when the score functions $\boldsymbol{u}_\beta$ lie within the space spanned by the polynomials $h_j$ not contained in $\boldsymbol{h}$. This happens when $\boldsymbol{u}_\beta$ contains polynomials, i.e., for distributions $g$ for which MLE and MME coincide. Finally, note that with the MLE $\hat{\boldsymbol{\beta}}$ and with the efficient score of Equation (4.16) the statistic $\hat{\boldsymbol{U}}_j = (1/\sqrt{n})\sum_{i=1}^n \boldsymbol{v}(x;\hat{\boldsymbol{\beta}}) = (1/\sqrt{n})\sum_{i=1}^n \boldsymbol{h}(x;\hat{\boldsymbol{\beta}})$, which would also have been the result if the ordinary score function $\boldsymbol{v} = \boldsymbol{h}$ were considered. Thus, numerically it makes no difference in this case. Moreover, with the choice $\boldsymbol{v} = \boldsymbol{h}$ Theorem 4.2 gives exactly the same variance–covariance matrix of Equation (4.18).

Although the efficient score arises naturally when MLE is considered, it has a much broader validity. For instance, Hall and Mathiason (1990) showed that the asymptotic null distribution of the efficient score statistic $\boldsymbol{V}(\hat{\boldsymbol{\beta}})$ is the same for any $n^{1/2}$-consistent estimator $\hat{\boldsymbol{\beta}}$.

#### 4.2.2.3 The Generalised Score Test

The name "generalised score test" was used by Boos (1992) to name a quadratic goodness-of-fit test which looks very similar to a score test, but which also works with more general estimation equations. This class of tests also works in a semiparametric framework in which the likelihood function is not specified. When applied to smooth alternatives, they are referred to as "generalised smooth tests" by Javitz (1975) and Rayner et al. (2009).

Theorem 4.2 that follows shortly states the asymptotic null distribution of statistics of the form

$$\hat{\boldsymbol{V}} = \boldsymbol{V}(\hat{\boldsymbol{\beta}}) = \frac{1}{\sqrt{n}}\sum_{i=1}^{n} \boldsymbol{v}(X_i;\hat{\boldsymbol{\beta}}),$$

where $\boldsymbol{v}$ is a $k$-dimensional vector-valued function which satisfies the same regularity conditions as imposed on the influence functions (Section 5.1.3). Because the exposition in the next few paragraphs is fairly general and technical, it may be skipped by readers who are only interested in the applications.

We need the following additional notation for any two vector-valued functions $\boldsymbol{r}$ and $\boldsymbol{s}$,

$$\boldsymbol{\Sigma}_{rs} = \mathrm{Cov}_0\{\boldsymbol{r}, \boldsymbol{s}\} = <\boldsymbol{r}, \boldsymbol{s}>_g$$
$$\boldsymbol{\Sigma}_{r\beta} = \mathrm{Cov}_0\{\boldsymbol{r}, \boldsymbol{u}_\beta\} = <\boldsymbol{b}, \boldsymbol{u}_\beta>_g .$$

We also use the convention $\boldsymbol{\Sigma}_r = \boldsymbol{\Sigma}_{rr}$.

**Theorem 4.2.** *Let $X_1, \ldots, X_n$ denote a sample of i.i.d. observations which have, under the null hypothesis, density function $g(x; \boldsymbol{\beta})$. Suppose the p-dimensional nuisance parameter is estimated by means of a locally asymptotic linear estimator $\hat{\boldsymbol{\beta}}$ which is determined by estimation function $\boldsymbol{b}$. Let $\hat{\boldsymbol{V}} = \boldsymbol{V}(\hat{\boldsymbol{\beta}}) = (1/\sqrt{n})\sum_{i=1}^n \boldsymbol{v}(X_i; \hat{\boldsymbol{\beta}})$, where $\boldsymbol{v}$ is a k-dimensional vector-valued function with $\mathrm{E}_0\{\boldsymbol{v}(X; \boldsymbol{\beta})\} = \boldsymbol{0}$ and finite $\mathrm{E}_0\{\dot{\boldsymbol{v}}(X; \boldsymbol{\beta})\dot{\boldsymbol{v}}^t(X; \boldsymbol{\beta})\}$ under the null hypothesis. Then, the asymptotic null distribution of $\hat{\boldsymbol{V}}$ is a zero-mean multivariate normal distribution with variance-covariance matrix*

$$\boldsymbol{\Sigma}_{\hat{v}} = \boldsymbol{\Sigma}_v + \boldsymbol{\Sigma}_{v\beta}\boldsymbol{\Sigma}_{b\beta}^{-1}\boldsymbol{\Sigma}_{bb}(\boldsymbol{\Sigma}_{b\beta}^{-1})^t\boldsymbol{\Sigma}_{\beta v} - \boldsymbol{\Sigma}_{vb}(\boldsymbol{\Sigma}_{b\beta}^{-1})^t\boldsymbol{\Sigma}_{\beta v} - \boldsymbol{\Sigma}_{v\beta}\boldsymbol{\Sigma}_{b\beta}^{-1}\boldsymbol{\Sigma}_{bv}. \tag{4.19}$$

The proof of this theorem is a direct consequence of a more general theorem which is stated and proved in Section A.8 of Appendix A, where the asymptotic distribution of $\hat{\boldsymbol{V}}$ is studied under sequences of local alternatives.

The next theorem gives the generalised smooth test statistic and its asymptotic null distribution. The theorem is based on the Neyman model, but using similar arguments as in the proof of Theorem 4.1, it can be shown that the same result holds for the Barton model too.

**Theorem 4.3.** *Let $\{h_j(.; \boldsymbol{\beta})\}$ denote a set of orthonormal functions w.r.t. density $g(.; \boldsymbol{\beta})$ and let $\boldsymbol{h}^t = (h_1, \ldots, h_k)$. Consider the statistic*

$$\hat{\boldsymbol{V}}_k = \frac{1}{\sqrt{n}}\sum_{i=1}^n \boldsymbol{h}(X_i; \hat{\boldsymbol{\beta}}),$$

*and assume that the regularity conditions of Theorem 4.2 apply to $\boldsymbol{h}$ and $\hat{\boldsymbol{\beta}}$. The score (smooth) test statistic for testing $H_0 : \boldsymbol{\theta} = \boldsymbol{0}$ in the order k Neyman smooth model is then given by*

$$T_k = \hat{\boldsymbol{V}}^t \boldsymbol{\Sigma}_{\hat{v}}^- \hat{\boldsymbol{V}}, \tag{4.20}$$

*where* $\boldsymbol{\Sigma}_{\hat{v}}^{-}$ *is the generalised inverse of the variance–covariance matrix* $\boldsymbol{\Sigma}_{\hat{v}}$ *as given in Equation (4.19) with* $\boldsymbol{v}$ *replaced by* $\boldsymbol{h}$. *Under* $H_0$, *as* $n \to \infty$,

$$T_k \xrightarrow{d} \chi^2_r,$$

*where* $r$ *is the rank of* $\boldsymbol{\Sigma}_{\hat{v}}$.

At first sight Theorems 4.2 and 4.3 may look quite complicated, but in particular important cases, they give quite simple results. We have seen already the efficient score test, and in the next paragraphs we give another interesting illustration of the generalised score test.

When MME is used for nuisance parameter estimation, the estimation equations are $\boldsymbol{b} = \boldsymbol{h}_N = \boldsymbol{0}$, where $\boldsymbol{h}_N$ denotes the vector function which is built from the first $p$ polynomials $h_j$ ($j = 1, \ldots, p$). Earlier in this section we have argued that with these estimation functions, it makes no sense to include the first $p$ polynomials in the goodness-of-fit test, or, equivalently, the first $p$ $\theta$ parameters may be removed from the smooth alternative. Let $\boldsymbol{h}_T$ denote the vector function containing the polynomials $h_j$, $j = p+1, \ldots, k$. When working in a likelihood framework, as before $\boldsymbol{h}_T$ is the score function for the $\theta$ parameters. Hence, $\boldsymbol{v} = \boldsymbol{h}_T$ seems a natural choice. Theorem 4.2 now gives

$$\boldsymbol{\Sigma}_{\hat{v}} = \boldsymbol{I}_{k-p} + \boldsymbol{\Sigma}_{h_T\beta}\boldsymbol{\Sigma}^{-1}_{h_N\beta}(\boldsymbol{\Sigma}^{-1}_{h_N\beta})^t\boldsymbol{\Sigma}_{\beta h_T}.$$

Although a test based on the choice $\boldsymbol{v} = \boldsymbol{h}_T$ makes sense, it is a naive construction. It has been shown that power may be gained if $\boldsymbol{v}$ is still taken as the efficient score function of Equation (4.16) (with $\boldsymbol{h}$ limited to the $(k-p)$-dimensional $\boldsymbol{h}_T$). For this particular construction Theorem 4.2 gives

$$\boldsymbol{\Sigma}_{\hat{v}} = \boldsymbol{I}_{k-p} - \boldsymbol{\Sigma}_{h_T\beta}\boldsymbol{\Sigma}^{-1}_{\beta\beta}\boldsymbol{\Sigma}_{\beta h_T}. \tag{4.21}$$

When MLE and MME coincide, the $p$ score functions in $\boldsymbol{u}_\beta$ are polynomial. When the polynomials are of orders $1, \ldots, p$, the covariance matrix $\boldsymbol{\Sigma}_{\hat{v}}$ reduces to the identity matrix.

*Example 4.5 (Cultivars data).* To demonstrate a smooth test for a composite null hypothesis we consider the cultivars data and we test the null hypothesis that the data come from a normal distribution. We apply a $k = 6$ order smooth test and use MLE for the estimation of the mean and the variance, but for the normal distribution the results would have been same with MME. It is known that the convergence to the asymptotic null distribution is rather slow, and so it is recommended that the $p$-values are approximated by means of simulation, e.g., by the bootstrap (see Appendix B.2). However, because a normal distribution is location-scale invariant, the null distribution does not depend on the true mean and the variance, and therefore the simulations should be performed only once. With the `smooth.test` function comes a database of simulated null distributions for location-scale families, and $p$-values are thus rapidly obtained.

```
> smooth.test(cultivars,order=6,distr="norm",method="MLE")
Smooth goodness-of-fit test
Null hypothesis:  norm  against  6 th order alternative
Nuisance parameter estimation:  MLE
Parameter estimates:  2.067579 33.94631  ( MEAN VAR )

Smooth test statistic S_k =  3.247817 p-value =  0.18
     3 th component V_k =  0.394757  p-value =  0.64
     4 th component V_k =  1.055294  p-value =  0.08
     5 th component V_k =  0.476652  p-value =  0.53
     6 th component V_k = -1.323307  p-value =  0.02

All p-values are obtained by referring to a simulated null
distribution based on 10,000 runs
```

The first two components are exactly zero because MME and MLE coincide. We read $p = 0.18$ for $k =$ 6th-order smooth test, and conclude at the 5% level of significance that there is no reason to suspect the data from not being normally distributed. Despite this nonsignificance, the output also shows that the 6th individual component test gives $p = 0.02$.

In the next section we see methods that may be used to select the order $k$ of the smooth test, and in Section 4.5 we show how the rescaling method of Henze and Klar (see Section 4.2.1) can be applied in the presence of nuisance parameters. All these methods to come will be illustrated in Section 4.6.

## 4.3 Adaptive Smooth Tests

### *4.3.1 Consistency, Dilution Effects and Order Selection*

A well known and important problem with the smooth tests is that the order $k$ must be fixed before looking at the data. This is essential for the distribution theory to hold. It is, however, easy to understand that a "bad" choice of $k$ may result in a smooth test with unfavourable power characteristics for particular alternatives. To explain this problem we first consider the smooth tests for a simple null hypothesis and we use the representation of a density function in a Hilbert space. As before consider the linear expansion of Equation (4.4), resulting in the expansion $f(x) = (1 + \sum_{j=1}^{\infty} \theta_j h_j(x))g(x)$ of the true density $f$. Any density $f$ which satisfies some regularity conditions, has a representation in an infinite-dimensional Hilbert space which is spanned by the orthonormal basis functions $h_j$. In Section 4.1.1 it was shown that the $\theta$

parameters are the solutions to minimising Pearson's $\phi^2$ measure, resulting in $\theta_j = <h_j, (f/g)>_g = <h_j, f>$. In a Hilbert space the $\theta_j h_j$ are also recognised as the orthogonal projections of the relative density $f/g$ onto the basis function $h_j$. Hence, $\tilde{f}_k(x) = (1 + \sum_{j=1}^k \theta_j h_j(x))g(x)$ is the orthogonal projection of $f/g$ onto the subspace $\mathcal{P}_k \subset L_2(\mathcal{S})$ spanned by $\{h_0, \ldots, h_k\}$, and which in a geometric sense is interpretable as the relative density within $\mathcal{P}_k$ that has the smallest distance to $f/g$. Furthermore, a statistical interpretation of $\theta_j$ may come from

$$\theta_j = <h_j, f/g>_g = \int_{\mathcal{S}} h_j(x) f(x) dx = \mathrm{E}_f\{h_j(X)\} = \sqrt{n}\,\mathrm{E}_f\{U_j\},$$

where $U_j$ is the score statistic of Section 4.2.1. Using this result, and relying on Theorem 4.1, the following lemma follows almost immediately (the proof is omitted).

**Lemma 4.5.** *Let $X_1, \ldots, X_n$ denote i.i.d. random variables with density function $f$, let $S$ denote any finite nonempty subset of $\{1, 2, \ldots\}$ and let $\boldsymbol{U}_S^t = (U_j)_{j \in S}$. Then, as $n \to \infty$,*

$$\boldsymbol{U}_S - \sqrt{n}\boldsymbol{\theta}_S \xrightarrow{d} MVN(\boldsymbol{0}, \boldsymbol{\Sigma}_S),$$

*where $\boldsymbol{\theta}_S^t = (\theta_j)_{j \in S}$ and $\boldsymbol{\Sigma}_S = \mathrm{Var}_f\{\boldsymbol{U}_S\}$ of which the $j, k$th element is given by $\sigma_{jk} = \mathrm{Cov}_f\{h_j(X), h_k(X)\}$.*

This lemma, together with the projection interpretation of the $\theta$ parameters is important for the understanding of the power characteristics of the order $k$ smooth test and its components. We immediately give some important consequences, but first we make some additional assumptions about $f$. We suppose that $f$ belongs to $\{g \in L_2(\mathcal{S}) : 0 < \sum_{j \in S} \mathrm{Var}_g\{U_j\} < \infty\}$. This restriction can sometimes be dropped if instead of $\boldsymbol{U}_S$ the empirically scaled score statistics of Henze and Klar were used. However, in the light of our current discussion this would only make things unnecessarily more complex. Here are some important consequences.

1. Let $\mathcal{P}_S$ denote the Hilbert subspace spanned by $h_j$ $(j \in S)$. First we consider a single-component test based on $V_j = U_j/\sigma_j$ with $\sigma_j^2$ the asymptotic variance of $U_j$ under the null hypothesis$(j \in S)$. When the orthogonal projection of $f/g$ onto $\mathcal{P}_S$ gives $\theta_j \neq 0$, Lemma 4.5 implies that $\mathrm{E}_f\{V_j\}$ grows unboundedly in probability. Hence, if $c_\alpha$ is the (asymptotic) critical point of $V_j^2$ at the $\alpha$ level, $\Pr_f\{V_j^2 > c_\alpha\} \to 1$ as $n \to \infty$. This shows that $V_j$ gives a consistent test against the alternative $f$, and the necessary condition is that the $j$th component of the orthogonal projection of $f/g$ on $h_j$ should be nonzero.
2. Leaving the one degree-of-freedom situation and considering the partial sum test statistic $T_S = \sum_{j \in S} V_j^2$, Lemma 4.5 again shows that $T_S$ grows unboundedly, even if there is only one $j$ for which $\theta_j \neq 0$. Thus also $T_S$

gives a consistent test against the alternative $f$ as long as at least one of the components of the projection of $f/g$ on $\mathcal{P}_S$ is different from zero. Note that this reasoning only holds if the size of $S$ is kept finite.

3. In contrast to the two previous situations, it is clear that an order $k$ smooth test has no power (i.e., power equals significance level) if all $\theta_j = 0$ ($j \leq k$).
4. The two previous points were based on asymptotic arguments leading to a trivial asymptotic power equal to one. This makes power comparison difficult. A classical theoretical framework used to compare nontrivial powers is to consider sequences of local alternatives, which are indexed by the sample size $n$ and which converge to the hypothesised density $g$ at a convenient rate so that the power is kept away from the significance level and from one. Such arguments can be made formal if we rely on the results of Appendix A.8, but instead we give a rather intuitive argumentation.
We suppose that Lemma 4.5 remains approximately valid for large but finite sample sizes $n$. Imagine that exactly one $\theta_j \neq 0$ ($j \in S$), let $S = \{1, \ldots, k\}$, and let $c_{k,\alpha}$ denote the $\alpha$-level critical value of the asymptotic $\chi^2_k$ null distribution of $T_S = T_k$. Thus, for large $n$ we have approximately $\Pr_g\{T_k > c_{k,\alpha}\} = \alpha$. Under the alternative $f$, Lemma 4.5 implies that $T_k$ has approximately a noncentral $\chi^2$ distribution with $k$ degrees-of-freedom and noncentrality parameter $c = n\theta_j^2$. Thus, for a given sample size $n$ and $\theta_j$, the noncentrality parameter remains constant, but the degrees of freedom increase with order $k$. To illustrate the effect of increasing $k$ when the order $j$ for which $\theta_j \neq 0$ is rather small, say $j = 2$, the approximate powers for $n\theta_1^2$ ranging from 1 to 10 are presented in Table 4.2. These powers suggest that (1) the power increases with the noncentrality parameter; (2) there is no power (power equals significance level) when $k < j = 2$; (3) for each value of the noncentrality parameter the test with order $k = 2$ has the largest power; and (4) for a given noncentrality parameter, the power decreases when $k > j = 2$ increases. The latter effect is called the *dilution effect.* It illustrates that $k$ should be chosen appropriately: not too small and not too large.

The discussion of the previous paragraph makes clear that it is very important to specify $k$ appropriately so that a dilution effect is avoided. There

**Table 4.2** The approximate powers of the partial sum test $T_k$ (with $k$ ranging from 1 to 6) under alternatives with noncentrality parameter $n\theta_2^2$ ranging from 1 to 10

| | Noncentrality Parameter | | | | | | | | | |
|---|---|---|---|---|---|---|---|---|---|---|
| $k$ | 1 | 2 | 3 | 4 | 5 | 6 | 7 | 8 | 9 | 10 |
| 1 | 0.05 | 0.05 | 0.05 | 0.05 | 0.05 | 0.05 | 0.05 | 0.05 | 0.05 | 0.05 |
| 2 | 0.13 | 0.23 | 0.32 | 0.42 | 0.50 | 0.58 | 0.65 | 0.72 | 0.77 | 0.82 |
| 3 | 0.12 | 0.19 | 0.27 | 0.36 | 0.44 | 0.52 | 0.59 | 0.65 | 0.71 | 0.76 |
| 4 | 0.11 | 0.17 | 0.24 | 0.32 | 0.40 | 0.47 | 0.54 | 0.61 | 0.66 | 0.72 |
| 5 | 0.10 | 0.16 | 0.22 | 0.29 | 0.36 | 0.43 | 0.50 | 0.56 | 0.62 | 0.68 |
| 6 | 0.09 | 0.15 | 0.21 | 0.27 | 0.34 | 0.40 | 0.47 | 0.53 | 0.59 | 0.64 |

are two important practical situations. In the first, the user is particularly interested in rejecting the null hypothesis when the true distribution $f$ belongs to a specific restricted class of alternatives. For instance, when testing for normality as a pretest to a $t$-test for equality of means, the user is often more interested in detecting skewness than in detecting any other type of (lower order) deviation, for it is generally known that $t$-tests are quite sensitive to skewed deviations from the normal distribution. This sensible argumentation advocates the use of an order $k = 3$ partial sum test. On the other hand, there are many situations in which the user wants to detect almost any kind of deviation from the hypothesised distribution. In these situations, $k$ should be large, but, as illustrated above, when $k$ is too large as compared to the noncentrality parameter or when $k$ is much larger than the largest order $j$ for which $\theta_j \neq 0$, the dilution effect kicks in and substantial power may be lost. This order $j$ defined here is sometimes called the *effective order* of the alternative $f$ (Rayner and Best (1989)). Unfortunately, in this last situation, the user has a priori no idea about the effective order.

The solution consists of making the smooth test *adaptive* in the sense that the order, or, more generally, the subset $S$, is "estimated" from the sample observations. This data-driven process makes, however, the order $k$ or the indices in $S$ random variables, and therefore the distribution theory of the resulting adaptive smooth tests is affected. In the next section we make a distinction between "order selection" and "subset selection". The latter concerns selecting the subset $S \subseteq \{1, 2, \ldots, m\}$, where $m$ denotes the maximum order that may be considered, whereas the former restricts the subsets to be of the form $S = \{1, \ldots, k\}$ $(k \leq m)$. Another important distinction to be made is the size of $m$. If $m$ is finite, we say that subsets are within a *finite horizon*. Sometimes $m$ is allowed to grow with the sample size $n$, denoted as $m_n$. In this last situation, only the order selection has a sound asymptotic theory.

### 4.3.2 Order Selection Within a Finite Horizon

Ledwina (1994) was the first to note that the order selection problem is basically a model selection problem. Indeed, smooth tests are actually parametric score tests within the framework of a flexible and particularly constructed family of smooth alternatives of order $k$. Thus, estimating the $\theta$ parameters and selecting the order $k$ in the Neyman density (Equation (4.1)) can be considered as a type of nonparametric density estimation. Nonparametric density estimation based on a Barton-type expansion was already proposed in 1962 by Cencov (1962). Also the Neyman model has been studied well before 1994 as a basis for nonparametric density estimation, but it has never been as popular as the Barton representation. See Section 2.8 for a brief introduction to nonparametric density estimation. Looking at the problem from the point of view of model selection, Ledwina (1994) suggested that the order $k$ could be selected based on a model selection criterion such as the Bayesian information criterion (BIC) Schwarz (1978). Although different model selection criteria exist, she argued that BIC should be chosen because it is consistent in the sense that it asymptotically selects with probability one the most parsimonious model among the models that are closest to the true model in terms of the Kullback–Leibler divergence. In this 1994 paper, only testing a simple null hypothesis was discussed. Later, Kallenberg and Ledwina (1997) and Inglot et al. (1997) extended the data-driven smooth test framework to testing composite hypotheses when the nuisance parameters are estimated by a $\sqrt{n}$-consistent estimator (details follow). These two last papers actually concern order selection within an infinite horizon, but here we restrict the discussion to a finite $m$. In the next paragraphs we give the main results from these papers.

First we suppose that $\boldsymbol{\beta}$ is completely known (simple null hypothesis). Let $l(\boldsymbol{\theta}_k; \boldsymbol{\beta})$ denote the log-likelihood function based on the Neyman model including terms up to order $k$. When $\hat{\boldsymbol{\theta}}_k$ denotes the MLE of $\boldsymbol{\theta}_k$, then $l(\hat{\boldsymbol{\theta}}_k; \boldsymbol{\beta})$ is the maximised log-likelihood. For a model of order $k$, the BIC is defined as

$$\text{BIC}_n(k; \boldsymbol{\beta}) = l(\hat{\boldsymbol{\theta}}_k; \boldsymbol{\beta}) - \frac{1}{2} k \log(n), \tag{4.22}$$

where the last term is interpreted as a penalty term which makes the BIC favouring lower-dimensional models as the sample size increases. The order selection rule specifies the selected order as

$$K = K_n(\boldsymbol{\beta}) = \min\left\{k : 1 \leq k \leq m, \text{BIC}_n(k; \boldsymbol{\beta}) \geq \text{BIC}_n(j; \boldsymbol{\beta}), j = 1, \ldots, m\right\}. \tag{4.23}$$

Because it can be quite tedious to find the MLE $\hat{\boldsymbol{\theta}}$ (see, e.g., Buckland (1992) and Efron and Tibshirani (1996)), a similar but simpler selection rule has been proposed by Kallenberg and Ledwina (1997),

$$\begin{aligned} K2 = K2_n(\boldsymbol{\beta}) = \min\big\{k : 1 \leq k \leq m, \boldsymbol{U}_k^t \boldsymbol{U}_k - k \log(n) \geq \\ \boldsymbol{U}_j^t \boldsymbol{U}_j - j \log(n), j = 1, \ldots, m\big\}, \end{aligned} \tag{4.24}$$

where $\boldsymbol{U}_k = \boldsymbol{U}_k(\boldsymbol{\beta})$ is the vector of score statistics $(1/\sqrt{n}) \sum_{i=1}^n h_j(X_i; \boldsymbol{\beta})$, $j = 1, \ldots, k$. This simplification results from the fact that the maximised log-likelihood is equal to the log-likelihood ratio statistic for testing $\boldsymbol{\theta} = \boldsymbol{0}$ versus $\boldsymbol{\theta} \neq \boldsymbol{0}$ which is locally equivalent to $\frac{1}{2}$ times the score test statistic $\boldsymbol{U}_k^t \boldsymbol{U}_k$. See, for example, Javitz (1975) for more details on this equivalence. The data-driven smooth test statistic is now defined as the usual score test statistic with $k$ replaced by $K$ or $K2$,

$$T_{K2} = \boldsymbol{U}_{K2}^t \boldsymbol{U}_{K2} = \sum_{j=1}^{K2} \left( \frac{1}{\sqrt{n}} \sum_{i=1}^n h_j(X_i; \boldsymbol{\beta}) \right)^2 .$$

When the nuisance parameter $\boldsymbol{\beta}$ is estimated by a $\sqrt{n}$-consistent estimator, say $\hat{\boldsymbol{\beta}}$, Kallenberg and Ledwina (1997) and Inglot et al. (1997) suggested to replace $\boldsymbol{\beta}$ in the selection rules (4.23) and (4.24) with $\hat{\boldsymbol{\beta}}$, but to use the efficient score test statistic of Equation (4.20) with $\boldsymbol{\Sigma}_{\hat{v}}$ given in Equation (4.18); i.e.,

$$T_{K2} = \hat{\boldsymbol{U}}_{K2}^t \left( \boldsymbol{I}_{K2} - \boldsymbol{\Sigma}_{h\beta}\boldsymbol{\Sigma}_{\beta\beta}^{-1}\boldsymbol{\Sigma}_{\beta h} \right)^{-1} \hat{\boldsymbol{U}}_{K2}, \tag{4.25}$$

where the $\boldsymbol{\Sigma}_{\beta h}$ matrix refers to the first $K2$ components of $\boldsymbol{h}$. Later, Janic-Wróblewska (2004) noted that when nuisance parameters are estimated, it would be better to replace the score statistics in the $K2$ selection rule (Equation (4.24)) by the efficient score statistic,

$$\hat{\boldsymbol{U}}_{K2} = \frac{1}{\sqrt{n}} \sum_{i=1}^{n} \left\{ \boldsymbol{h}_{K2}(X_i; \hat{\boldsymbol{\beta}}) - \boldsymbol{\Sigma}_{h\beta}\boldsymbol{\Sigma}_{\beta\beta}^{-1}\boldsymbol{u}_{\beta}(X_i) \right\}$$

(see Section 4.2.2 for details on the notation). This slightly different selection rule is denoted as $\tilde{K2}$. Note that $\tilde{K2} \equiv K2$ when MLE is used. Note also that the $K2$ and $\tilde{K2}$ selection rules are based on sum of squared statistics, and not on the corresponding smooth test statistics that also involve the asymptotic variance of $\boldsymbol{U}$. While studying smooth tests for location-scale distributions, Janic-Wróblewska and Ledwina (2009) suggested using the modified selection rule

$$\begin{aligned} K1 = K1_n(\boldsymbol{\beta}) = \min\{k : 1 \leq k \leq m : T_k - k\log(n) \geq \\ T_j - j\log(n), j = 1, \ldots, m\}, \end{aligned}$$

in which $T_k$ is now the order $k$ efficient score test statistic (4.20).

For all four order selection rules, the following lemma holds true (Ledwina (1994), Kallenberg and Ledwina (1997), Inglot et al. (1997), Janic-Wróblewska (2004) and Janic-Wróblewska and Ledwina (2009)). We refer to these papers for some technical regularity conditions for the lemma and following theorem to hold.

**Lemma 4.6.** *Let $\hat{\boldsymbol{\beta}}$ be a $\sqrt{n}$ consistent estimator of $\boldsymbol{\beta}$. Under $H_0$, as $n \to \infty$,*

$$Pr\left\{ O_n(\hat{\boldsymbol{\beta}}) = 1 \right\} \to 1,$$

*where $O_n$ denotes any of $K_n$, $K2_n$, $\tilde{K2}_n$ and $K1_n$.*

This lemma says that asymptotically always the first-order model is selected under the null hypothesis. This result, together with Theorem 4.3 immediately gives the following theorem.

**Theorem 4.4.** *Assume the conditions of Lemma 4.6 and Theorem 4.3. Under $H_0$, as $n \to \infty$,*

$$T_{O_n}(\hat{\boldsymbol{\beta}}) \xrightarrow{d} \chi_1^2,$$

*where $O_n$ denotes any of $K_n$, $K2_n$ or $\tilde{K}2_n$, and where $T$ is the efficient score statistic.*

Note that Lemma 4.6 and Theorem 4.4 also apply when $\boldsymbol{\beta}$ is fixed in a simple null hypothesis. Up to now we have worked within the Neyman model (4.1) for which it was assumed that the normalisation constant $C(\boldsymbol{\theta}, \boldsymbol{\beta})$ exists. When $\mathcal{S}$ is a closed set, this will usually be the case, but when $\mathcal{S}$ is open, this will often be problematic. For instance, when testing for a normal distribution the natural set of orthonormal functions is the Hermite polynomials which are unfortunately not bounded. To avoid such problems, many papers on data-driven smooth tests work within a slightly different type of Neyman model,

$$g_k(x; \boldsymbol{\theta}, \boldsymbol{\beta}) = C(\boldsymbol{\theta}, \boldsymbol{\beta}) \exp\left(\sum_{j=1}^{k} \theta_j \phi_j(G(x; \boldsymbol{\beta}))\right) g(x; \boldsymbol{\beta}), \tag{4.26}$$

where $G$ denotes the CDF of $g$ and $\{\phi_j\}$ is now a set of bounded orthonormal functions in $L_2([0,1])$. Typical examples include the cosine basis or the Legendre polynomials. This representation basically results from applying a probability integral transformation (PIT) prior to the analysis (see Section 2.4 for PIT). We refer to Equation (4.26) as the Neyman–PIT model. The consequences of using the Neyman–PIT rather than the usual Neyman model are minor. The orthonormal basis functions satisfy

$$h_j(x; \boldsymbol{\beta}) = \phi_j(G(x; \boldsymbol{\beta})). \tag{4.27}$$

All definitions and theoretical results on the distribution theory of the smooth tests remain valid, but the expression of the asymptotic variance–covariance matrix of the score or efficient score statistic becomes more complicated, because, as (4.27) shows, the nuisance parameter now enters in $h_j$ through the CDF $G$. More important from a practical point of view some of the nice moment interpretations of the components are lost because the $\phi_j$ may refer to the $j$ the moment of $G(X)$, but it is not always straightforward to translate this to a moment interpretation of $X$ itself. When the Neyman–PIT model is used, the Hilbert space should be redefined too. In particular, the Hilbert space is spanned by the functions $\phi_j \circ G$ $(j = 0, \ldots)$, which agrees with Equation (4.27), and it is denoted by $L_2([0,1] \circ \mathcal{S})$. Furthermore, the subspace $\mathcal{P}$ now denotes the the subspace spanned by the first $k$ functions $\phi_j \circ G$ $(j \leq k)$.

Before we go to the next section, it is important to say something about the consistency of the data-driven tests discussed so far. In Section 4.3.1 we have seen that a smooth test of order $k$ is consistent if the projected relative density $\tilde{f}_k/g$ has at least one $\theta_j \neq 0$ $(j \leq k)$. The following theorem is a consequence of Theorem 2.6 (and Remark 2.7) of Inglot et al. (1997) and Theorem 3 of Janic-Wróblewska (2004).

**Theorem 4.5.** *Assume the conditions of Lemma 4.6 and Theorem 4.3 apply. Let $\mathcal{F}_m$ denote the set of density functions $f$ for which the orthogonal projections of $f/g$ onto $\mathcal{P}_m$ have at least one $\theta_j \neq 0$ ($j \leq m$). Suppose that $f \in \mathcal{F}_m$, and let $k_m$ denote the smallest $j$ for which $\theta_j \neq 0$ occurs. Let $O_n$ denote any of $K_n$, $K2_n$, $\tilde{K2}_n$, or $K1_n$. Then, (1) as $n \to \infty$, $Pr_f\left\{O_n(\hat{\beta}) \geq k_m\right\} \to 1$, and (2) the data-driven smooth test based on $T_{O_n(\hat{\beta})}$ is consistent against $f$.*

This theorem illustrates the limitation of the finite horizon restriction: the data-driven smooth tests with fixed and finite $m$ are only consistent against alternatives $f \in \mathcal{F}_m$. A natural extension is to allow $m$ to become infinitely large.

### *4.3.3 Order Selection Within an Infinite Horizon*

The extension of data-driven smooth tests with finite $m$ to the situation where $m$ is allowed to grow unboundedly with the sample size $n$ was first proposed by Kallenberg and Ledwina (1995a) and Kallenberg and Ledwina (1995b) for the case of testing a simple null hypothesis, and later, among other, by Kallenberg and Ledwina (1997), Inglot et al. (1997), Janic-Wróblewska (2004), and Janic-Wróblewska and Ledwina (2009) for the composite case. We denote $m = m_n$ to stress the dependence of $m$ on the sample size $n$. These papers study the asymptotic behaviour of the order selection rules and the data-driven smooth test statistics when $m$ is replaced with $m_n$. To get nice asymptotic results restrictions must be placed on the rate at which $m_n$ grows with $n$. These restrictions depend on (1) the system of orthonormal functions used, and (2) the method of nuisance parameter estimation. To avoid too many technicalities, we only give some details on the conditions for testing a simple null hypothesis. Later in this section we only summarise some of the main results for testing a composite null hypothesis. We refer to the papers mentioned above for more technical details.

Before we continue we mention that all theoretical results on this type of data-driven test are derived for the Neyman–PIT model. This necessity may be seen from an assumption that must be made on the convergence rate of $m_n$. Let

$$V_m = \max_{1 \leq j \leq m} \sup_{z \in [0,1]} |\phi_j(z)|. \tag{4.28}$$

It is assumed that, as $n \to \infty$,

$$m_n V_{m_n} \sqrt{\frac{\log(n)}{n}} \to 0. \tag{4.29}$$

Thus, when in Equation (4.28) orthonormal polynomials over an unbounded support $\mathcal{S}$ would be considered, $V_m$ would be infinite, and thus condition (4.29) could not be met.

The next theorem summarises some of the main results for testing the simple null hypothesis (Theorems 3.2, 3.4, and 4.1 of Kallenberg and Ledwina (1995a)).

**Theorem 4.6.** *Let $m_n \to \infty$ as $n \to \infty$, and assume that condition (4.29) holds. Let $O_n$ denote $K_n$ or $K2_n$. Let $\mathcal{F}_m$ be the set of density functions $f$ for which the orthogonal projections of $f/g$ onto $\mathcal{P}_m$ have at least one $\theta_j \neq 0$ $(j \leq m)$. Let $f \in \mathcal{F}_m$ $(m > 0)$ and let $k_m$ denote the smallest $m$ for which $f \in \mathcal{F}_m$. Then, as $n \to \infty$, (1) $Pr_g\{O_n = 1\} \to 1$; (2) $T_{O_n} \xrightarrow{d} \chi_1^2$ (where convergence is w.r.t. G); (3) $Pr_f\{O_n \geq k_m\} \to 1$ whenever $f \in \mathcal{F}_{k_m}$.*

This theorem states that the asymptotic null distribution of $T_{O_n}$ is again simply $\chi_1^2$, and the data-driven test is now omnibus consistent; i.e. it is consistent against essentially all fixed alternatives $f \neq g$.

Before leaving the simple null hypothesis case, we consider two important examples: $\{\phi_j\}$ is the system of Legendre polynomials or the cosine basis. We find $V_k = \sqrt{2k+1}$ and $V_k = \sqrt{2}$, respectively. This gives $m_n = o((n/\log(n))^{1/3})$ and $m_n = o((n/\log(n))^{1/2})$, respectively.

For the composite case, the rates of convergence of $m_n$ depend not only on the method of nuisance parameter estimation and on the the system of orthonormal functions, but also on the hypothesised distribution . A summary of the some appropriate rates for $m_n$ is presented in Table 4.3 (Inglot et al. (1997) and Janic-Wróblewska (2004)).

Despite the nice theoretical results of these data-driven tests, it is not obvious how the convergence rates of $m_n$ should be translated to a realistic situation in which the sample size $n$ is always finite. It seems that the only practical solution is to choose $m_n$ according to some empirical guidelines which are typically derived from simulation studies. Most of these studies suggest that there is no need to choose $m_n$ large; e.g., for sample sizes $n \leq 100$, $5 \leq m_n \leq 10$ seems appropriate. These simulation studies also show that the powers do not change dramatically with the choice of $m_n \geq 5$.

### 4.3.4 Subset Selection Within a Finite Horizon

Although the data-driven tests of the previous section have good theoretical properties such as omnibus consistency, the last paragraph explained that in

**Table 4.3** Some selected rates of convergence of $m_n$ for several hypothesised density functions $g$. Let $\varepsilon > 0$ and $c < 27/(2\pi^2(6+\pi^2))$

| | Legendre | | cosine | |
|---|---|---|---|---|
| $g$ | MLE | MME | MLE | MME |
| Normal | $o((n/\log(n))^{1/9})$ | $o((n/\log(n))^{1/9})$ | $o(n^{1/6-\varepsilon})$ | $o(n^{1/6-\varepsilon})$ |
| Exponential | | | $o((n/\log(n))^{1/4})$ | $o((n/\log(n))^{1/4})$ |
| Extreme value | | $o((n/\log(n))^{1/9})$ | | |
| Logistic | | $o(n^c)$ | | |

practice the maximal order $m_n$ is always finite and that its dependence on the sample size is something that empirically should be determined. In this section we again focus on model selection within a finite horizon, but now the selected models are not restricted to index sets of the form $\{1, \ldots, k\}$, instead the index set $S$ may be any nonempty subset of $\{1, \ldots, m\}$ where $m < \infty$ is fixed. The methodology in this section may thus be considered as an extension of order selection within a finite horizon. Besides BIC, we also discuss AIC (Akaike's Information Criterion) as a model selection rule. The methods described in this and the next sections are mainly based on Claeskens and Hjort (2004). They focused, however, on testing a simple null hypothesis, and the composite case is only briefly addressed in their Section 6. Because composite hypothesis testing is of more practical importance, we have extended some of their results to fit better in this chapter.

To make the exposition slightly more general we assume that (1) $\hat{\boldsymbol{\beta}}$ is an asymptotically linear estimator; (2) $\hat{T}_{n,S}$ denotes any of the smooth test statistics based on $h_j$, $j \in S$. Moreover, $\hat{T}_{n,S}$ may even represent $\hat{\boldsymbol{V}}^t\hat{\boldsymbol{V}}$, which is the squared norm of $\hat{\boldsymbol{V}}$ and does not take $\boldsymbol{\Sigma}_v = \text{Var}\left\{\hat{\boldsymbol{V}}\right\}$ into account. The limit distribution of $\hat{T}_{n,S}$ can be directly derived from Theorems 4.2 and 4.3 for a fixed subset $S$. The asymptotic null distribution of $\hat{T}_{n,M_n}$ where the index set $M_n$ is determined by a data-driven selection rule, is provided in this section.

A general BIC-type selection rule can be formulated as

$$M_n = \left\{R \subseteq S : R \neq \phi \text{ and } \hat{T}_{n,R} - |R|\log(n) \geq \hat{T}_{n,Q} - |Q|\log(n), \forall Q \subseteq S\right\},$$

where $|R|$ denotes the cardinality of the set $R$. In analogy with the notation of the previous section, we call $M_n$ one of $S2_n(\hat{\boldsymbol{\beta}})$, $\tilde{S2}_n(\hat{\boldsymbol{\beta}})$ or $S1_n(\hat{\boldsymbol{\beta}})$ for $\hat{T}_{n,R}$ being $\hat{U}_R^t\hat{U}_R$, $\hat{V}_R^t\hat{V}_R$, or $\hat{V}_R^t\boldsymbol{\Sigma}_{\hat{v}}^{-1}\hat{V}_R$, respectively.

The next lemma shows that when the null hypothesis is true, the BIC-type selection rules always asymptotically select a model with exactly one term. This lemma is important for finding the asymptotic null distribution.

**Lemma 4.7.** *Under $H_0$, as $n \to \infty$,*

$$Pr_g\left\{|M_n| > 1\right\} \to 0.$$

*Proof.* Let $S \neq \phi$, $j \leq m$, and $j \notin S$. We show that $\{j\}$ always "wins" from $S \cup \{j\}$ according to the $M_n$ selection rule. This happens when $\hat{T}_{n,R} - |R|\log(n)$ is the largest for $R = \{j\}$.

$$\begin{aligned}
&\left(\hat{T}_{n,\{j\}} - \log(n)\right) - \left(\hat{T}_{n,\{j\}\cup S} - (1+|S|)\log(n)\right) \\
&= |S|\log(n) - \left(\hat{T}_{n,\{j\}\cup S} - \hat{T}_{n,\{j\}}\right).
\end{aligned}$$

Clearly this difference goes to infinity with probability one. This completes the proof. □

The following theorem now follows immediately.

**Theorem 4.7.** *Suppose that for all nonempty $R \subseteq S$, $\hat{T}_{n,R} \xrightarrow{d} T_R$, where $T_R$ represents a random variable with a nondegenerate distribution. Under $H_0$, as $n \to \infty$,*

$$\hat{T}_{n,M_n} \xrightarrow{d} \max_{j \in S} T_j.$$

Although the theorem gives the asymptotic null distribution, it is still not always easy to apply this in practice.

*Example 4.6 (BIC subset selection with MLE nuisance parameter estimation).* After the presentation of Theorems 4.2 and 4.3 in Section 4.2.2, we have discussed three special cases. The most traditional case is where the nuisance parameter $\boldsymbol{\beta}$ is estimated by means of MLE, say $\hat{\boldsymbol{\beta}}$. In this situation the score and the efficient score statistics coincide; i.e., $\hat{\boldsymbol{V}} = \hat{\boldsymbol{U}} = (1/\sqrt{n}) \sum_{i=1}^{n} \boldsymbol{h}(X_i; \hat{\boldsymbol{\beta}})$, which is under $H_0$ asymptotically zero-mean multivariate normally distributed with covariance matrix $\boldsymbol{\Sigma}_{\hat{v}} = \boldsymbol{I} - \boldsymbol{\Sigma}_{h\beta}\boldsymbol{\Sigma}_{\beta\beta}^{-1}\boldsymbol{\Sigma}_{\beta h}$.

To get an easy expression for $T_j$ in Theorem 4.7, we define $\boldsymbol{V} = \boldsymbol{N} - \boldsymbol{\Sigma}_{h\beta}\boldsymbol{\Sigma}_{\beta\beta}^{-1}\boldsymbol{B}$, where $\boldsymbol{N}$ and $\boldsymbol{B}$ are jointly zero-mean multivariate normal with variance–covariance matrix

$$\begin{bmatrix} \boldsymbol{I} & \boldsymbol{\Sigma}_{h\beta} \\ \boldsymbol{\Sigma}_{\beta h} & \boldsymbol{\Sigma}_{\beta\beta} \end{bmatrix}.$$

Thus, as $n \to \infty$, $\hat{\boldsymbol{V}} \xrightarrow{d} \boldsymbol{V}$, and $\hat{T}_{n,M_n} \xrightarrow{d} \max_{j \in S} T_j$, where $T_j = \boldsymbol{V}^t \boldsymbol{\Sigma}_{\hat{v}} \boldsymbol{V}$. We now have an representation of $T_j$ in terms of $\boldsymbol{V}$, but it is still not possible to simulate $T_j$ because the true value of the nuisance parameter $\boldsymbol{\beta}$ is generally unknown. Fortunately, in two particular and important cases we have a simplification.

1. When $g$ belongs to the exponential family with polynomial sufficient statistics, we know that MLE and MME coincide and that the corresponding estimation functions can typically be formulated in terms of the first few orthonormal polynomials $h_1, \ldots, h_p$. Hence $\boldsymbol{\Sigma}_{h\beta} = \boldsymbol{0}$ and thus $\boldsymbol{V} = \boldsymbol{N}$, $\boldsymbol{\Sigma}_{\hat{v}} = \boldsymbol{I}$, $T_j = N_j^2$, and the asymptotic null distribution of the data-driven test statistic becomes $\max_{j \in S} N_j^2$, where the $N_j$ are i.i.d. standard normal. This is very easy to simulate.
2. When the hypothesised $g$ is a location-scale invariant distribution all the covariance matrices become independent of $\boldsymbol{\beta}$ and can therefore be specified without any further knowledge of $\boldsymbol{\beta}$. Examples include the normal and the logistic distribution. For example, for a two-parameter logistic distribution we find

$$\boldsymbol{\Sigma}_{\hat{v}} = \begin{bmatrix} 1-\frac{9}{\pi^2} & 0 & \frac{\sqrt{21}}{2\pi^2} & 0 \\ 0 & 1-\frac{45}{12+4\pi^2} & 0 & \frac{3\sqrt{5}}{6+2\pi^2} \\ \frac{\sqrt{21}}{2\pi^2} & 0 & 1-\frac{7}{12\pi^2} & 0 \\ 0 & \frac{3\sqrt{5}}{6+2\pi^2} & 0 & 1-\frac{1}{3+\pi^2} \end{bmatrix}.$$

See Thas and Rayner (2009) for more details on smooth tests for the logistic distribution.

The AIC is defined as

$$\text{AIC}_n(S;\boldsymbol{\beta}) = 2l(\hat{\boldsymbol{\theta}}_S;\boldsymbol{\beta}) - 2|S|,$$

but again it is more convenient to use an alternative definition that avoids the use of the maximised likelihood. Without using a different notation, we adopt

$$\text{AIC}_n(S;\boldsymbol{\beta}) = T_{n,S}(\boldsymbol{\beta}) - 2|S|.$$

The subset selection rule has general form

$$M_n(\boldsymbol{\beta}) = \{R \subseteq S : R \neq \phi \text{ and } \text{AIC}_n(R;\boldsymbol{\beta}) \geq \text{AIC}_n(Q;\boldsymbol{\beta}), \forall Q \subseteq S\}.$$

To study the asymptotic null distribution of the adaptive smooth test with AIC-selected index set $M_n$, we first need to know the asymptotic behaviour of the AIC criterion for a fixed nonempty set $S$. Because $\text{AIC}_n(S;\boldsymbol{\beta})$ only depends on the data through the statistic $T_{n,S}(\boldsymbol{\beta})$ we conclude that $\text{AIC}_n(S;\hat{\boldsymbol{\beta}})$ converges in distribution to $\text{AIC}(S;\boldsymbol{\beta}) = T_S - 2|S|$, where $T_S$ is a random variable with the same distribution as the asymptotic distribution of $T_{n,S}(\hat{\boldsymbol{\beta}})$.

**Theorem 4.8.** *Under $H_0$, as $n \to \infty$,*

$$\hat{T}_{n,M_n} \xrightarrow{d} \sum_{R\subseteq S} \left(I(R=M)\, T_R\right),$$

*where $M = \{R \subseteq S : R \neq \phi$ and $AIC(R;\boldsymbol{\beta}) \geq AIC(Q;\boldsymbol{\beta}), \forall Q \subseteq S\}$.*

*Proof.* The proof is straightforward. Write

$$\begin{aligned} \hat{T}_{n,M_n} &= \sum_{R\subseteq S} \hat{T}_{n,R}\text{I}\left(\text{AIC}_n(R;\hat{\boldsymbol{\beta}}) \text{ larger than all other } \text{AIC}_n(Q;\hat{\boldsymbol{\beta}}), Q \subseteq S\right) \\ &\xrightarrow{d} \sum_{R\subseteq S} T_R\text{I}\left(\text{AIC}(R;\boldsymbol{\beta}) \text{ larger than all other } \text{AIC}(Q;\boldsymbol{\beta}), Q \subseteq S\right) \\ &= \sum_{R\subseteq S} \left(\text{I}(R=M)\, T_R\right). \end{aligned}$$

□

The asymptotic null distribution can again be simulated, and, as before, the complexity depends on the asymptotic representation of $T_R$ which simplifies when $g$ belongs to the exponential family or when $g$ is location-scale invariant. Finally, we refer to Inglot and Ledwina (2006) who went one step further. Their data-driven selection rule not only selects the order, but it also makes a data-driven choice of the order selection criterion (AIC or BIC).

### 4.3.5 Improved Density Estimates

The methods for selecting terms as described in the previous sections clearly rely on the close relation between smooth goodness-of-fit testing and nonparametric density estimation using orthogonal series expansions. The order and subset selection criteria are indeed all model section criteria that are applied to the smooth alternatives, which are basically orthogonal series expansions. From this point of view the adaptive tests may be considered as testing after model selection. It also suggests that at the rejection of the null hypothesis, the selected model may be considered as an appropriate nonparametric density estimate of the true distribution.

When introducing the orthogonal series estimators in Section 2.8.2, we limited the discussion to estimators of the form (here we use the order selection technique)

$$\hat{f}(x) = g(x)\left\{1 + \sum_{j\in S}\hat{\theta}_j h_j(x)\right\},$$

where $h_j \in L_2(\mathcal{S}, G)$. In this chapter we actually went one step further by using a composite carrier density $g(.;\boldsymbol{\beta})$ that is indexed by the nuisance parameter $\boldsymbol{\beta}$. The nonparametric density estimator thus becomes

$$\hat{f}_{M_n}(x) = g(x;\hat{\boldsymbol{\beta}})\left\{1 + \sum_{j\in M_n}\hat{\theta}_j h_j(x;\hat{\boldsymbol{\beta}})\right\}, \tag{4.30}$$

where $M_n$ is any of the subset selection criteria presented in the previous section, and where ($j \in M_n$),

$$\hat{\theta}_j = \frac{1}{n}\sum_{i=1}^{n} h_j(X_i;\hat{\boldsymbol{\beta}}).$$

In the present context we refer to (4.30) as the *improved density estimator*. Because (4.30) is a Barton representation it is not necessarily a bona fide density. This can be corrected using the methods described in Section 2.8.2. Finally, we refer to Chapter 10 of Rayner et al. (2009) for a more detailed exposition on improved density estimates.

Finally note that the improved density estimate contains basically the same information as comparison density estimated by the same orthogonal series estimator,

$$\frac{\hat{f}_{M_n}(x)}{g(x;\hat{\boldsymbol{\beta}})} = 1 + \sum_{j \in M_n} \hat{\theta}_j h_j(x;\hat{\boldsymbol{\beta}}). \tag{4.31}$$

## 4.4 Smooth Tests for Discrete Distributions

### 4.4.1 Introduction

The smooth testing framework as described in the previous sections was completely developed for continuous distributions. In this section we discuss how smooth tests can be constructed for discrete distributions. Because most of the theory is very parallel to what has been given in detail in Sections 4.1 and 4.2 of this chapter, and Section 1.3 on the Pearson $\chi^2$ test, we can keep the discussion brief. For notational comfort we start again with the simple null hypothesis case, and extend this later to the composite null hypothesis situation. As in Section 1.3, we restrict our exposition to *pure* discrete distributions; i.e., we do not explicitly consider categorised or grouped continuous distributions. For more details on the latter we refer to Chapters 5 and 7 of Rayner et al. (2009).

Using the notation of Section 1.3, the null hypothesis of interest is $H_0 : \boldsymbol{\pi} = \boldsymbol{\pi}_0$.

### 4.4.2 The Simple Null Hypothesis Case

Rayner and Best (1989) showed how the smooth tests for discrete distributions arise naturally as a score test for testing $H_0 : \boldsymbol{\theta} = \mathbf{0}$ in an order $k$ smooth family of alternatives, which is now given by

$$\pi_{ki} = C(\boldsymbol{\theta}) \exp\left(\sum_{j=1}^{k} \theta_j h_{ij}\right) \pi_{0i} \quad i = 1, \ldots, m, \tag{4.32}$$

where $\{\boldsymbol{h}_j\}$, with $\boldsymbol{h}_j^t = (h_{1j}, \ldots, h_{mj})$, is a set of orthonormal vectors in the $m$-dimensional vector space with inner product defined by $< \boldsymbol{p}, \boldsymbol{q} >_{\pi_0} = \sum_{i=1}^{m} p_i q_i \pi_{0i}$. Let $V(\boldsymbol{\pi}_0)$ denote this vector space of vectors $\boldsymbol{h}$ for which $< \boldsymbol{h}, \boldsymbol{h} >_{\pi_0}$ is finite. The orthonormality condition thus implies

$$\sum_{i=1}^{m} h_{ij} h_{il} \pi_{0i} = \delta_{jl}. \tag{4.33}$$

It is convenient to write restriction (4.33) in matrix notation. Let $\boldsymbol{H}^t$ denote the $m \times k$ matrix with $(i,j)$th element equal to $h_{ij}$; i.e., $\boldsymbol{H}^t = (\boldsymbol{h}_1, \ldots, \boldsymbol{h}_k)$, and let $\boldsymbol{D}_\pi = \text{diag}(\boldsymbol{\pi}_0)$. Then (4.33) is equivalent to

$$\boldsymbol{H}\boldsymbol{D}_{\pi_0}\boldsymbol{H}^t = \boldsymbol{I},$$

with $\boldsymbol{I}$ the $k \times k$ identity matrix.

For a given distribution $\boldsymbol{\pi}_0$, the orthonormal vectors $\{\boldsymbol{h}_j : j = 0, \ldots, k\}$ are usually easy to find. We always impose the restriction $h_{i0} = 1$ for $i = 1, \ldots, m$.

The score test statistic and its asymptotic null distribution are presented in the next theorem.

**Theorem 4.9.** *Let $Y_1, \ldots, Y_n$ denote a sample of i.i.d. observations that take values in $\{1, \ldots, m\}$ and which have under the null hypothesis distribution function $\pi_{0i} = \text{Pr}_0\{Y = i\}$ $(i = 1, \ldots, m)$. Let $\boldsymbol{N}^t = (N_1, \ldots, N_m)$ denote the vector of counts $N_i$ of sample observations $Y$ equal to $i$. Finally, let $\{\boldsymbol{h}_j\}$, with $\boldsymbol{h}_j^t = (h_{1j}, \ldots, h_{mj})$, be a set of orthonormal vectors in the $m$-dimensional vector space $V(\boldsymbol{\pi}_0)$.*

*(1) The score test statistic for testing $H_0 : \boldsymbol{\theta} = \boldsymbol{0}$ in the order $k$ smooth model (4.32) is given by*

$$T_k = \sum_{j=1}^{k} U_j^2, \tag{4.34}$$

*where $U_j = (1/\sqrt{n}) \sum_{i=1}^{m} N_i h_{ij}$ $(j = 1, \ldots, k)$.*

*(2) Let $\boldsymbol{U}^t = (U_1, \ldots, U_k)$ and let $\boldsymbol{I}$ denote the $k \times k$ identity matrix. Under the null hypothesis, as $n \to \infty$,*

$$\boldsymbol{U} \xrightarrow{d} MVN(\boldsymbol{0}, \boldsymbol{I}) \text{ and } T_k \xrightarrow{d} \chi_k^2. \tag{4.35}$$

The proof of the theorem is very similar to the proof of Theorem 4.1 and we therefore omit it here.

The next theorem shows a nice relation between the smooth test statistic and the Pearson $\chi^2$ statistic. In particular, it demonstrates that Pearson's $\chi^2$ is basically a smooth test statistic, and it can therefore also be decomposed into $m - 1$ components. Appendix A.7 contains the proof.

**Theorem 4.10.** *Consider the notation of Theorem 4.9. If $k = m - 1$, then*

$$T_k = \sum_{j=1}^{m-1} U_j^2 = \sum_{i=1}^{m} \frac{(N_i - n\pi_{0i})^2}{n\pi_{0i}}. \tag{4.36}$$

### *4.4.3 The Composite Null Hypothesis Case*

When nuisance parameters are involved, both the hypothesised distribution $\boldsymbol{\pi}_0$ and the orthonormal vectors in $\boldsymbol{H}(\boldsymbol{\beta}) = \boldsymbol{H}$ depend on the $p$-dimensional vector $\boldsymbol{\beta}$. The smooth test statistics and their asymptotic null distributions may again be found in a similar fashion as in Section 4.2.2. Also the results on the Pearson $\chi^2$ test in the composite case are useful in proving the results presented here. Particularly, the proof of Theorem 1.2 is very useful.

**Theorem 4.11.** *Assume that the conditions of Theorem 4.9 hold, and write the score vector* $\boldsymbol{U}(\boldsymbol{\beta}) = (1/\sqrt{n})\boldsymbol{H}(\boldsymbol{\beta})\boldsymbol{N} = \sqrt{n}\boldsymbol{H}(\boldsymbol{\beta})\,(\hat{p} - \boldsymbol{\pi}_0(\boldsymbol{\beta}))$.

*(1) Suppose that* $\hat{\boldsymbol{\beta}}$ *is a BAN estimator of* $\boldsymbol{\beta}$. *The order* $k$ $(1 < k < m)$ *smooth test statistic is given by*

$$T_k = \boldsymbol{U}^t(\hat{\boldsymbol{\beta}})\hat{\boldsymbol{\Sigma}}^{-1}\boldsymbol{U}(\hat{\boldsymbol{\beta}}), \tag{4.37}$$

*where*

$$\hat{\boldsymbol{\Sigma}}^{-1} = \hat{\boldsymbol{D}}_{\pi_0} - \hat{\boldsymbol{\pi}}_0\hat{\boldsymbol{\pi}}_0^t - \hat{\boldsymbol{D}}_{\pi_0}^{1/2}\hat{\boldsymbol{A}}(\hat{\boldsymbol{A}}^t\hat{\boldsymbol{A}})^{-1}\hat{\boldsymbol{A}}^t\hat{\boldsymbol{D}}_{\pi_0}^{1/2}, \tag{4.38}$$

*and the* $\hat{\ }$ *notation is used to indicate that the nuisance parameter is replaced by* $\hat{\boldsymbol{\beta}}$. *Under* $H_0$, *as* $n \to \infty$,

$$T_k \xrightarrow{d} \chi^2_{k-p-1}. \tag{4.39}$$

*(2) Suppose that* $\hat{\boldsymbol{\beta}}$ *is a* $\sqrt{n}$*-consistent estimator of* $\boldsymbol{\beta}$. *Let*

$$u_{\beta_j i}(\boldsymbol{\beta}) = \frac{1}{\pi_{0i}(\boldsymbol{\beta})}\frac{\partial \pi_{0i}(\boldsymbol{\beta})}{\partial \boldsymbol{\beta}_j} \quad (i = 1, \ldots, m; j = 1, \ldots, p),$$

*and* $\boldsymbol{u}^t_{\beta_j} = (u_{\beta_j 1}, \ldots, u_{\beta_j m})$. *Similarly,* $\boldsymbol{u}^t_{\beta i} = (u_{\beta_1 i}, \ldots, u_{\beta_p i})$. *Let* $\boldsymbol{u}_\beta$ *denote the* $p \times m$ *matrix with* $i$*th row equal to* $\boldsymbol{u}^t_{\beta i}$ $(i = 1, \ldots, p)$. *Let* $\boldsymbol{\Sigma}_{h\beta}$ *be a* $k \times p$ *matrix with* $(i, j)$*th element equal to* $< \boldsymbol{h}_i, \boldsymbol{u}_{\beta_j} >_{\pi_0}$ $(i = 1, \ldots, k,$ $j = 1, \ldots, p)$, *and the* $p \times p$ *matrix* $\boldsymbol{\Sigma}_{\beta\beta}$ *has* $(i, j)$*th element given by* $< \boldsymbol{u}_{\beta_i}, \boldsymbol{u}_{\beta_j} >_{\pi_0}$ $(i, j = 1, \ldots, p)$. *The efficient score statistic is then given by*

$$\boldsymbol{V}(\boldsymbol{\beta}) = (V_1, \ldots, V_k)^t = \frac{1}{\sqrt{n}}\left(\boldsymbol{H}(\boldsymbol{\beta}) - \boldsymbol{\Sigma}_{h\beta}\boldsymbol{\Sigma}^{-1}_{\beta\beta}\boldsymbol{U}(\boldsymbol{\beta})\right)\boldsymbol{N}. \tag{4.40}$$

*Using the notation* $\hat{\boldsymbol{V}} = \boldsymbol{V}(\hat{\boldsymbol{\beta}})$ *(with also all* $\boldsymbol{\beta}$ *in the covariance matrices replaced by* $\hat{\boldsymbol{\beta}}$*), we find, under* $H_0$, *as* $n \to \infty$,

$$T_k = \hat{\boldsymbol{V}}^t\hat{\boldsymbol{\Sigma}}^{-1}\hat{\boldsymbol{V}} \xrightarrow{d} \chi^2_{k-p}, \tag{4.41}$$

*where* $\hat{\boldsymbol{\Sigma}}$ *is the matrix* $\boldsymbol{\Sigma} = \boldsymbol{I} - \boldsymbol{\Sigma}_{h\beta}\boldsymbol{\Sigma}^{-1}_{\beta\beta}\boldsymbol{\Sigma}_{\beta h}$ *with all* $\boldsymbol{\beta}$ *replaced by* $\hat{\boldsymbol{\beta}}$.

*Example 4.7 (Pulse rate).* To illustrate the smooth test for a discrete distribution, we test the null hypothesis that the pulse rate data of Section 1.2.3 comes from a Poisson distribution. Note that for the Poisson distribution the MLE and MME coincide. The null hypothesis is tested by means of a smooth test of order $k = 6$, and the first component is exactly zero by the estimation process.

The R-code and the resulting output is shown below. For the computation of the $p$-values the asymptotic $\chi^2$ approximation is chosen.

```
> smooth.test(pulse,order=6,distr="pois",method="MLE",B=NULL)
Smooth goodness-of-fit test
Null hypothesis:  pois  against  6 th order alternative
Nuisance parameter estimation:  MLE
Parameter estimates:  82.3  ( lambda )

Smooth test statistic S_k =  20.9846  p-value =  0.0008155
     2 th component V_k = -0.246051  p-value =  0.8056427
     3 th component V_k =  3.041709  p-value =  0.0023524
     4 th component V_k =  3.242072  p-value =  0.0011866
     5 th component V_k =  0.662461  p-value =  0.5076757
     6 th component V_k = -0.849816  p-value =  0.3954270

All p-values are obtained by the asymptotical chi-square
   approximation
```

From the output we read that the $p$-value of the order $k$ smooth test equals $p = 0.0008 < 0.05$, and therefore we conclude at the 5% level of significance that the observations do not come from a Poisson distribution. A closer look at the individual components may shed some light on how the distribution differs from the Poisson distribution. Here the third- and the fourth-order components show very large values. This suggests that the pulse rate distribution has a different skewness and a different kurtosis from a Poisson distribution with mean equal to 82.3.

It is interesting to compare this conclusion with the exploratory analysis that we have presented in Section 3.3.3 by plotting the comparison distribution. This plot showed that there were too many counts observed around the a pulse rate of 80, and too few counts to the immediate left and right of this pulse rate. In other words, the plot suggested that the mode of the distribution does not correspond to what was expected for a Poisson distribution. Moving the mode of a distribution, but keeping the mean equal to 82.3 does indeed have an immediate effect on the skewness and the kurtosis.

## 4.5 A Semiparametric Framework

### 4.5.1 The Semiparametric Hypotheses

It has become clear by now that smooth tests within a finite horizon are not omnibus consistent, for they are not sensitive to deviations of the higher-order moments of the hypothesised distribution $g$. By restricting the order

$k < \infty$ it actually looks as if the statistician is only interested in the first $k$ moments of $g$. This may be formalised by adopting a semiparametric null hypothesis.

We restrict the discussion to continuous densities $f \in \mathcal{F} = \{f \in L_2(\mathcal{S}) : \int_{\mathcal{S}} x^j f(x)dx < \infty, j = 1, \ldots, k\}$. The set of densities with the first $k$ moments equal to those of $g(.;\boldsymbol{\beta})$ is defined as

$$\mathcal{F}_0 = \{f \in \mathcal{F} : \mathrm{E}_f\{h_j(X;\boldsymbol{\beta})\} = 0, j = 1, \ldots, k\},$$

where $\{h_j\}$ is the set of orthonormal polynomials w.r.t. density $g$. In this context the distribution $g$ only plays the role of a hypothesised-moment generating density. The semiparametric hypotheses may now be formulated as

$$H_0 : f \in \mathcal{F}_0 \text{ and } H_1 : f \in \mathcal{F} \setminus \mathcal{F}_0.$$

To get a deeper insight and a correct interpretation of the meaning of the parameter $\boldsymbol{\beta}$, we look at it from a Hilbert space perspective. In Section 4.1.1 we have shown that the Barton model corresponds to the representation of the relative density $f/g$ in a Hilbert space $L_2(\mathcal{S}; G)$ spanned by the orthonormal basis functions $h_j$. The full parametric null hypothesis ($\boldsymbol{\theta} = \mathbf{0}$) corresponds to $< f/g, h_j >_g = 0$ for all $j = 1, \ldots$; i.e., the relative density is orthogonal to all basis functions $h_j$. Although

$$< f/g, h_j >_g = \int_{\mathcal{S}} \frac{f(x)}{g(x;\boldsymbol{\beta})} h_j(x;\boldsymbol{\beta}) g(x;\boldsymbol{\beta}) dx = \mathrm{E}_f\{h_j(X;\boldsymbol{\beta})\}$$

is expressed in term of expectations as those in $\mathcal{F}_0$, the Hilbert space $L_2(\mathcal{S}; G)$ is not suited for the semiparametric hypothesis. The reason is that inner product $< .,. >_g$ of $L_2(\mathcal{S}; G)$ depends explicitly on the density $g$ which is only meaningful under the full parametric null hypothesis. Consider instead the space $L_2(\mathcal{S}; F)$. In this space we have

$$\mathrm{E}_f\{h_j(X;\boldsymbol{\beta})\} = < h_j, 1 >_f,$$

which does not depend on $g$. Hence, $\mathcal{F}_0$ is the set of functions $f$ so that in the space $L_2(\mathcal{S}; F)$ the identity function 1 is orthogonal to the linear subspace spanned by $h_1, \ldots, h_k$. This subspace is denoted by $\mathcal{P}_k = \mathrm{span}(h_1, \ldots, h_k)$, or by $\mathcal{P}_k(\boldsymbol{\beta})$ to stress the dependence on the parameter $\boldsymbol{\beta}$. Note that in $L_2(\mathcal{S}; F)$, the functions $h_j$ do not necessarily form an orthogonal basis.

### 4.5.2 Semiparametric Tests

When the full parametric null hypothesis is replaced by a semiparametric null hypothesis, do we need to construct different statistical tests, or can we still work with, e.g., the smooth tests discussed in this chapter? We give an answer to this question in this section, but first we mention that there is a vast literature on semiparametric inference, which is, however, predominantly about efficient estimation. We refer to Bickel et al. (2006) for a good treatment on semiparametric hypothesis testing, but we de not follow their method of test construction here.

Consider a test statistic of the form $T_n = \hat{\boldsymbol{\theta}}^t \hat{\boldsymbol{\Sigma}}^{-1} \hat{\boldsymbol{\theta}}$ where $\hat{\boldsymbol{\theta}}$ is a $\sqrt{n}$-consistent estimator of $\boldsymbol{\theta}$ and where $\hat{\boldsymbol{\Sigma}}$ is a $\sqrt{n}$-consistent estimator of $\text{Var}\left\{\hat{\boldsymbol{\theta}}\right\}$ under certain conditions specified below. The statistic $T_n$ has clearly an appropriate form for the testing problem at hand. We want the test to be asymptotically unbiased under the semiparametric null hypothesis, and consistent against the alternatives to the semiparametric null hypothesis. Both "unbiasedness" and "consistency" are defined w.r.t. the distributions of the observations under the semiparametric null hypothesis and alternative hypothesis, respectively. Thus, for each $\alpha \in (0, 1)$ there exists a $c_\alpha$ so that the test is

(1) *Asymptotically unbiased*:

$$\lim_{n\to\infty} \sup_{f\in\mathcal{F}_0} \Pr_f \{T_n > c_\alpha\} \leq \alpha;$$

(2) *Consistent*:

$$\lim_{n\to\infty} \inf_{f\in\mathcal{F}\setminus\mathcal{F}_0} \Pr_f \{T_n > c_\alpha\} = 1.$$

A sufficient condition for (1) to hold is that $T_n$ has asymptotically the same null distribution for all $f \in \mathcal{F}_0$. It usually holds that $\hat{\boldsymbol{\theta}}$ has asymptotically a zero mean multivariate normal distribution for all $f$ under the semiparametric null hypothesis, so that it remains to be assured that $\hat{\boldsymbol{\Sigma}}$ is $\sqrt{n}$-consistent for all $f \in \mathcal{F}_0$. The consistency property (2) may often simplified to the condition that

$$\lim_{n\to\infty} \sup_{f\in\mathcal{F}\setminus\mathcal{F}_0} |\hat{\boldsymbol{\Sigma}}| < \infty \tag{4.42}$$

with probability one.

In almost all of this chapter, except for Section 4.2.1.2, the estimator of the variance of $\hat{\boldsymbol{\theta}} = \hat{\boldsymbol{U}}$ has been constructed so that it is consistent under the full parametric null hypothesis. This is a consequence of the smooth test being essentially a score test, which is always constructed by imposing the (full parametric) null hypothesis. However, in Section 4.2.1.2, where the diagnostic property of the component tests in the simple null case was discussed, we referred to the work of Henze and Klar (Henze (1997), Henze and Klar (1996), Klar (2000)) who showed that the covariance matrix $\text{Var}\{\boldsymbol{U}\}$ should be estimated by its empirical covariance matrix estimator, which in the simple null hypothesis case works well under quite mild regularity conditions. In the following subsections we elaborate briefly on some rather recent developments in the area of semiparametric goodness-of-fit testing related to the smooth tests.

### 4.5.3 A Distance Function

Suppose first that $\boldsymbol{\beta}$ is known. As before, we construct a goodness-of-fit test statistic on a distance function. However, in a semiparametric framework we cannot use a distance function between the true $f$ and the hypothesised $g$, because the latter implies more restrictions than expressed by the semiparametric null hypothesis. Instead we use a distance function between $f$ and the set $\mathcal{F}_0$. Because the restrictions of $\mathcal{F}_0$ indicate that the identity function 1 is orthogonal to the subspace $\mathcal{P}_k$, a meaningful distance function exists in (1) projecting 1 orthogonally onto $\mathcal{P}_0$; (2) calculating the length of this projection. Details follow.

1. Although the $h_j$ that span $\mathcal{P}_k$ are not orthogonal in $L_2(\mathcal{S}, F)$, they are linearly independent. Using the notation $\boldsymbol{h}_1^t = (h_1, \ldots, h_k)$ to denote a vector-valued function, the orthogonal projection of 1 onto $\mathcal{P}_k$ is therefore given by
$$< 1, \boldsymbol{h}_1 >_f < \boldsymbol{h}_1, \boldsymbol{h}_1 >_f^{-1} \boldsymbol{h}_1$$
(see Section 2.5 for details on orthogonal projections). Note that the $j$th element in $< 1, \boldsymbol{h}_1 >_f$ equals $< 1, h_j >_f = \theta_j$ when the Barton model representation is considered.
2. Next, we calculate the squared length of the projection. Simple algebra results in the squared length
$$d_k^2(\boldsymbol{\beta}) = \boldsymbol{\theta}_k^t \boldsymbol{C}^{-1} \boldsymbol{\theta}_k, \tag{4.43}$$
where $\boldsymbol{C} = < \boldsymbol{h}_1, \boldsymbol{h}_1 >_f$, which is a $k \times k$ matrix with $(i, j)$th element equal to $< h_i, h_j >_f$.

Clearly, when $f \in \mathcal{F}_0$, there exists a $\boldsymbol{\beta}$ so that $d_k^2(\boldsymbol{\beta}) = 0$.

### 4.5.4 Interpretation and Estimation of the Nuisance Parameter

In the full parametric setting the parameter $\boldsymbol{\beta}$ has an unambiguous interpretation as it simply appears as a parameter in a well-defined density function $g$. Now, however, $g$ only serves as a hypothesised moment generating density function, and the nuisance parameter appears in the moment restrictions $\mathrm{E}_f\{h_j(X; \boldsymbol{\beta})\}$. In the Hilbert space, $\boldsymbol{\beta}$ determines the position of the subspace $\mathcal{P}_k(\boldsymbol{\beta})$.

From the construction of the distance function $d_k^2$, we could find a definition of $\boldsymbol{\beta}$,
$$\boldsymbol{\beta} = \mathrm{ArgMin}_{\boldsymbol{b}} d_k^2(\boldsymbol{b}).$$

The parameter $\boldsymbol{\beta}$ is thus defined so that it makes the subspace $\mathcal{P}_0$ as orthogonal to 1 as possible. Or, in other words, $\boldsymbol{\beta}$ places the subspace $\mathcal{P}_0(\boldsymbol{\beta})$ so that in some sense the first $k$ moments of $f$ come as close as possible to the hypothesised moments. If $f \in \mathcal{F}_0$, then $d_k^2(\boldsymbol{\beta}) = 0$, but even when $f \notin \mathcal{F}_0$, the parameter $\boldsymbol{\beta}$ is still well defined!

The above discussion includes a hint regarding nuisance parameter estimation. First, because $g$ has no meaning as a density function in the semiparametric setting, it is obvious that MLE does not exist here. The minimum distance approach, however, suggests another simple estimation method: find $\tilde{\boldsymbol{\beta}}$ that minimizes some estimator of the squared distance function. Equation (4.43) suggests that such an estimator is given by

$$\tilde{\boldsymbol{\theta}}^t \tilde{\boldsymbol{C}}^{-1} \tilde{\boldsymbol{\theta}}, \tag{4.44}$$

where the $j$th element of $\tilde{\boldsymbol{\theta}}$ equals $\tilde{\theta}_j = (1/\sqrt{n}) U_j(\tilde{\boldsymbol{\beta}})$ and $\tilde{\boldsymbol{C}}$ is a $\sqrt{n}$-consistent estimator of $\boldsymbol{C}$. Since $\boldsymbol{C}$ depends on the unknown $f$, we consider the empirical estimator which as $(i,j)$th element equal to $(1/n)\sum_{l=1}^n h_i(X_l;\tilde{\boldsymbol{\beta}}) h_j(X_l;\tilde{\boldsymbol{\beta}})$. Note that $\boldsymbol{C}$ has the interpretation of the variance–covariance matrix of $\tilde{\boldsymbol{\theta}}$ calculated under the semiparametric null hypotheses.

## 4.5.5 The Quadratic Inference Function

In the previous subsections we have described how a semiparametric null hypothesis is expressed in terms of $k$ moment restrictions. Within a Hilbert space we have defined a quadratic distance function which measures how far $f$ is from $\mathcal{F}_0$ for a given nuisance parameter $\boldsymbol{\beta}$. This parameter is well defined in the semiparametric setting as the minimiser of the distance function. By replacing the distance function by an estimator, we immediately arrived at an estimation method for the nuisance parameter. This method was first proposed by Qu et al. (2000) in a more general setting. They refer to the statistic in (4.44) as the quadratic inference function (QIF), which we further denote by $\text{QIF}_k(\boldsymbol{\theta})$. The estimator of $\boldsymbol{\beta}$ which is defined as the minimiser of (4.44) is therefore referred to as the minimum quadratic influence function estimator (MQIFE).

We have used the QIF as an inference function to find an estimator of the nuisance parameter $\boldsymbol{\beta}$. Qu and coworkers showed that the MQIFE is consistent, even when $f \notin \mathcal{F}_0$. Because $\text{QIF}_k(\tilde{\boldsymbol{\beta}})$ is an estimator of the minimised squared distance function, they further proposed using this statistic as a goodness-of-fit test statistic. In particular, under the semiparametric null hypothesis, as $n \to \infty$,

$$\text{QIF}_k(\tilde{\boldsymbol{\beta}}) = \tilde{\boldsymbol{\theta}}^t \tilde{\boldsymbol{C}}^{-1} \tilde{\boldsymbol{\theta}}_k \xrightarrow{d} \chi^2_{k-p}. \tag{4.45}$$

They also showed that the MQIFE $\tilde{\boldsymbol{\beta}}$ is asymptotically normally distributed.

We have performed an extensive simulation study in which we have studied goodness-of-fit tests based on QIF. These results are not published, merely because of the poor results. First, the convergence to the asymptotic $\chi^2$ approximations is very slow ($n > 1000$ is still not satisfactory). Second, on using the semiparametric bootstrap method of Bickel and Ren (2001), which is described in Appendix B.3, we still found biased test results. Moreover, poor powers were found.

### *4.5.6 Relation with the Empirically Rescaled Smooth Tests*

Earlier in this chapter we already mentioned briefly that the smooth tests of Henze and Klar were actually developed in a semiparametric setting (see Section 4.2.1). Their test statistic is of the same form as the QIF statistic (4.45), except that the nuisance parameter $\boldsymbol{\beta}$ is not estimated as the MQIFE, but rather as the MME (in this section denoted by $\hat{\boldsymbol{\beta}}$). MME forces the first $p$ moments of $g(x; \hat{\boldsymbol{\beta}})$ to coincide with the corresponding sample moments, implying the first $p$ components of $\hat{\boldsymbol{\theta}} = (1/\sqrt{n})\boldsymbol{U}_k(\hat{\boldsymbol{\beta}})$ to be zero. These zero elements are removed from $\hat{\boldsymbol{\theta}}$, and their statistic becomes

$$T_{k-p} = \hat{\boldsymbol{\theta}}^t \hat{\boldsymbol{C}}^{-1} \hat{\boldsymbol{\theta}}, \tag{4.46}$$

where $\hat{\boldsymbol{\theta}}$ is a vector with $k-p$ nonzero elements $\hat{\theta}_j = (1/\sqrt{n})U_j(\hat{\boldsymbol{\beta}})$, $j = p+1, \ldots, k$ and $\hat{\boldsymbol{C}}$ is the $(k-p) \times (k-p)$ empirical variance–covariance matrix estimator, but now with the MME $\hat{\boldsymbol{\beta}}$. They considered the components scaled by using the appropriate diagonal element of $\hat{\boldsymbol{C}}$ as the basis of component tests that have the *diagnostic property.*

The MME-based generalised smooth test statistic (4.46) measures thus the distance between the $p+1$ up to the $k$th sample moments and the corresponding moments of $g(x; \hat{\boldsymbol{\beta}})$ which fits exactly the first $p$ sample moments. Or, similarly, given that the first $p$ moments of $g(x; \hat{\boldsymbol{\beta}})$ agree with the sample observations, (4.46) measures how far the other $k-p$ sample moments deviate from the hypothesised.

When the MQIFE $\tilde{\boldsymbol{\beta}}$ is used instead, the QIF test statistic $\text{QIF}_k(\tilde{\boldsymbol{\beta}})$ measures how close the first $k$ moments of $g$ can be brought to their sample counterparts, and thus $\text{QIF}_k(\tilde{\boldsymbol{\beta}})$ avoids in some sense the conditioning on the equality of the first $p$ moments of $g$. The QIF approach treats all $k$ moments evenly.

The theory of Klar (2000) is quite general, but the empirical covariance matrix $\hat{\boldsymbol{C}}$ may only be used when MME and MLE coincide for the hypothesised distribution $g$. When MLE and MME are different, this estimator does not correctly account for the estimation of the nuisance parameters. In this case Klar (2000) suggested to express $\text{Var}_f\left\{\hat{\boldsymbol{\theta}}\right\}$ ($f \in \mathcal{F}_0$) in terms of the moments of $f$, and subsequently replace these moments by their empirical counterparts, and use this estimator instead of $\hat{\boldsymbol{C}}$. Another solution, which involves the nuisance estimation equations explicitly, was proposed by Thas and Rayner (2009) and is also illustrated in Rayner et al. (2009).

## 4.6 Example

We illustrate now the methods of the previous sections on the PCB data. The data have been used before in Section 2.1.1 to demonstrate the construction and the interpretation of the comparison distribution. There it was concluded that the density of PCB concentrations is slightly larger than expected for a normal distribution around concentrations of 200, and slightly smaller than expected for concentrations of about 270. This conclusion was of course formulated in terms of the relative density, but it is often more informative to formulate the conclusion in other terms. For instance, this relative density interpretation, together with the accompanying nonparametric density estimation shown in the top panel of Figure 3.15, suggests that the PCB distribution may perhaps be bimodal.

In this section we test the composite null hypothesis that the PCB concentration data come from a normal distribution. We test this hypothesis first with a traditional smooth test based on the efficient scores. Because the normal distribution belongs to the exponential family, and MME and MLE coincide, it does not matter which $\sqrt{n}$-consistent estimation scheme we choose. The output below shows the R-code and the results of two smooth tests with fixed orders $k = 6$ and $k = 7$. All $p$-values are obtained from the asymptotic $\chi^2$ approximation, but the results based on the simulated null distribution give the same conclusions.

```
> smooth.test(PCB,distr="norm",method="MLE",order=3,B=NULL)
Smooth goodness-of-fit test
Null hypothesis:  norm  against  3 th order alternative
Nuisance parameter estimation:  MLE
Parameter estimates:  210 72.26383  ( MEAN VAR )

Smooth test statistic S_k =  5.436919  p-value =  0.01971542
      3 th component V_k =  2.331720  p-value =  0.01971542

All p-values are obtained by the asymptotical chi-square
approximation

> smooth.test(PCB,distr="norm",method="MLE",order=6,B=NULL)
Smooth goodness-of-fit test
Null hypothesis:  norm  against  6 th order alternative
Nuisance parameter estimation:  MLE
Parameter estimates:  210 72.26383  ( MEAN VAR )
```

```
Smooth test statistic S_k =  10.18261  p-value =  0.03746153
     3 th component V_k =  2.331720  p-value =  0.01971542
     4 th component V_k =  2.030241  p-value =  0.042332
     5 th component V_k =  0.434342  p-value =  0.6640404
     6 th component V_k = -0.659661  p-value =  0.5094708


All p-values are obtained by the asymptotical chi-square
approximation

> smooth.test(PCB,distr="norm",method="MLE",order=7,B=NULL)
Smooth goodness-of-fit test
Null hypothesis:  norm  against  7 th order alternative
Nuisance parameter estimation:  MLE
Parameter estimates:  210 72.26383  ( MEAN VAR )

Smooth test statistic S_k =  10.59477  p-value =  0.06003358
     3 th component V_k =  2.331720  p-value =  0.01971542
     4 th component V_k =  2.030241  p-value =  0.042332
     5 th component V_k =  0.434342  p-value =  0.6640404
     6 th component V_k = -0.659661  p-value =  0.5094708
     7 th component V_k = -0.641999  p-value =  0.5208738

All p-values are obtained by the asymptotical chi-square
approximation
$statistics
```

We present the tests with three different orders for demonstrating the dilution effect as explained in Section 4.3.1. Our statistical analyses show that the smooth tests with $k = 3$ and with $k = 6$ give $p$-values of 0.020 and 0.037, respectively. Thus they both reject the null hypothesis of normality at the 5% level of significance. However, if $k = 7$ were chosen, then the smooth test would have $p$-value equal to 0.060 which does not imply the rejection of the null hypothesis. The reason may be found by looking at the $p$-values of the individual component tests. The third- and the fourth-order component tests have small $p$-values, but as the order increases, the $p$-values increase too. This is a typical illustration of the dilution effect.

We previously used the $p$-values of the individual component tests, but in Section 4.2.1 we argued extensively that the components should be rescaled to recover their full diagnostic property. Later, in Section 4.5.6, we explained the method of Henze and Klar in the presence of nuisance parameters. The following R-code and output pe concern these rescaled component tests (using the rescale=T option in the smooth.test function).

```
> smooth.test(PCB,distr="norm",method="MLE",order=6,rescale=T,
+ B=1000)
Smooth goodness-of-fit test with Henze and Klar rescaling of
the components
Null hypothesis:  norm  against  6 th order alternative
Nuisance parameter estimation:  MLE
Parameter estimates:  210 72.26383  ( MEAN VAR )

Smooth test statistic S_k =  10.18261  p-value =  0.024
  3 th rescaled component V_k =  1.493205  p-value =  0.135
  4 th rescaled component V_k =  1.212814  p-value =  0.276
  5 th rescaled component V_k =  0.350246  p-value =  0.779
  6 th rescaled component V_k = -0.974392  p-value =  0.290

All p-values are obtained by the bootstrap with  1000  runs
```

This output first shows the simulated $p$-value of order $k$ smooth test: $p = 0.024$. The next lines show the empirically rescaled components and the $p$-values. Whereas we previously concluded that the third- and the fourth-order component tests gave significant results, we must now conclude that they are not significant. This may look like a contradiction. There are two possible explanations. The first is that the skewness and the kurtosis of the PCB concentration distribution agree with those of the normal distribution, and that it was falsely suggested by the nonrescaled component tests due to an incorrect standardisation of the components. A second explanation might be that the use of the empirical variance estimator in the rescaled component test introduces additional variance, which further implies a loss in power. Thus maybe the large $p$-values of the rescaled component tests are a consequence of a smaller power. Which one of the two arguments is correct is still not clear at this point.

There is also still another problem left unanswered. Which analysis should we trust: the smooth test with $k < 7$ or with $k = 7$? To avoid the problem of choosing the order $k$ in an arbitrary way, as we have done here, we can also apply one of the adaptive smooth tests of Section 4.3. In particular, we apply the BIC-based data-driven test as described in Section 4.3.2. The BIC criterion is given in (4.22), the order selection rule in (4.24), and the test statistic in (4.25). The R-code and output follow.

```
> smooth.test(PCB,distr="norm",method="MLE",
+ adaptive=c("BIC","order"),max.order=7,plot=T,B=10000)
Adaptive Smooth goodness-of-fit test
Null hypothesis:  norm  against  7 th order alternative
Nuisance parameter estimation:  MLE
Parameter estimates:  210 72.26383  ( MEAN VAR )
Order selection rule: BIC
```

```
Adaptive smooth test statistic S_k = 5.436919
p-value = 0.0325
Selected order = 3

All p-values are obtained by the bootstrap with  10000  runs
```

The adaptive smooth tests are invoked by the smooth.test function with the adaptive option specifying the selection rule (BIC). The specification "order" means that BIC is used to select the order of the test. If "subset" were used instead, then BIC would be used to select a subset model. The option max.order specifies the maximal order of the model that can be chosen. Although the theory says that this data-driven test statistic has asymptotically a $\chi_1^2$ null distribution, empirical studies have indicated that the convergence is rather slow. We have therefore computed the bootstrap $p$-values based on 10,000 simulation runs.

The $p$-value of this data-driven smooth test is 0.0325. Based on this adaptive test we decide to reject the null hypothesis of normality at the 5% level of significance. The BIC criterion selected only the third-order term. Although the test statistic that was used here is not properly scaled to guarantee the diagnostic property, we may at least have trust in the overall conclusion: rejection of the null hypothesis of normality. With this argument in mind, the large $p$-value of the rescaled test is likely to be a consequence of the smaller powers of rescaled tests. When the diagnostic property of smooth tests is not present, it is often instructive to plot the improved density estimate and use this graphical representation as a basis for formulating conclusions. This improved density estimate is plotted by the smooth.test function by setting the argument plot=T. The graph is presented in the left panel of Figure 4.1. In

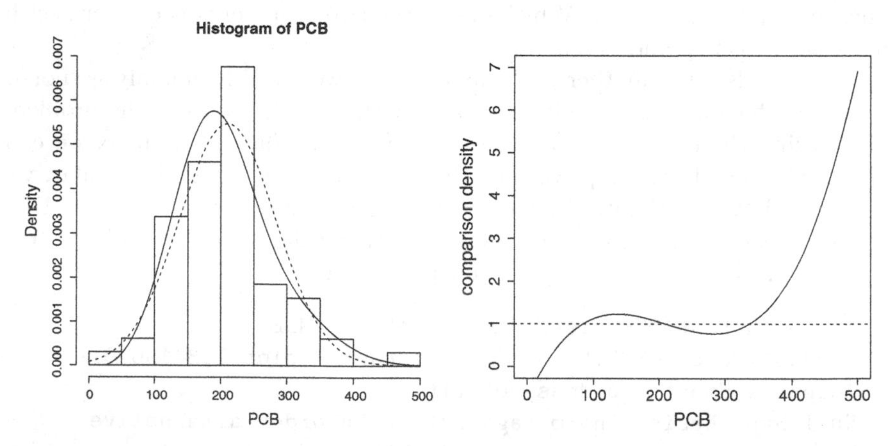


**Fig. 4.1** The left panel shows the histogram of the PCB data, the fitted normal density (dashed line), and the improved density estimate (full line); the right panel shows the comparison density

this quite simple example, for which only the first nonzero term is selected, the improved density estimate of course also shows the skewness of the PCB distribution. In situations for which several terms are selected and for which the diagnostic property does not work, it may be safer to use the improved density estimate for formulating conclusions. Improved density estimates can be plotted together with confidence intervals. The right panel of Figure 4.1 shows the comparison density, which contains the same information as the improved density.

## 4.7 Some Practical Guidelines for Smooth Tests

In general smooth goodness-of-fit tests have many good properties. We name here the most important.

- Smooth tests are easy to compute.
- Smooth tests are available for many distributions.
- Many simulation studies have indicated that (data-driven) smooth tests have good power for detecting many important alternatives. For most practical applications, it is sufficient to chose $k = 4$ for small sample sizes ($n < 50$), or $k = 6$ for larger datasets ($n \approx 100$). For the data-driven tests, there seems to be little need to take the maximal order larger than 7 for small datasets, and 10 for larger datasets.
- Although the smooth test statistic has an asymptotic $\chi^2$ distribution, we recommend using simulations to compute $p$-values (see Appendix B.2 for details on the parametric bootstrap).
- For many distributions the smooth test statistic decomposes into components (this happens, e.g., for the normal, exponential, Poisson, . . .). These components possess limited diagnostic power, in the sense that if the $j$th component is large, the statistic suggests that the data are inconsistent with the hypothesised distribution in at least one moment of order $\leq 2j$. Such conclusions must however be taken with great care, particularly when there are large inconsistencies in more than one moment. Rescaling the components by using an empirical variance estimator only works in situations where (1) there are no nuisance parameters, or (2) the hypothesised distribution belongs to a restricted, though important class of distributions. Also for these rescaled components one should be careful in the interpretation, because simulations studies have shown that large samples are needed for the method to work well.
- The remark given in the previous paragraph suggest the following practical guideline: when looking at the individual components, always start with the lowest-order component, and stop interpreting them as soon as a large component is encountered.
- Because the smooth tests can be interpreted as tests for testing that the parameters in an orthogonal series estimator of the comparison density

are all zero, the plot of the comparison density or the improved density estimate may be helpful in seeking a deeper understanding of how the true and the hypothesised distributions are different. This is particularly helpful when the diagnostic property of the components is in doubt.

- For some distributions (e.g., the logistic and extreme value distributions) the smooth test statistic does not naturally decompose into its components. For these distributions the MLE and MME do not coincide, and we suggest using MME here instead, and to use a generalised smooth test. With this construction, it is still informative to look at the individual components. The rescaling technique with the empirical covariance matrix requires a different estimator of the covariance matrix; see Thas and Rayner (2009).

# Chapter 5
# Methods Based on the Empirical Distribution Function

In this chapter a very wide class of statistical tests based on the empirical distribution function (EDF) is introduced. Among these tests we find some old tests, as the Kolmogorov–Smirnov test, but also in recent years new tests have still been added to this class. A discussion on the EDF and empirical processes has been given in Sections 2.1 and 2.2. Sections 5.1 and 5.2 are devoted to the Kolmogorov–Smirnov and the Cramér–von Mises type tests, respectively. In Section 5.3 we generate the class of EDF tests so that also more recent tests based on the empirical quantile function or the empirical characteristic function fit into the framework. We show that many of these tests are closely related to the class of smooth tests. Practical guidelines are provided in Section 5.6.

## 5.1 The Kolmogorov–Smirnov Test

### *5.1.1 Definition*

In Section 2.1.2 we have argued that a distance or divergence function between the hypothesised distribution function and the EDF is a natural quantity to assess the quality of fit. In this section we discuss one of the traditional goodness-of-fit tests, the Kolmogorov–Smirnov (KS) test, which originates from the work of Kolmogorov (1933) and Smirnov (1939). For testing the null hypothesis $H_0 : F = G$ versus $H_1 : F \neq G$, the KS test statistic is given by

$$D_n = \sqrt{n} \sup_{x \in \mathcal{S}} \left| \hat{F}_n(x) - G(x) \right| = \sup_{x \in \mathcal{S}} \left| \mathbb{B}_n(x) \right| . \tag{5.1}$$

Note that $D_n$ is of the form of (2.2) with $d$ the supremum function. Thus, $D_n$ is the largest absolute deviation between the hypothesised distribution $G$ and the EDF. This difference may also be written as

O. Thas, *Comparing Distributions,* Springer Series in Statistics,
DOI 10.1007/978-0-387-92710-7_5, 

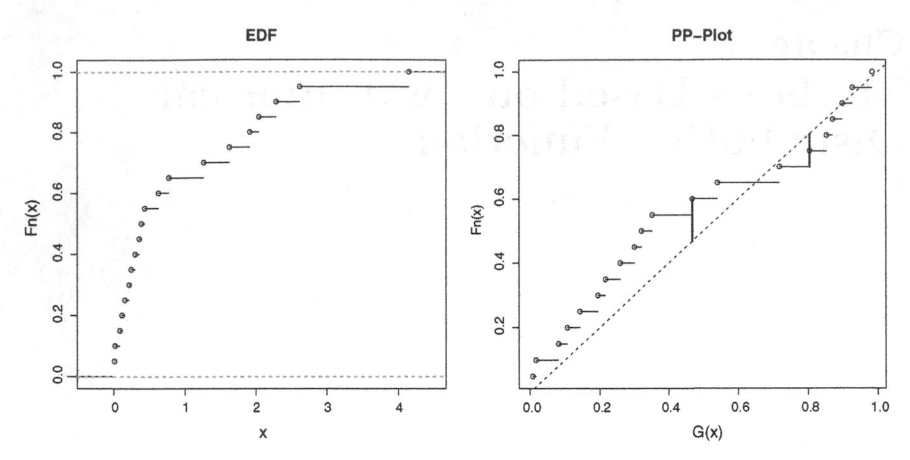


**Fig. 5.1** The EDF of a sample of 20 observations (left panel) and its PP plot w.r.t. a standard exponential distribution (right panel). The two thick vertical lines in the right panel show the $D_n^+$ (left line) and $D_n^-$ (right line) statistics

$$\hat{F}_n(x) - G(x) = \hat{F}_n(G^{-1}(p)) - p \text{ where } p = G(x).$$

The KS statistic may thus also be read from the sample PP plot. This is illustrated in Figure 5.1. One way to look at this relation is to say that the KS test is a formal test procedure which comes with the PP plot.

Closely related to the KS statistic, are the statistics studied by Smirnov (1939),

$$D_n^+ = \sqrt{n} \sup_{x \in \mathcal{S}} \left( \hat{F}_n(x) - G(x) \right) = \sup_{x \in \mathcal{S}} \mathbb{B}_n(x)$$

$$D_n^- = \sqrt{n} \sup_{x \in \mathcal{S}} \left( G(x) - \hat{F}_n(x) \right) = \sup_{x \in \mathcal{S}} \left( -\mathbb{B}_n(x) \right).$$

They represent the largest positive ($D_n^+$) and the largest negative ($D_n^-$) deviations (see Figure 5.1). The KS statistic may also be defined as $D_n = \max(D_n^+, D_n^-)$. The $D_n^-$ and $D_n^+$ statistics are used in directional tests. Because $D_n^+$ is only large when $\hat{F}_n(x) > G(x)$, it is used to test $H_0 : F = G$ versus $H_1 : F > G$. Similarly, $D_n^-$ is used when the alternative hypothesis is $H_1 : F > G$. The alternatives formulated in terms of $F < G$ and $F > G$ reflect *stochastic orderings* of $F$ and $G$.

To understand the meaning of stochastic orderings, suppose the random variables $X$ and $Y$ have CDFs $F$ and $G$, respectively. When $F > G$, then we say that $X$ is stochastically smaller than $Y$, which means that for any $z$, $\Pr\{X < z\} > \Pr\{Y < z\}$. Thus $X$ takes on smaller values with a larger probability, or, equivalently, it is more likely that $X$ takes on smaller values. Stochastic ordering can also be easily detected in a PP plot. Suppose $F(x) < G(x)$ for all $x \in \mathcal{S}$, and let $u = G(x)$. Then, $F(G^{-1}(u)) < u$, for all $u \in [0, 1]$.

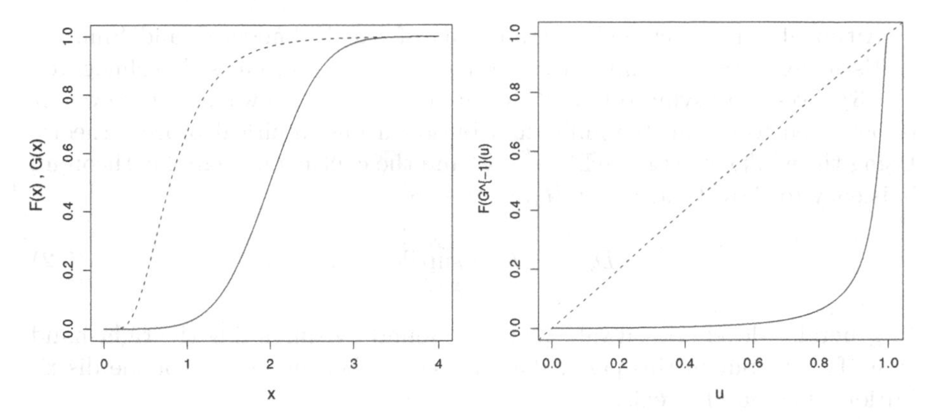


**Fig. 5.2** An example of stochastic ordining of the type $F < G$. In the left panel the two CDFs are shown ($F$: full line; $G$ dashed line), and in the right-hand panel the population PP plot is shown

This latter expression relates directly to the population PP plot (3.5). This is illustrated in Figure 5.2 in which in the left panel $F(x)$ and $G(x)$ are plotted, and in the right panel the corresponding population PP plot is shown. The PP plot is thus completely situated under the diagonal reference line.

The computation of a supremum of a nondifferentiable function typically requires the evaluation of many points. However, because $\hat{F}_n$ is a step function, and because $G$ is a monotone increasing function, the statistic $D_n^+$ simplifies to

$$D_n^+ = \max_{1 \leq i \leq n} \left( \frac{i}{n} - G(X_{(i)}) \right).$$

In a similar way we find

$$D_n^- = \max_{1 \leq i \leq n} \left( G(X_{(i)}) - \frac{i-1}{n} \right)$$

(the use of $(i-1)/n$ becomes clear from Figure 5.1). The calculation of $D_n$ requires thus only the evaluation of $D_n^+$ and $D_n^-$ in the $n$ sample observations.

## *5.1.2 Null Distribution*

The statistics $D_n$, $D_n^+$ and $D_n^-$ have the advantage of being distribution free; i.e., for any hypothesised distribution $G$, the null distributions of these statistics are the same, even for finite sample sizes. It is therefore most convenient to present the results for the uniform distribution. Its exact null distribution has been tabulated by Massey (1951) for sample sizes up to 35.

Although the exact null distributions of the Kolmogorov and Smirnov statistics exist, their asymptotic counterparts are more often used. Kolmogorov (1933) gives the asymptotic null distribution of $D_n$. Nowadays, however, it is preferred to obtain the limit distribution using empirical process theory. Using the weak convergence $\mathbb{B}_n \xrightarrow{w} \mathbb{B}$ and the continuous mapping theorem, it is easy to show that, under $H_0$, as $n \to \infty$,

$$D_n \xrightarrow{d} D = \sup_{x \in \mathcal{S}} |\mathbb{B}(x)| \,. \tag{5.2}$$

In general, $\alpha$-level critical values may be found by simulating the right-hand side of (5.2), but in this particular case an explicit expression of the distribution function of $D$ exists,

$$F_D(d) = 1 - 2\sum_{j=1}^{\infty}(-1)^{j+1}\exp(-2j^2d^2).$$

Although the proof of this result is beyond the scope of this book, it is quite simple by using properties of sample paths of a Brownian bridge. An accessible proof may be found in, e.g., Shorack and Wellner (1986).

*Example 5.1 (Pseudo-random generator data).* The 100,000 numbers generated with the runif function in R are used to test the null hypothesis that the pseudo-random generator in R samples from a uniform distribution over $[0, 1]$. Because the unform distribution is completely specified, this is a simple null hypothesis, and we may apply the KS test.

```
> ks.test(PRG,"punif",min=0,max=1)

        One-sample Kolmogorov-Smirnov test

data:  PRG
D = 0.0029, p-value = 0.349
alternative hypothesis: two.sided
```

In the output we see the calculated test statistic. The ks.test function in R, however, computes $\sup_{x\in\mathcal{S}}\left|\hat{F}_n(x) - G(x)\right|$, and so we have to multiply 0.0029 by $\sqrt{n} = \sqrt{100000}$ to find $D_n = 0.917$. The output also gives the corresponding $p$-value, $p = 0.349$. The ks.test function always uses the asymptotic null distribution for the one-sample KS test, which is definitely allowed here on our very large dataset. Because $p = 0.349 > 0.05$, we conclude at the 5% level of significance to accept the null hypothesis. So, we may conclude that the runif function gives uniformly distributed numbers.

Although we have not discussed the power properties of the KS test so far, it is interesting to note here that we have applied the KS test to a very large dataset with 100,000 observations. With such a large sample size it is expected that the test has a large power so that even a rather small deviation from the uniform distribution should result in a rejection of the null hypothesis. The fact that this has not happened here convinces us even more that the pseudo-random generator produces good numbers.

In Appendix B.1 we show how the null distribution of the KS test can be simulated by using simulations of approximations of the Brownian bridge.

### *5.1.3 Presence of Nuisance Parameters*

We have started the section on the KS test by looking at the problem of the one-sample *simple* null hypothesis in which the hypothesised distribution $G$ is completely specified. In most practical situations, however, the distribution $G$ is only specified up to some $p$-dimensional nuisance parameter vector $\boldsymbol{\beta}^t = (\beta_1, \ldots, \beta_p)$. All test statistics for the simple null hypothesis are typically also used for testing thc composite null hypothesis. The only adaptation is the replacement of $G(x)$ by $G(x; \hat{\boldsymbol{\beta}}_n)$, where $\hat{\boldsymbol{\beta}}_n$ is an estimator of $\boldsymbol{\beta}$ (more technical conditions are given later). An important consequence is that the (asymptotic) null distribution of the test statistic changes and often becomes more complicated. We show next how the distribution theory of the KS test changes under nuisance parameter estimation. Because the KS test, as well as all other EDF tests presented in this chapter, is based on the empirical process, we first show how the the empirical process behaves.

To make the dependence on the parameter $\boldsymbol{\beta}$ more explicit the notations of the empirical and Gaussian processes are slightly changed. Let $\mathbb{B}_n(x) = \mathbb{B}_n(x; \boldsymbol{\beta}) = \sqrt{n}(\hat{F}_n(x) - G(x; \boldsymbol{\beta}))$, and $\mathbb{B}(x) = \mathbb{B}(x; \boldsymbol{\beta})$ denote the limiting zero-mean Gaussian process with covariance function

$$\begin{aligned} c(x, y) = c(x, y; \boldsymbol{\beta}) &= \text{Cov}\left\{\mathbb{B}(x; \boldsymbol{\beta}), \mathbb{B}(y; \boldsymbol{\beta})\right\} \\ &= G(x \wedge y; \boldsymbol{\beta}) - G(x; \boldsymbol{\beta})G(y; \boldsymbol{\beta}). \end{aligned}$$

Note that this covariance function is exactly the covariance function given in Equation (2.5), except that here the dependence on $\boldsymbol{\beta}$ is made explicit. When the nuisance parameters are estimated these estimators are plugged into the empirical process, resulting in the *estimated empirical process* $\hat{\mathbb{B}}_n(x) = \mathbb{B}_n(x; \hat{\boldsymbol{\beta}}_n)$. To find the asymptotic behaviour of $\hat{\mathbb{B}}_n(x)$ some assumptions on the distribution $G$ and on the estimation method are required. A complete proof can be found in, e.g., Theorem 4.1 in Babu and Rao (2004) or Theorem 19.23 in van der Vaart (1998).

Before we can give the limit process of $\hat{\mathbb{B}}_n$, we need some more notation. Let $\boldsymbol{h}(x; \boldsymbol{\beta}) = \partial G(x; \boldsymbol{\beta})/\partial \boldsymbol{\beta}$, $\boldsymbol{\Psi}(x; \boldsymbol{\beta}) = \int_{-\infty}^{x} \boldsymbol{\psi}(z; \boldsymbol{\beta}) dG(z; \boldsymbol{\beta})$, and $\boldsymbol{\Sigma}_\psi = \text{Var}\left\{\boldsymbol{\psi}(X; \boldsymbol{\beta})\right\}$.

A self-contained proof of the following theorem can be found in Babu and Rao (2004) (Theorem 4.1).

**Theorem 5.1.** *Given a locally asymptotically linear estimator $\hat{\boldsymbol{\beta}}_n$, the estimated empirical process $\hat{\mathbb{B}}_n$ converges weakly to a zero-mean Gaussian process $\hat{\mathbb{B}}$ with covariance function*

$$\begin{aligned} c(x,y) &= \text{Cov}\left\{\hat{\mathbb{B}}(x), \hat{\mathbb{B}}(y)\right\} \\ &= G(x \wedge y; \boldsymbol{\beta}) - G(x;\boldsymbol{\beta})G(y;\boldsymbol{\beta}) \\ &\quad - \boldsymbol{\Psi}^t(x;\boldsymbol{\beta})\boldsymbol{h}(y;\boldsymbol{\beta}) - \boldsymbol{\Psi}^t(y;\boldsymbol{\beta})\boldsymbol{h}(x;\boldsymbol{\beta}) \\ &\quad + \boldsymbol{h}^t(x;\boldsymbol{\beta})\boldsymbol{\Sigma}_\psi \boldsymbol{h}(y;\boldsymbol{\beta}). \end{aligned} \tag{5.3}$$

There are two very important consequences of the weak convergence of $\hat{\mathbb{B}}_n$ to $\hat{\mathbb{B}}$. The first is that we now can find the asymptotic null distribution of the KS test statistic. In particular, under $H_0$, as $n \to \infty$,

$$\hat{D}_n = D_n(\hat{\boldsymbol{\beta}}_n) = \sqrt{n} \sup_{x \in \mathcal{S}} |\hat{F}_n(x) - G(x;\hat{\boldsymbol{\beta}}_n)| = \sup_{x \in \mathcal{S}} |\hat{\mathbb{B}}_n| \xrightarrow{d} \sup_{x \in \mathcal{S}} |\hat{\mathbb{B}}|.$$

The second implication, however, is that this limit distribution depends on the unknown parameter $\boldsymbol{\beta}$ and on the hypothesised distribution $G$. So the KS test for the composite null hypothesis is no longer distribution free, not even in an asymptotic sense. Consequently, the asymptotic null distribution cannot be used directly to perform the KS test. Fortunately, there exist solutions that circumvent this problem.

For location-scale invariant distributions it has been shown that the asymptotic null distribution of $\hat{D}_n$ reduces to a form which still depends on the distribution $G$, but not anymore on the unknown parameter $\boldsymbol{\beta}^t = (\mu, \sigma)$, where $\mu$ and $\sigma$ denote the location and scale parameter, respectively. A location-scale invariant distribution is a distribution with a density that satisfies $g(x;\mu,\sigma) = g((x-\mu)/\sigma, 0, 1)/\sigma$. This independence of the nuisance parameters is a direct consequence of a simplification of the covariance function $c(x,y)$ when $G$ is a location-scale distribution.

A well-known family of location-scale invariant distributions is the normal distribution. For this distribution, it was already recognised by Lilliefors (1967) that the asymptotic null distribution of $\hat{D}_n$ does not depend on the parameters. He was the first one to tabulate the distribution of $\hat{D}_n$ for the normal distribution. The test is often named after Lilliefors . For small $p$-values Dallal and Wilkinson (1986) give a method to approximate the asymptotic distribution. For larger $p$-values, Stephens (1974) gave approximations.

Another solution to get an approximation of the asymptotic null distribution of $\hat{D}_n$ is to apply the bootstrap. Babu and Rao (2004) showed that the parametric bootstrap gives asymptotically the correct critical values. For the nonparametric bootstrap, however, a bias correction is needed. In Appendix

B.2 more details on the practical implementation of the parametric bootstrap are given.

*Example 5.2 (PCB concentration data).* In the PCB concentration data it is of interest to test for normality. The R-function ks.test may not be used for this purpose, because the null distribution used in this function is only correct for known mean and variance. The Lilliefors corrected KS test is available via the lillie.test function in the nortest R package, which is also made available in the cd package. It makes use of the approximations of Dallal and Wilkinson (1986) and Stephens (1974).

```
> lillie.test(PCB)

        Lilliefors (Kolmogorov-Smirnov) normality test

data:  PCB
D = 0.1093, p-value = 0.0521
```

Because $p = 0.0521$ we cannot reject the null hypothesis of normality at the 5% level of significance. However, the $p$-value is only nearly larger than the nominal 5% level. So the conclusion should be made with care. Maybe one or more outliers are causing the small $p$-value, or maybe the true distribution is not the normal distribution, but this was not detected due to a small sample size.

We now test the same null hypothesis, but using the bootstrap approximation.

```
> ksboot.test(PCB,distr="pnorm",B=10000)

        Bootstrap One-sample Kolomogorov Smirnov Test for the
        normal distribution

data:  PCB
D = 0.1093, number of bootstrap runs = 10000, p-value = 0.051
```

## 5.2 Tests as Integrals of Empirical Processes

### *5.2.1 The Anderson–Darling Statistics*

The KS statistic is only one example of a statistic of the form $T_n = c(n)d(\hat{F}_n, G)$. Yet another important class of statistics was introduced by Anderson and Darling (1952),

$$T_n = \int_S w(G(x))\mathbb{B}_n^2(x)dG(x), \tag{5.4}$$

where $w(.)$ is a weight function. When $w(u) = 1$ (for all $0 \leq u \leq 1$), the Anderson–Darling (AD) statistic reduces to the statistic which is today known as the Cramér–von Mises (CvM) statistic, which has its origin in the work of Cramér (1928), von Mises (1931), and von Mises (1947). Although the AD statistic is basically a class of statistics indexed by a weight function, there is actually only one particular weight function $w(u) \neq 1$ which is very popular, $w(u) = 1/(u(1-u))$. The test with this choice of $w$ is generally known as *the* AD test in the literature, probably because it was this particular weight function that was suggested by Anderson and Darling (1954). They advocate this choice because $w(u) = 1/(u(1-u))$ has a variance stabilising effect; i.e., $\sqrt{w(u)}\mathbb{B}_n(u)$ has constant variance equal to 1. To make a clear distinction between the AD and CvM statistics, we use the notation $A_n$ and $W_n$, respectively.

Although at first sight it may seem difficult to calculate $T_n$ for a given dataset, fortunately for the two most popular weight functions there exist simple formulae. Let $U_i = G(X_i)$, and let $U_{(i)}$ denote the $i$th order statistic of $U_1, \ldots, U_n$. Then,

$$\begin{aligned} A_n &= -n - \frac{1}{n}\sum_{i=1}^{n}(2i-1)\left(\log U_{(i)} + \log(1 - U_{(n+1-i)})\right) \\ &= -n - \frac{1}{n}\sum_{i=1}^{n}\left((2i-1)\log U_{(i)} + (2n+1-2i)\log(1-U_{(i)})\right) \\ W_n &= \sum_{i=1}^{n}\left(U_{(i)} - \frac{2i-1}{n}\right)^2 + \frac{1}{12n}. \end{aligned}$$

When a composite null hypotheses $H_0 : F(x) = G(x;\boldsymbol{\beta})$ has to be tested, we proceed as with the KS test. First, the nuisance parameter $\boldsymbol{\beta}$ has to be estimated, and we assume that the estimator $\hat{\boldsymbol{\beta}}_n$ is asymptotically linear. The AD and CvM statistics may now be calculated using $G(.;\hat{\boldsymbol{\beta}}_n)$ instead of $G(.)$, and they are denoted by $\hat{A}_n$ and $\hat{W}_n$, respectively, or by $\hat{T}_n$ in general. As with the KS test, this has an effect on the asymptotic null distribution. This is discussed in the Section 5.2.3.

### 5.2.2 Principal Components Decomposition of the Test Statistic

In Section 2.2.3 we have introduced a decomposition of a Gaussian process. Here we show a similar decomposition, but applied to the empirical process

$\mathbb{B}_n$, and we show that by substituting this decomposition into the definition of the AD test statistic, we obtain a decomposition of the AD test statistic into interpretable components that are related to the components of smooth test statistics. The decomposition was proposed by Durbin and Knott (1972) and Durbin et al. (1975).

The central idea is quite simple: consider the construction of the Kac–Siegert principal components of a Gaussian process (see Equation (2.8)), and replace the process with the empirical process. We illustrate this program by applying it to the Anderson–Darling and the Cramér–von Mises statistics for testing uniformity (simple null hypothesis). For the integral statistics of the form (5.4), the process to be considered is $\mathbb{P}_n(x) = \sqrt{w(x)}\mathbb{B}_n(x)$, because the test statistic is then simply the integral of the squared process; i.e., $T_n = \int_0^1 \mathbb{P}_n^2(x)dx$. When nuisance parameters are to be estimated, $\boldsymbol{\beta}$ is replaced by its estimator, and the process is denoted by $\hat{\mathbb{P}}_n$.

#### 5.2.2.1 Principal Components Decomposition of the Cramér–von Mises Statistic (Simple Null)

For the CvM statistic the weight function is $w(x) = 1$, and we thus have to consider the empirical process $\mathbb{P}_n = \mathbb{B}_n$ for which the covariance function is $c(x, y) = x \wedge y - xy$ when testing for uniformity. For this covariance function, $\{\lambda_j\}$ and $\{l_j\}$ are the eigenvalues and eigenfunctions. It can be shown that $(j = 1, 2 \ldots)$

$$\lambda_j = \frac{1}{j^2\pi^2} \text{ and } l_j(x) = \sqrt{2}\sin(j\pi x).$$

A Kac–Siegert type of decomposition of the empirical process now looks like (cfr. Equation (2.7))

$$\mathbb{P}_n(x) = \sum_{j=1}^{\infty} \sqrt{\lambda_j} l_j(x) Z_{nj},$$

where

$$Z_{jn} = \frac{1}{\sqrt{\lambda_j}} \int_0^1 \mathbb{P}_n(x) l_j(x) dx. \tag{5.5}$$

These components can be simplified

$$\begin{aligned} Z_{jn} &= \frac{1}{\sqrt{\lambda_j}} \int_0^1 \mathbb{P}_n(x) l_j(x) dx \\ &= \sqrt{2} j\pi \int_0^1 \left(\sqrt{n}(\hat{F}_n(x) - x)\right) \sin(j\pi x) dx \\ &= \sqrt{2n} j\pi \left[\int_0^1 \hat{F}_n(x) \sin(j\pi x) dx - \int_0^1 x \sin(j\pi x) dx\right] \end{aligned}$$

$$
\begin{aligned}
&= \sqrt{2nj\pi} \int_0^1 \hat{F}_n(x) \sin(j\pi x) dx \\
&= \sqrt{2n} \int_0^1 \cos(j\pi x) d\hat{F}_n(x) \\
&= \sqrt{2n} \frac{1}{n} \sum_{i=1}^n \cos(j\pi X_i) \\
&= \sqrt{\frac{2}{n}} \sum_{i=1}^n \cos(j\pi X_i).
\end{aligned}
$$

It is easy to verify that under the null hypothesis these components are indeed asymptotically standard normally distributed and that they are asymptotically independent.

The principal component decomposition of the CvM test statistic is obtained as follows.

$$
\begin{aligned}
T_n = W_n &= \int_0^1 \mathbb{B}_n^2(x) dx \\
&= \int_0^1 \left[ \sum_{j=1}^\infty \sqrt{\lambda_j} l_j(x) Z_{nj} \right]^2 dx \\
&= \sum_{j=1}^\infty \sum_{m=1}^\infty \sqrt{\lambda_j \lambda_m} Z_{nj} Z_{nm} \int_0^1 l_j(x) l_m(x) dx \\
&= \sum_{j=1}^\infty \lambda_j Z_{nj}^2 \\
&= \sum_{j=1}^\infty \frac{1}{j^2 \pi^2} Z_{nj}^2. \qquad (5.6)
\end{aligned}
$$

Hence, $T_n$ has a representation as an infinite weighted sum of asymptotically independent squared components. The component $Z_{nj}$ is called the $j$th order component. Note that this decomposition is similar to the decomposition of smooth test statistics when the eigenfunctions $\{l_j\}$ are used for the construction. There are two important distinctions: (1) the integral statistic has an infinite number of components, but (2) they have a decreasing weight with the order $j$. To be more precise, the weights $1/(j^2\pi^2) \to 0$ as the order $j \to \infty$. This decreasing weight property is necessary to make $T_n$ have a proper limiting distribution.

Just as with smooth tests it is very informative to have a closer look at the interpretation of the components. In particular they may give us some more insight into the behaviour of the test under alternatives. The question of under which alternatives the test statistic $T_n$ becomes big now translates into the question of under which alternatives the components $Z_{nj}$ are expected to be very different from zero. Moreover, because the weights of components decrease rapidly with the order $j$, it is particularly important to understand the lower-order components. For the CvM statistic we see that $Z_{nj} = \sqrt{2/n}\sum_{i=1}^{n}\cos(j\pi X_i)$, which is exactly the $j$th component $U_j$ of the smooth test statistic for uniformity introduced in Section 4.2.1. Based on the discussion given there, we may conclude that the CvM test will have larger power for *slowly oscillating* alternatives than for *fast oscillating* alternatives. This is an important difference with the order $k$ smooth test, because for the latter the power drops to $\alpha$ when the alternative is oscillates so fast that it has only nonzero expectations of components of order smaller than $k$. The CvM test, on the other hand, is omnibus consistent.

#### 5.2.2.2 Principal Components Decomposition of the Anderson–Darling Statistic (Simple Null)

Because for the AD statistic the weight function is $w(x) = 1/(x(1-x))$, we need the covariance function of the process

$$\mathbb{P}_n(x) = \frac{\mathbb{B}_n(x)}{\sqrt{x(1-x)}},$$

which is

$$c(x,y) = \frac{x \wedge y - xy}{\sqrt{x(1-x)y(1-y)}}.$$

The components are again of the form (5.5), but now are the eigenvalues and the eigenfunctions given by

$$\lambda_j = \frac{1}{j(j+1)} \text{ and } l_j(x) = 2\sqrt{\frac{1}{j(j+1)}}\sqrt{x(1-x)}\frac{d}{dx}L_j(x), \tag{5.7}$$

where the $L_j$ denote the orthonormal Legendre polynomials. After similar calculations as in the previous section we find

$$T_n = A_n = \sum_{j=1}^{\infty}\frac{1}{j(j+1)}Z_{nj}^2, \tag{5.8}$$

where the components are

$$Z_{nj} = -\frac{1}{\sqrt{n}}\sum_{i=1}^{n}L_j(X_i).$$

Thus, just as with the CvM test we see here that the AD test statistic is a weighted sum of squared components, and these components are exactly those of the traditional smooth test of Section 4.2.1 for testing for uniformity.

The weights are equal to $1/(j(j+1))$, which suggests that the AD test will have a particularly large power against alternatives with deviations in the lower-order moments.

If we adopt the moment interpretation of the components, as we did in Chapter 4, we may conclude that the AD test is particularly sensitive to deviations in the lower-order moments, but asymptotically it also has power against alternatives that have differences in the higher-order moments. It is important to note, however, that the components of the AD statistic are always in terms of the Legendre polynomials, whatever the hypothesised distribution $G$ is. This may be explained by the relation

$$A_n = n \int_{\mathcal{S}} \frac{(\hat{F}_n(x) - G(x))^2}{G(x)(1-G(x))} dG(x) = n \int_0^1 \frac{(\hat{F}_n(G^{-1}(u)) - u)^2}{u(1-u)} du,$$

which shows that the AD test actually tests for uniformity after the PIT is applied. This makes the interpretation of the components less clear as compared to the case of the smooth tests for which the polynomials and the hypothesised distribution go hand in hand.

#### 5.2.2.3 Principal Components Decompositions for Composite Null Hypotheses

Before moving onwards to the composite case we say a little word about the relation between the eigenfunctions that appear in the Kac–Siegert expansion and the functions defining the components. We have used the notation $\{l_j\}$ for the eigenfunctions. The components of the AD and CvM statistics are, however, not of the form $(1/\sqrt{n})\sum_{i=1}^n l_j(X_i)$, but of the form $(1/\sqrt{n})\sum_{i=1}^n l_j^a(X_i)$, where $l_j^a$ is associated with $l_j$. For example, for the CvM test the eigenfunctions are sine functions, but the components are in terms of cosine functions. The exact association between the $l_j$ and $l_j^a$ comes from (5.5) which involves the $l_j$, and which can be turned into the form $Z_{jn} = (1/\sqrt{n})\sum_{i=1}^n l_j^a(X_i)$ by integration by parts. In conclusion, we do not expect the eigenfunctions to coincide with the orthogonal functions used for the construction of smooth tests, but rather the $l_j^a$ should.

As with smooth test statistics that do not always decompose naturally into asymptotically independent terms when estimated nuisance parameters are plugged in, this is also the case for the integral test statistics. We start in this section with the general form of the principal component decomposition of statistics of the form $\hat{T}_n = \int_0^1 \hat{\mathbb{P}}_n^2(x)dx$, which includes both the AD and CvM statistics. The resulting asymptotically independent components are, however, not necessarily nicely interpretable. We prefer the components to be in terms of the orthonormal functions $\{l_j^a\}$ which appear in the tests for the simple null hypothesis. We further show how the components can be transformed so that they are expressed in terms of $\{l_j^a\}$. At the end of the section we show how this decomposition relates to the smooth tests. We restrict the discussion to MLE. We do not intend to prove all results rigorously. Instead we rather give a sequence of heuristic arguments. For example, because there are infinitely many eigenfunctions and eigenvalues, we need to use infinite-dimensional matrices, but we do not focus on such technical issues. It is sufficient to read these matrices as large-dimensional. Similarly, when writing a sum with an index $j$ going from 1 to $\infty$, we simply write $\sum_j$ so that this may just as well be read as, say, $\sum_{j=1}^M$ with large $M$. This approach was also used in the seminal paper of Durbin et al. (1975).

Suppose that $c(x,y)$ is the covariance function of the process $\hat{\mathbb{P}}_n$, and denote the corresponding eigenfunctions and eigenvalues by $\{k_j\}$ and $\{\kappa_j\}$. We work in the Hilbert space $L_2(\mathcal{S},G)$. With this notation, the covariance function can be written as

$$c(x,y) = \sum_j \kappa_j k_j(x) k_j(y)$$

(cfr. (2.6)). The test statistic $\hat{T}_n = \int_0^1 \hat{\mathbb{P}}_n^2(x)dx$ can then be equivalently represented by its principal components decomposition; i.e.,

$$\hat{T}_n = \sum_j \kappa_j Z_{nj}^2, \tag{5.9}$$

where the components are given by $Z_{nj} = (1/\sqrt{\kappa_j}) \int_{\mathcal{S}} \hat{\mathbb{P}}_n(x) k_j(x) dG(x)$, which can be further simplified by means of partial integration as we did for the AD and CvM tests in the two previous sections. This generally results in components of the form $Z_{nj} = (1/\sqrt{n}) \sum_{i=1}^n k_j^a(X_i;\hat{\boldsymbol{\beta}})$, where the function set $\{k_j^a\}$ is associated with the eigenfunctions $\{k_j\}$. The components $Z_{nj}$ are asymptotically i.i.d. standard normal, and their interpretation depends on the $k_j^a$ functions. Durbin et al. (1975) showed that for a given process $\hat{\mathbb{P}}_n$, it is always true that $\kappa_j \leq \lambda_j$; i.e., the eigenvalues of the estimated process $\hat{\mathbb{P}}_n$ are never larger than those of the process $\mathbb{P}_n$ used to test the simple null hypothesis. This property is similar to the loss of degrees of freedom property of $\chi^2$-type statistics.

In general it is hard to find the eigenvalues and the eigenfunctions of $c(x,y)$, and, moreover, it is not guaranteed that the form of the $k_j^a$ allows simple interpretations. We therefore prefer components in terms of the $l_j^a$ orthonormal functions that appear in the decomposition of the test statistic when no nuisance parameters are estimated. Note that in the case of no nuisance parameters, the functions $k_j^a$ and $l_j^a$ coincide. The most interesting $l_j^a$ functions are those that also appear in the smooth tests of Chapter 4, therefore we also often use $h_j$ instead of $l_j^a$.

Before the main theorem is stated we introduce some notation. Let $\boldsymbol{h}^t(x)=(h_1(x),\ldots)$, $\boldsymbol{k}_a^t(x) = (k_1^a(x),\ldots)$, and let $\boldsymbol{\Sigma}_{hG\beta} = <\boldsymbol{h}\circ G, \boldsymbol{u}_\beta>_g$ denote the matrix with $(i,j)$th element equal to $\int_{\mathcal{S}} h_i(G(x;\boldsymbol{\beta}))(\partial \log g(x;\boldsymbol{\beta})/\partial\beta_j)$

$dG(x;\boldsymbol{\beta})$. The difference between the latter and the matrix $\boldsymbol{\Sigma}_{h\beta}$ that appears in the efficient score is that here we have $h \circ G(x) = h(G(x))$ instead of simply $h(x)$ because of the PIT. In analogy with the construction of the efficient score (4.16), we now need

$$\boldsymbol{v}(x;\boldsymbol{\beta}) = h_j(G(x;\boldsymbol{\beta})) - \boldsymbol{\Sigma}_{hG\beta}\boldsymbol{\Sigma}_{\beta\beta}^{-1}\boldsymbol{u}_\beta(x).$$

On using Theorem 4.2, the variance–covariance matrix of $\boldsymbol{h}(G(X;\hat{\boldsymbol{\beta}}))$ and $\boldsymbol{v}(x;\hat{\boldsymbol{\beta}})$ coincide and are equal to $\boldsymbol{\Sigma}_{\hat{v}} = \boldsymbol{I} - \boldsymbol{\Sigma}_{hG\beta}\boldsymbol{\Sigma}_{\beta\beta}^{-1}\boldsymbol{\Sigma}_{\beta Gh}$. We may also write $\boldsymbol{\Sigma}_{\hat{v}} =< \boldsymbol{v}, \boldsymbol{v} >_g$. Finally, using the vector functions $\boldsymbol{v}$ and $\boldsymbol{k}$ we define the statistics $\hat{\boldsymbol{V}} = (1/\sqrt{n})\sum_{i=1}^n \boldsymbol{v}(X_i;\hat{\boldsymbol{\beta}})$ and $\hat{\boldsymbol{K}}_a = (1/\sqrt{n})\sum_{i=1}^n \boldsymbol{k}_a(X_i;\hat{\boldsymbol{\beta}})$. Note that because of the of MLE, $\hat{\boldsymbol{V}}$ further reduces to $(1/\sqrt{n})\sum_{i=1}^n \boldsymbol{h}$ $(G(X_i;\hat{\boldsymbol{\beta}}))$. With this notation we may write $T_n$ of (5.9) as $T_n = \hat{\boldsymbol{K}}_a^t \boldsymbol{\Gamma} \hat{\boldsymbol{K}}_a$ with $\boldsymbol{\Gamma}$ a diagonal matrix with elements $\kappa_1, \kappa_2, \ldots$.

**Theorem 5.2.** *(1) The following equality holds,*

$$\hat{T}_n = \hat{\boldsymbol{K}}_a^t \boldsymbol{\Gamma} \hat{\boldsymbol{K}}_a = \left(\boldsymbol{\Sigma}_{\hat{v}}^{-1/2}\hat{\boldsymbol{V}}\right)^t \boldsymbol{\Gamma} \left(\boldsymbol{\Sigma}_{\hat{v}}^{-1/2}\hat{\boldsymbol{V}}\right) = \sum_j \kappa_j \hat{Q}_j^2, \tag{5.10}$$

*where the components are given by*

$$\hat{Q}_j = \sum_m \boldsymbol{\Sigma}_{\hat{v}j,m}^{-1/2}\hat{V}_m,$$

*with $\boldsymbol{\Sigma}_{\hat{v}j,m}^{-1/2}$ denoting the $(j,m)$th element of $\boldsymbol{\Sigma}_{\hat{v}}^{-1}$, which equals $< \boldsymbol{v}, \boldsymbol{k}_a >_g^{-1}$.*
*(2) The eigenvalues $\{\kappa_j\}$ can be calculated as*

$$\kappa_j = \boldsymbol{a}_j^t \left[\int_S \int_S c(x,y)\boldsymbol{v}(x)\boldsymbol{v}^t(y)dG(x)dG(y)\right]\boldsymbol{a}_j, \tag{5.11}$$

*where $\boldsymbol{a}_j$ is the $j$th column of the transformation matrix $\boldsymbol{\Sigma}_{\hat{v}}^{-1/2}$.*

The heuristic proof of the theorem is given in Appendix A.9.
From the theorem we learn the following.

- The difference between the test statistics in the simple and the composite null hypotheses cases is very similar to the difference that we observed in the order $k$ smooth test statistics in the previous chapter. As before, the $T_n$ statistic in (5.10) is a weighted sum of squared components. To see the link with the smooth test, write the order $k$ smooth test statistic in (4.20) as

$$T_k = \hat{\boldsymbol{V}}^t \boldsymbol{\Sigma}_{\hat{v}}^{-1}\hat{\boldsymbol{V}} = \left(\boldsymbol{\Sigma}_{\hat{v}}^{-1/2}\hat{\boldsymbol{V}}\right)^t \left(\boldsymbol{\Sigma}_{\hat{v}}^{-1/2}\hat{\boldsymbol{V}}\right),$$

which is indeed the unweighted and truncated version of the integral statistic. There is, however, one further important difference: smooth tests are constructed starting from polynomials that are orthonormal w.r.t. the hypothesised density, whereas the $h_j$ functions that appear in the integral statistics are all orthonormal w.r.t. the uniform distribution. In $\boldsymbol{\Sigma}_{\hat{v}}$ used in the smooth tests we find the matrix $\boldsymbol{\Sigma}_{h\beta}$, and in the integral tests this matrix is replaced by $\boldsymbol{\Sigma}_{hG\beta}$, which accounts for the PIT.

- The components $\hat{Q}_j = \sum_m \boldsymbol{\Sigma}_{\hat{v}j,m}^{-1/2} \hat{V}_m$ are linear combinations of the components $\hat{V}_m$ which are in turn defined in terms of the $h_m$ orthonormal functions. The interpretation of a single $\hat{Q}_j$ is thus based on the interpretation of several $h_m$ functions. The interpretation gets simpler the fewer of these $h_m$ functions get a large weight.
- In view of the previous remark, we would like that the $\hat{Q}_j$ component depend only on a single $h_m$ function. This happens in the important case that $\boldsymbol{\Sigma}_{\hat{v}}^{-1/2}$ is a diagonal matrix. This occurs if the elements of $\boldsymbol{\Sigma}_{hG\beta} = \, < \boldsymbol{h} \circ G, \boldsymbol{u}_\beta >_g$ are all zero. To our knowledge this does unfortunately not happen in any practical relevant case. Later we show generalisations of the EDF integral statistics that have neater forms (see Section 5.3).
- In the previous chapter we have argued that even when the smooth test statistic in the presence of nuisance parameters does not decompose into the $\hat{V}_j$ components, it is still informative to look at these components, or even apply the rescaled component tests (Sections 4.5.6 and 4.6). Because the components $\hat{V}_j$ are here expressed in terms of $l_j \circ G$, the interpretation is not very simple.

### *5.2.3 Null Distribution*

For testing a simple null hypothesis the AD and CvM tests are nonparametric tests in all of its meanings: the test statistics have, even for finite sample sizes, a null distribution which is independent of the hypothesised distribution $G$. However, this does not mean that the exact distribution is easy to obtain. The exact distribution of the CvM statistic has received much attention in the statistical literature already for more than 50 years, and still the exact distribution is only tabulated for $n = 1, \dots, 7$. Approximations to the exact distribution of the CvM statistic have also been heavily investigated. The best approximation up to now is given by Csörgö and Faraway (1996). It has a firm theoretical ground, it is based on a one-term correction to the asymptotic distribution function, and it gives quite good approximations, even for sample sizes as small as $n = 7$. Although their solution gives good results and it is easier as compared to most other approximations, it still requires substantial computation effort. Another type of approximation was suggested by Pearson and Stephens (1962), Tiku (1965), and Zhang and Wu (2001). They proposed to consider a flexible parameterised family of distributions, and find the parameter values so that the distribution matches to the first three or four moments of the exact distribution of the CvM statistic. The rationale is that (1) the exact first few moments are known, and (2) it is hoped that the mimicking distribution approximates the true exact distribution sufficiently closely, particularly in the tails. In this sense the methods of Zhang and Wu (2001) seem to give very acceptable approximations. Yet another approximation method was suggested by Stephens (1970). Based on an extensive empirical simulation study, he suggested to use a modified test statistic,

$$W_n^{\star} = \frac{W_n - 0.4/n + 0.6/n^2}{1 + 1/n}, \tag{5.12}$$

where the coefficients were estimated using simple regression techniques. When $W_n^{\star}$ is used the percentage points of the asymptotic null distribution of $W_n$ apply. Despite the simplicity of this approach, it works remarkably well.

Less is known about the exact null distribution of the Anderson–Darling statistic. Lewis (1961) gave the exact distribution when $n = 1$, but, to our knowledge at least, there are no exact results for larger sample sizes. Even the exact moments of $A_n$ are not known, and, therefore, the moment based approximation methods cannot be applied here. Fortunately, simulation studies have indicated that the distribution of the $A_n$ statistic converges very rapidly to its asymptotic distribution. For instance, D'Agostino and Stephens (1986) (p. 104) said that for sample sizes as small as $n = 3$ the asymptotic approximation is quite good. Lewis (1961), who estimated percentage points for sample sizes $n \leq 8$, is more conservative and recommends the asymptotic distribution only for $n > 8$.

The asymptotic null distributions of $T_n$ and $\hat{T}_n$ are again found by using the weak convergence of the empirical process or the estimated empirical process, for the simple and composite null hypothesis, respectively.

**Theorem 5.3.** *If $\int_0^1 t(1-t)w(t)dt < \infty$, then, under the simple null hypothesis with $G$ the uniform distribution, as $n \to \infty$,*

$$T_n \xrightarrow{d} \int_0^1 w(t)\mathbb{B}^2(t)dt.$$

Although this result gives a theoretically correct representation of the asymptotic null distribution, it is not convenient to get percentage points quickly. Expressions for the CDF of the CvM and AD statistics were obtained by Anderson and Darling (1952). They first found the characteristic functions, which, by inversion, results in the CDF. The CDF, however, contain an infinite sum which makes the exact evaluation difficult. Fortunately, using only the very first few terms already gives quite good approximations. An immediate and interesting conclusion which emerges directly from the form of the characteristic functions, is that the CvM and AD statistics are asymptotically equivalent in distribution to the random variables

$$W = \sum_{j=1}^{\infty} \frac{1}{j^2\pi^2} Z_j^2 \quad \text{(CvM)}$$

$$A = \sum_{j=1}^{\infty} \frac{1}{j(j+1)} Z_j^2 \quad \text{(AD)},$$

where $Z_1, Z_2, \ldots$ are i.i.d. standard normal random variables. These represent infinite weighted sums of independent chi-squared random variates, and the weights decrease quadratically with the index $j$. The same representation follows also immediately from (5.6) and (5.8).

*Example 5.3 (Pseudo-random generator data).* We analyse the PRG data again with the CvM and AD tests which are available in the EDF.test function of the cd package. The CvM test is implemented as the approximation of Stephens (1970) based on the modified statistic given in Equation (5.12).

```
> EDF.test(PRG,B=NA,distr="unif",type="AD",pars=c(0,1))

        Anderson-Darling Test for the uniform distribution

data:  PRG
T = 0.9829, number of bootstrap runs = NA, p-value =
0.25

Warning message:
The p-value is only a lower bound. in: EDF.test(PRG, B = NA,
    distr = "unif", type = "AD", pars = c(0,

> EDF.test(PRG,B=NA,distr="unif",type="CvM",pars=c(0,1))

        Cramer-von Mises Test for the uniform distribution

data:  PRG
T = 0.1517, number of bootstrap runs = NA, p-value =
0.25

Warning message:
The p-value is only a lower bound. in: EDF.test(PRG, B = NA,
    distr = "unif", type = "CvM", pars = c(0,
```

Both tests confirm that the 100,000 generated numbers may be considered as a sample from a uniform distribution. Note that the output contains a warning saying that the reported $p$-values are only an upper bound. The reason is that no approximations for $p$-value calculation are implemented in the EDF.test function. If a $p$-value is needed, the AD and CvM tests can be performed as bootstrap tests. This is illustrated with the AD test.

```
> EDF.test(PRG,B=100,distr="unif",type="AD",pars=c(0,1))
          Anderson-Darling Test for the uniform distribution

data:  PRG
T = 0.9829, number of bootstrap runs = 100, p-value = 0.35
```

When testing a composite null hypothesis, the nuisance parameter $\boldsymbol{\beta}$ must be estimated.

**Theorem 5.4.** *Assume that $\hat{\boldsymbol{\beta}}_n$ is locally asymptotically linear and $\int_0^1 t(1-t) w(t)dt < \infty$. Then, under the composite null hypothesis, as $n \to \infty$,*

$$\hat{T}_n \xrightarrow{d} \int_0^1 w(t)\hat{\mathbb{B}}^2(t)dt.$$

As noted in Section 5.1.3, this distribution generally depends on the hypothesised distribution $G$, as well as on the unknown nuisance parameters $\boldsymbol{\beta}$. Consequently, the distribution cannot be tabulated, and percentage points and $p$-values must be approximated using the bootstrap (see Appendix B). However, when $G$ is a location-scale invariant distribution, only the dependence on $G$ remains. For these distributions the percentage points may be obtained by simulation. For many popular distributions (normal, exponential, etc.), the asymptotic distributions of the AD and CvM tests have been tabulated. As for the simple null hypothesis case, there is no simple analytic expression for the asymptotic distribution function, and approximations are available. For the normal distribution, we mention the work of Stephens (1971, 1974) and Stephens (1976) (summarised in D'Agostino and Stephens (1986) (p. 122)), who suggested to use the modified statistics

$$W_n^\star = W_n(1 + 0.5/n) \text{ and } A_n^\star = A_n(1 + 0.75/n + 2.25/n^2). \tag{5.13}$$

*Example 5.4 (PCB concentration data).* In Section 5.1 the PCB data were analysed with the KS test, which resulted in a $p$-value only nearly larger than $\alpha = 0.05$. And the analysis of the PCB data in Section 4.6 revealed that only the low-order components of the smooth test gave significant results. Here we redo the analysis with the AD and CvM tests.

Many of the EDF tests are available in the cd package through the EDF.test function. For testing composite normality, the AD and CvM tests make use of the approximations of D'Agostino and Stephens (1986) based on the modified statistics given in Equation (5.13).

```
> EDF.test(PCB,B=NA,distr="norm",type="AD")

          Anderson-Darling Test for the normal distribution
```

```
data:  PCB
T = 0.7506, number of bootstrap runs = NA, p-value = 0.05076

Warning message:
The p-value is the D'Agostino-Stephens approximation

> EDF.test(PCB,B=NA,distr="norm",type="CvM")

        Cramer-von Mises Test for the normal distribution

data:  PCB
T = 0.134, number of bootstrap runs = NA, p-value = 0.03893

Warning message:
The p-value is the D'Agostino-Stephens approximation
```

This result of the AD test is very close to the analysis with the KS test, but with the CvM test we have to reject the null hypothesis and conclude that the data are not normally distributed. Both tests can also be performed by using the bootstrap. The approximated $p$-values are rather close to the nominal significance level, therefore we take a quite large number of bootstrap runs.

```
> EDF.test(PCB,B=20000,distr="norm",type="AD")

        Anderson-Darling Test for the normal distribution

data:  PCB
T = 0.7506, number of bootstrap runs = 1000, p-value = 0.051

> EDF.test(PCB,B=20000,distr="norm",type="CvM")

        Cramer-von Mises Test for the normal distribution

data:  PCB
T = 0.134, number of bootstrap runs = 1000, p-value = 0.04095
```

These results confirm the conclusions from the previous analyses. Finally, we refer to the analysis of these data presented in Section 4.6, where we did the analysis by means of order $k$ smooth tests. There the significance depended strongly on the choice of the order $k$. This problem does not play any role here. The choice of the order is replaced by the weighting scheme that is completely determined by the weight function $w$ in the definition of $T_n$, and by the score function $\boldsymbol{u}_\beta$.

## 5.2.4 The Watson Test

### 5.2.4.1 The Test Statistic

Watson (1961) proposed a test that can test goodness-of-fit of distributions on a circle. An example of circular data is, e.g., the measurement of the wind direction. When measuring on a circle, there is no natural origin. For the wind direction data, one typically takes the north-direction as origin, but one could just as well have chosen any other direction. A test for goodness-of-fit for such data should of course be invariant to the choice of the origin. The Watson test statistic is defined as

$$U_n = n\int_{\mathcal{S}}\left\{\hat{F}_n(x) - G(x) - \int_{\mathcal{S}}(\hat{F}_n(y) - G(y))dG(y)\right\}^2 dG(x) \tag{5.14}$$

$$= n\int_{\mathcal{S}}\int_{\mathcal{S}}\left\{\left(\hat{F}_n(x) - \hat{F}_n(y)\right) - (G(x) - G(y))\right\}^2 dG(x)dG(y). \tag{5.15}$$

Although the form in (5.14) is usually used to study the theoretical properties of the test, it is (5.15) that clearly shows that $U_n$ is independent of the choice of origin. It can be interpreted as an average of the differences of the empirical probability that an observation is in the interval $[x, y]$ and the corresponding hypothesised probability.

Although the test was originally constructed for testing goodness-of-fit on the circle, it can just as well be used to test goodness-of-fit on the real line.

When testing for uniformity, the computational form is given by

$$U_n = \sum_{i=1}^{n}\left(X_{(i)} - \frac{2i-1}{2n}\right)^2 - n\left(\bar{X} - \frac{1}{2}\right) + \frac{1}{12n}.$$

As the AD and CvM statistics, also the Watson (W) statistic has a representation in terms of an empirical process. Let

$$\mathbb{P}_n(x) = \mathbb{B}_n(x) - \int_0^1 \mathbb{B}_n(y)dy.$$

Then $U_n = \int_0^1 \mathbb{P}_n^2(x)dx$.

In the following subsections we provide some more details on the decomposition and the asymptotic null distribution for simple null hypothesis. In general the theory for circular distributions is more complicated, and therefore we omit a discussion on composite null hypotheses and on how the decomposition in the latter case relates to components of smooth tests. For more information we refer to Wouters et al., who established the link between the Watson and smooth test statistics.

#### 5.2.4.2 Principal Components Decomposition of the Watson Statistic (Simple Null)

The principal component decomposition was given by Shorack and Wellner (1986). The covariance function of the process $\mathbb{P}_n$ is given by

$$c(x,y) = x \wedge y - (x+y)/2 + (x-y)/2 + 1/12.$$

It has eigenvalues $\lambda_{2j-1} = \lambda_{2j} = 1/(4\pi^2 j^2)$, and eigenfunctions $k_{2j-1}(x) = \sqrt{2}\sin 2j\pi x$, and $k_{2j}(x) = \sqrt{2}\cos 2j\pi x$, $j = 1, 2, \ldots$. Every two consecutive eigenvalues of odd and even order are equal, thus the principal component decomposition may be written as

$$U_n = \sum_{j=1}^{\infty} \frac{1}{4\pi^2 j^2} \left(Y_{nj}^2 + Z_{nj}^2\right), \tag{5.16}$$

where the components are

$$Y_{nj} = \sqrt{\frac{2}{n}} \sum_{i=1}^{n} \cos(2j\pi X_i) \text{ and } Z_{nj} = \sqrt{\frac{2}{n}} \sum_{i=1}^{n} \sin(2j\pi X_i). \tag{5.17}$$

Note that due to the orthogonality of the sin and the cos terms in $Y_{nj}$ and $Z_{nj}$, the components have zero covariance. The term $Y_{nj}^2 + Z_{nj}^2$ can be interpreted as the resultant length of the $j$th empirical trigonometric moment of the circular distribution. The first component ($j = 1$) can be recognised as the Rayleigh test statistic (Rayleigh (1919)).

Another interesting interpretation of the components can be seen by applying some simple trigonometric calculus. Write

$$\begin{aligned}
Y_{nj}^2 + Z_{nj}^2 &= \left(\sqrt{\frac{2}{n}} \sum_{i=1}^{n} \cos(2j\pi X_i)\right)^2 + \left(\sqrt{\frac{2}{n}} \sum_{i=1}^{n} \sin(2j\pi X_i)\right)^2 \\
&= \frac{2}{n} \sum_{i=1}^{n} \sum_{m=1}^{n} \left(\cos(2j\pi X_i)\cos(2j\pi X_m) + \sin(2j\pi X_i)\sin(2j\pi X_m)\right) \\
&= \frac{2}{n} \sum_{i=1}^{n} \sum_{m=1}^{n} \cos\left(2j\pi(X_i - X_m)\right) \\
&= 2 + \frac{4}{n} \sum_{i<m}^{n} \cos\left(2j\pi(X_{(i)} - X_{(m)})\right).
\end{aligned}$$

This last expression clearly shows that this component gets large when there are relatively too many observations too close to one another, i.e., too many small $X_{(i)} - X_{(m)}$. This happens, for instance, when the distribution is too peaked, or when there are too many modi. Furthermore, the smaller the differences $X_{(i)} - X_{(m)}$ are, their contribution will be large for the more orders $j$.

#### 5.2.4.3 Null Distribution (Simple Null)

The first way to get at the asymptotic null distribution of the Watson test statistic is based on its principal component decomposition. Because the components $Y_{nj}$ and $Z_{nj}$ in (5.17) converge to independent standard normal variates, Equation (5.16) implies that $U_n$ converges to $U$, which has representation

$$U = \sum_{j=1}^{\infty} \frac{1}{4\pi^2 j^2} \left( Z_{2j-1}^2 + Z_{2j}^2 \right),$$

where the $Z_j$ $(j = 1, 2, \ldots)$ are i.i.d. standard normal. Thus we could also write

$$U = \sum_{j=1}^{\infty} \frac{1}{4\pi^2 j^2} X_j^2$$

with $X_j^2$ i.i.d. $\chi_2^2$.

The other solution is based on empirical process theory. Because the process $\mathbb{P}_n$ is a function of the empirical process $\mathbb{B}_n$, we find by the continuous mapping theorem, under $H_0$, as $n \to \infty$,

$$\mathbb{P}_n(x) = \mathbb{B}_n(x) - \int_0^1 \mathbb{B}_n(y)dy \xrightarrow{w} \mathbb{P} = \mathbb{B}(x) - \int_0^1 \mathbb{B}(y)dy,$$

and, similarly,

$$U_n = \int_0^1 \mathbb{P}_n^2(x)dx \xrightarrow{w} U = \int_0^1 \mathbb{P}^2(x)dx.$$

## 5.3 Generalisations of EDF Tests

In Section 2.1.2 we have explained the rationale of EDF tests and we gave the general form of an EDF test statistic, $T_n = c(n)d(\hat{F}_n, G)$. The distance or divergence functional $d$ satisfies the property that $d(F, G) = 0$ if and only if $F(x) = G(x)$ for all $x \in \mathcal{S}$, and $d(\hat{F}_n, G)$ actually serves as a plug-in estimator of $d(F, G)$. Although this formulation is quite general, all tests discussed so far have a $d$ functional that depends on $F$ and $G$ through $B(x) = (F(x) - G(x))w(x)$, where $w(x)$ is a weight function, and in terms of $B(x)$ we have

$$B(x) = 0 \text{ for all } x \in \mathcal{S} \Leftrightarrow F(x) = G(x) \text{ for all } x \in \mathcal{S}. \tag{5.18}$$

Thus the functional $d$ could just as well be denoted by $d(B)$. Some examples:

| | | |
|---|---|---|
| CvM | $w(x) = \sqrt{g(x)}$ | $d = \int B^2(x)dx$ |
| AD | $w(x) = \sqrt{\frac{g(x)}{G(x)(1-G(x))}}$ | $d = \int B^2(x)dx$ |
| KS | $w(x) = 1$ | $d = \sup_x \lvert B(x) \rvert$ |

Thus all $B(x)$ are explicitly functionals of $F$ and $G$. This led to the important role of the empirical process which appears when $F$ is replaced by the EDF. There is, however, no necessity to have $B$ explicitly depending on $F$ and $G$. The only necessity is condition (5.18). This opens the way to other definitions of $B$. For instance, a distribution is not only uniquely characterised by its CDF $F(x)$. Other characterisations are based on the quantile function (QF) $F^{-1}(x)$, or the characteristic function (CF) $\Phi_F(x) = \mathrm{E}_f\{\exp(ixY)\}$ ($Y$ with CDF $F$). This brings us to the following choices for $B$,

$$B(x) = F^{-1}(x) - G^{-1}(x)$$
$$B(x) = \Phi_F(x) - \Phi_G(x).$$

All these $B$ functionals satisfy condition (5.18), and empirical versions, say $B_n$, can be obtained by replacing $F^{-1}$ or $\Phi_F$ by their estimators. For each of these types more details are given in the next few sections.

## 5.3.1 Tests Based on the Empirical Quantile Function (EQF)

### 5.3.1.1 The Empirical Quantile Function

The choice $B(x) = F^{-1}(x) - G^{-1}(x)$ involves the quantile function $F^{-1}(x)$ which is the inverse of the CDF and which is defined as $(0 < p < 1)$

$$F^{-1}(p) = \inf\{y \in \mathcal{S} : p \leq F(y)\}. \tag{5.19}$$

Its empirical plug-in estimator, say $\hat{F}_n^{-1}(p)$, is obtained by replacing $F$ in Equation (5.19) by the EDF. Note that $\hat{F}_n^{-1}$ always equals a sample observation. This can be stressed by adopting an alternative definition of $\hat{F}_n^{-1}$,

$$\hat{F}_n^{-1}(u) = X_{(i)} \text{ if } \frac{i-1}{n} \leq u < \frac{i}{n} \text{ for some } 1 \leq i \leq n.$$

These definitions, as well as some basic theory were provided in Section 2.3.

#### 5.3.1.2 EQF Tests for the Simple Null Hypothesis

The study of the EQF tests is based on the empirical quantile process (EQP),

$$\mathbb{Q}_n(x) = \sqrt{n}\left(F_n^{-1}(x) - F^{-1}(x)\right),$$

$(0 \leq x \leq 1)$. It is indeed possible to express the test statistics in terms of the EQP. Theorem 2.4 says that the EQP converges weakly to a weighted Brownian bridge; i.e., $\mathbb{Q}_n \xrightarrow{w} -\mathbb{B}/f(F^{-1}(x))$ as $n \to \infty$.

In analogy with the AD and CvM statistics, a natural distance measure between two distributions may be defined as

$$d(F,G) = \int_0^1 B^2(x)dx = \int_0^1 \left(F^{-1}(x) - G^{-1}(x)\right)^2 dx,$$

which is known as the (squared) Wasserstein distance; i.e., $d(F,G) = W^2(F,G)$. del Barrio et al. (1999) and del Bario et al. (2000) gave some asymptotic properties of the test based on

$$W_n = nW^2(\hat{F}_n, G) = n\int_0^1 \left(\hat{F}_n^{-1}(x) - G^{-1}(x)\right)^2 dx.$$

de Wet (2002) extended these tests by considering a weighted Wasserstein distance; i.e.,

$$d(F,G) = \int_0^1 \left(F^{-1}(x) - G^{-1}(x)\right)^2 w(x)dx,$$

where $\int_0^1 w(x)dx = 1$, and the resulting test statistic is denoted by $W_n^w$. In the remainder of this section we discuss this weighted integral statistic. We refer to the corresponding tests as EQF tests. Sometimes the integrals in the definition of $d$ may be hard to compute analytically. In these cases an alternative statistic may be considered (Mason (1984)),

$$M_n^w = \sum_{i=1}^n w\left(\frac{i}{n+1}\right)\left(X_{(i)} - G^{-1}\left(\frac{i}{n+1}\right)\right)^2,$$

which is, under certain conditions, asymptotically equivalent to the integral tests.

For the proofs of the convergence of the integral tests based on the EQP, a stronger type of convergence is often required: strong approximations for weighted versions of $\mathbb{Q}_n$. We, however, do not go into detail here, but we refer the interested reader to, e.g., Csörgö (1983), Mason (1984), Csörgö et al. (1986), Einmahl and Mason (1988), Csörgö and Horváth (1993), and Csörgö et al. (1993). In these references conditions on the distributions and weight functions are given, so that $M_n^w$ and $W_n^w$ have limiting distribution

$$\int_0^1 \frac{\mathbb{B}^2(x)w^2(x)}{f^2(F^{-1}(x))}dx.$$

The relation between EQF statistics and Brownian bridges is similar to what we have seen for the EDF statistics. It further suggests that these statistics allow a principal component decomposition, but now the eigenfunctions and eigenvalues are related to the covariance function

$$c(s,t) = \frac{(s \wedge t - st)w(s)w(t)}{f(F^{-1}(s))f(F^{-1}(t))}.$$

In the previous paragraphs we have restricted the discussion to a completely specified hypothesised distribution $G$ (simple null case). Before going to the composite null situation we mention briefly two specific tests for uniformity; i.e., $G(x) = x$ for $x \in [0,1]$. We consider the EQF versions of the CvM and AD statistics. With $w(u) = 1$ the EQF–CvM statistic is found. We give here the form proposed by Durbin and Knott (1972),

$$W_n^q = \frac{n+1}{n}\sum_{i=1}^n \left(X_{(i)} - \frac{i}{n+1}\right)^2,$$

which has limiting null distribution $\int_0^1 \mathbb{B}^2(x)dx$, which is the same as for the EDF–CvM test. Kaigh (1992) discussed some of the characteristics of the EQF–AD statistic,

$$A_n^q = (n+1)(n+2)\sum_{i=1}^n \frac{\left(X_{(i)} - \frac{i}{n+1}\right)^2}{i(n-i+1)},$$

which has the same asymptotic null distribution as the EDF–AD statistic. Kaigh (1992) noted, however, that the convergence is very slow and that the EQF–AD test has quite poor powers for many interesting alternatives. He further gave the principal component decomposition of the EQF–CvM and AD statistics and he showed that they are both functions of the *spacings* of the sample observations. Spacings are defined as $D_i = X_{(i)} - X_{(i-1)}$, $i = 1, \ldots, n+1$, with the convention that $X_{(0)} \equiv 0$ and $X_{(n+1)} \equiv 1$. He showed

$$W_n^q = \sum_{j=1}^\infty \frac{1}{j^2\pi^2} Z_{nj}^2,$$

where

$$Z_{nj} = -\sqrt{n+2}\sum_{i=1}^{n+1} \sqrt{2}\cos\left(\frac{2i-1}{2n+2}j\pi\right) D_i.$$

The components $Z_{nj}$ are asymptotically normally distributed, and, under the null hypothesis, they are asymptotically independent and standard normally distributed. A similar decomposition holds for the EQF–AD statistic, but instead of a cosine basis of orthonormal functions, it makes use of a basis of *Hahn orthonormal polynomial vectors* $\{\boldsymbol{\pi}_{0,n}, \ldots, \boldsymbol{\pi}_{n,n}\}$, with $\boldsymbol{\pi}^t_{j,n} = (\pi_{j,n}(1), \ldots, \pi_{j,n}(n+1))$. With this notation, we get

$$W^q_n = \sum_{j=1}^{\infty} \frac{1}{j(j+1)} Z^2_{nj},$$

where now

$$Z_{nj} = -\sqrt{(n+1)(n+2)} \sum_{i=1}^{n+1} \pi_{j,n}(i) \left( D_i - \frac{1}{n+1} \right).$$

For both EQF statistics the components $Z_{nj}$ are thus functions of the spacings, whereas for their EDF counterparts the components depend directly on the (unordered) observations. On the other hand, EDF and EQF have the same eigenfunctions and the same asymptotic representation under the null hypothesis. Finally, Kaigh (1992) showed that the correlation between the components of the EDF and EQF statistics, say $Z^D_{nj}$ and $Z^Q_{nk}$, is zero unless $j = k$.

#### 5.3.1.3 EQF Tests for Location-Scale Distributions

Recent literature on statistics based on the Wasserstein distance has focused on goodness-of-fit tests for location-scale families. In Section 5.1.3 we have defined location-scale families as distributions with densities satisfying $g(x; \mu, \sigma) = g((x - \mu)/\sigma, 0, 1)/\sigma$. In terms of the quantile function this becomes

$$G^{-1}(u; \mu, \sigma) = \mu + \sigma G^{-1}(u; 0, 1). \tag{5.20}$$

The linear representation of Equation (5.20) is particularly appropriate to plug into the Wasserstein distance. del Barrio et al. (1999) were the first to notice this. They investigated $W_n$ statistics for testing normality.

In this location-scale setting, however, we encounter again two nuisance parameters, $\mu$ and $\sigma$, which have to be estimated from the data. Instead of considering MLE or MME, there seem to be advantages of using a *minimum distance estimator*. In the present context, the most natural minimum distance estimator is the one which minimises the Wasserstein distance between $G^{-1}$ and the empirical $\hat{F}^{-1}_n$. Let $S^2$ denote the sample variance.

The close connection between the test statistic and the method of parameter estimation allows the test statistic to be written as

$$\begin{aligned} W_n &= nW^2(\hat{F}_n, \hat{G}) \\ &= n \inf_{\mu,\sigma} \left\{ W^2(\hat{F}_n, G(.; \mu, \sigma)) : -\infty < \mu < +\infty, 0 < \sigma < +\infty \right\} \end{aligned}$$

$$= n \inf_{\mu,\sigma} \left\{ \int_0^1 \left( \hat{F}_n^{-1}(u) - \mu - \sigma G^{-1}(u;0,1) \right)^2 du \right\}$$

$$= n \left( S^2 - \left( \int_0^1 \hat{F}_n^{-1}(u) G^{-1}(u;0,1) du \right)^2 \right),$$

where the last step is obtained by recognising that $\mu = \hat{\mu} = \bar{X}$ and $\sigma^2 = \hat{\sigma}^2 = \int_0^1 \hat{F}_n^{-1}(u)G^{-1}(u;0,1)du$ determines the infimum. Hence, $W_n = n(S^2 - \hat{\sigma}^2)$ is not affected by the location. del Barrio et al. (1999) suggested to consider a scaled version of $W_n$,

$$R_n = \frac{W_n}{nS^2} = 1 - \frac{\hat{\sigma}^2}{S^2},$$

which is asymptotically also scale invariant. This test statistic can be interpreted as a measure which compares $S^2$, which is an unrestricted consistent estimator of $\sigma^2$, with a restricted estimator $\hat{\sigma}^2$ which is only a consistent estimator of $\sigma^2$ when $F$ belongs to the hypothesised location-scale family. It can be shown that $R_n$ is asymptotically equivalent to the Shapiro–Wilk, Shapiro–Francia and the de Wet–Venter statistics.

The statistic $R_n$ may be conveniently expressed in terms of the *estimated empirical quantile process*,

$$\hat{\mathbb{Q}}_n = \mathbb{Q}_n - < \mathbb{Q}_n, 1 > 1 - < \mathbb{Q}_n, F^{-1} > F^{-1},$$

which is similar to the estimated empirical process of Section 5.1.3, but here the location and scale parameters are estimated in a different way. Similarly, $\hat{\mathbb{Q}} = \mathbb{Q} - < \mathbb{Q}, 1 > 1 - < \mathbb{Q}, F^{-1} > F^{-1}$. With this notation, we find

$$R_n = \frac{1}{S^2} \int_0^1 \hat{\mathbb{Q}}_n^2(u) du. \tag{5.21}$$

It may be tempting to conclude that $R_n$ converges in distribution to

$$\frac{1}{\sigma^2} \int_0^1 \hat{\mathbb{Q}}^2(u) du,$$

but apparently there appear to be some technical caveats. In particular it is not always true that

$$\int_0^1 \mathbb{Q}_n^2(u) du \xrightarrow{d} \int_0^1 \frac{\mathbb{B}^2(u)}{g^2(G^{-1}(u))} du$$

as may be suspected from Theorem 2.4 and the continuous mapping theorem. For some distributions $g$ a slightly different convergence holds,

$$\int_0^1 \mathbb{Q}_n^2(u)du - a_n \xrightarrow{d} \int_0^1 \frac{\mathbb{B}^2(u) - \mathrm{E}_g\left\{\mathbb{B}^2(u)\right\}}{g^2(G^{-1}(u))}du, \tag{5.22}$$

where

$$a_n = \int_{1/n}^{1-1/n} \frac{u(1-u)}{g^2(G^{-1}(u))}du.$$

For a general discussion of statistics of the form (5.21), and generalisations of it, we refer to Del Barrio et al. (2005). In the next paragraphs, however, we treat only the case of the normal distribution.

We consider again the normal distribution as an example of a location-scale invariant distribution. Using the conventional notation $\Phi$ and $\phi$ for the CDF and density function, it can be shown that under $H_0$, as $n \to \infty$,

$$\begin{aligned} R_n - a_n &\xrightarrow{d} \int_0^1 \mathbb{Q}^2(u)du - \int_0^1 \mathrm{E}\left\{\mathbb{Q}^2(u)\right\}du \\ &= \int_0^1 \frac{\mathbb{B}^2(u) - \mathrm{E}_\phi\left\{\mathbb{B}^2(u)\right\}}{\phi^2(\Phi^{-1}(u))}du - \left[\int_0^1 \frac{\mathbb{B}(u)}{\phi(\Phi^{-1}(u))}du\right]^2 \\ &\quad - \left[\int_0^1 \frac{\mathbb{B}(u)\Phi^{-1}(u)}{\phi(\Phi^{-1}(u))}du\right]^2, \end{aligned}$$

where

$$a_n = \int_{1/n}^{1-1/n} \frac{u(1-u)}{\phi^2(\Phi^{-1}(u))}du.$$

Further insight into the limiting distribution is given when an orthonormal components decomposition of $\hat{\mathbb{Q}}^2$ is performed. del Barrio et al. (1999) and del Bario et al. (2000) showed that

$$R_n - a_n \xrightarrow{d} -\frac{3}{2} + \sum_{j=3}^{\infty} \frac{1}{j}\left(Z_j^2 - 1\right), \tag{5.23}$$

where the $Z_j$ are i.i.d. standard normal. The asymptotic representation of $R_n - a_n$ now becomes

$$-\frac{3}{2} + \sum_{j=3}^{\infty} \frac{1}{j}\left(Z_{nj}^2 - 1\right),$$

where

$$Z_{nj} = \sqrt{j} < \hat{\mathbb{Q}}_n, H_j(\Phi^{-1}) >$$

$H_j$ is the Hermite polynomial of degree $j-1$. These $Z_{nj}$ are asymptotically i.i.d. standard normal under the null hypothesis.

The decomposition presented in (5.23) looks very similar to what we have seen many times before, except that the summation only starts at the third order term. The first two terms vanish due to the estimation of the two nuisance parameters. Because this sum of (weighted) squared independent standard normal variates looks similar to a $\chi^2$ distributed random variable, this loss of the first two terms is referred to the *loss of degrees of freedom* property.

Because the EQF test for normality has a principal component decomposition into the same components as the smooth test for normality based on the Hermite polynomials, it is again very meaningful to augment a statistical data analysis using this EQF test with a study of the first few components.

Instead of using the Wasserstein distance similar EQF tests may be constructed based on a weighted Wasserstein distance. This is studied in detail by Csörgö (2002) and de Wet (2002). They give results for scale-invariant distributions, and de Wet also gives some further results on location-invariant distributions. In particular, de Wet shows that for a given location- or scale-invariant distribution, the weight function can be chosen in such a way that the resulting minimum distance estimator is asymptotically efficient. Moreover, the same weight functions result in the loss of one degree of freedom property. Finally, we note that the normal distribution is the only location-scale invariant distribution which gives a two degrees of freedom loss.

### *5.3.2 Tests Based on the Empirical Characteristic Function (ECF)*

Another rather new type of goodness-of-fit test is based on the ECF. In general a test statistic could be constructed starting from $B(t) = \Phi_F(t) - \Phi_G(t)$ and replacing $\Phi_F$ by its empirical estimator, $\Phi_n(t) = (1/n)\sum_{l=1}^{n} \exp(itX_l)$, which gives $B_n(t)$. Because $\Phi$ and $\Phi_n$ are functions with values in the set of complex numbers, we should take care in defining the distance measure $d$. For instance, $d$ could be defined on the real or the imaginary part only (see, e.g., Heathcore (1972) and Feigin and Heathcore (1977)), but here we prefer to define $d$ in terms of the modulus of $B$ which is denoted as $|B|$. Epps and Pulley (1983) were the first to consider tests based on

$$C_n = \int_{-\infty}^{+\infty} |B(t)|^2 w(t) dt \tag{5.24}$$

for testing normality. The idea of using the characteristic function, however, dates from earlier, but most previous suggestions had the argument $t$ of $B_n(t)$ fixed at a given value. More recent tests based on statistics like $C_n$ are proposed by Meintanis (2004a), Meintanis (2004b), and Matsui and Takemura (2005) for the logistic, the Laplace, and the Cauchy distribution, respectively. For the Cauchy, see also Gürtler and Henze (2000). Note that all four distributions mentioned here are location-scale families. Let $\delta$ and $\gamma$ denote the

**Table 5.1** The characteristic functions of some distributions and appropriate weight functions to use in the test statistics. All distributions are location-scale families with parameters $\delta$ and $\gamma$, except the symmetric stable distribution which has scale parameter $\gamma$ and shape parameter $\theta$

| Distribution | CF | Weight function |
|---|---|---|
| Cauchy | $\phi(t) = \exp(it\delta - \gamma\|t\|)$ | $w(t) = \exp(-a\|t\|)$ |
| Exponential | $\phi(t) = \frac{i\gamma}{t+i\gamma}$ | $w_1(t) = \exp(-a\|t\|)$ or $w_2(t) = \exp(-at^2)$ |
| Laplace | $\phi(t) = \frac{\exp(it\delta)}{1+\gamma^2 t^2}$ | $w_1(t) = \exp(-a\|t\|)$ or $w_2(t) = \exp(-at^2)$ |
| Logistic | $\phi(t) = \frac{\pi t}{\sinh(\pi t)}$ | $w(t) = \exp(-a\|t\|)$ |
| Normal | $\phi(t) = \exp(it\delta - \frac{1}{2}t^2\gamma^2)$ | $w(t) = \frac{a}{\sqrt{2\pi}}\exp(-\frac{1}{2}a^2t^2)$ |
| Symmetric stable $(\gamma, \theta)$ | $\phi(t) = \exp(-\gamma^\theta\|t\|^\theta)$ | $w(t) = \|t\|\exp(-a\|t\|)$ |

location and scale parameter, respectively. Table 5.1 presents the CFs of some important distributions.

For location-scale invariant distributions it is natural to construct the test statistic in terms of the standardised observations, say $Y_i = (X_i - \hat{\delta})/\hat{\gamma}$ ($i = 1, \ldots, n$), where the estimators $\hat{\delta}$ and $\hat{\sigma}$ are locally asymptotically linear. A desirable property of goodness-of-fit tests for location-scale families is to be location-scale invariant. This can be guaranteed when the estimator $\hat{\delta}$ is scale-invariant and $\hat{\sigma}$ is location-invariant and scale-equivariant. The method of moment estimators, among others, possess this property.

One could think of the simplest ECF test statistic of the form of $C_n$ by taking $w(x) = 1$. However, with this choice the resulting integral in (5.24) has no analytic solution and thus the statistic $C_n$ has no simple computational form. The weight function is therefore to be determined so that $C_n$ has an analytic form, but at the same time one should take care that $C_n$ has a proper limiting distribution (see further down). Table 5.1 shows the weight functions that have been proposed in the literature. They all have a tuning parameter $a > 0$, and so the behavior of the test can be modified by changing $a$. The effect of $a$ on the behaviour of $C_n$ is explained intuitively in the next paragraph.

Suppose, for instance, that the weight function has the form $w(t) = \exp(-a|t|)$, with $a > 0$. Thus, large values of $a$ will make $w(t)$ a fast decaying function so that $C_n$ is dominated by $B(t)$ with small $t$. When, on the other hand, we consider a small value of $a$, then $w(t)$ decreases slowly with $|t|$ so that $C_n$ will also be influenced by large values of $|t|$. To get a deeper understanding, consider the expansion

$$\Phi_G(t) = \sum_{k=1}^{\infty} \frac{(it)^k}{k!}\mu'_{Gk},$$

where $\mu'_{Gk}$ denotes the $k$th moment of distribution $G$ about 0. Because $\hat{\Phi}_n(t)$ is a consistent estimator of $\Phi_F$, we can consider $B_n(t)$ as a consistent estimator of

$$B(t) = \Phi_F(t) - \Phi_G(t) = \sum_{k=1}^{\infty} \frac{(it)^k}{k!} \left(\mu'_{Fk} - \mu'_{Gk}\right).$$

Thus, $B(t)$ is a weighted sum of moment differences between $F$ and $G$, and the weights are determined by $t^k$. When $B(t)$ and $w(t)$ are combined in the construction of the test statistic $C_n$, this representation of $B(t)$ shows that large $a$ will result in a test which is particularly sensitive to deviations in the lower-order moments, whereas small $a$ will make the test more sensitive to deviations in the larger-order moments.

Just as for the EDF and EQF tests, the asymptotic distribution theory of the ECF tests is based on empirical process theory. We do not give details here. Instead we only summarise the major steps leading to the limiting null distribution of $C_n$. We restrict the discussion to the case without nuisance parameters. The extension to estimated nuisance parameters is similar to what is presented in Section 5.1.3, which shows that the limiting distribution depends on the method of estimation.

First, it is recognised that $Z_n = |B_n|$ takes random elements in an appropriate Hilbert space $L_2$. Sometimes it is not simple to express $Z_n$ as $(1/\sqrt{n})\sum_{i=1}^n W_i(t)$, with $W_i \in L_2$, in which case one should find a $Z_n^\star = (1/\sqrt{n})\sum_{i=1}^n W_i^\star(t)$ which is a strong approximation of $Z_n$. Next, the central limit theorem in Hilbert spaces can be applied to obtain a weak convergence of $Z_n$ to a Gaussian process $Z$ of which the covariance function $c$ is determined by $W_i$ or $W_i^\star$. The asymptotic distribution of $C_n$ is found by using strong approximation results to cope with the weight function, and by applying the CMT. Details can be found in the previously mentioned references, particularly Gürtler and Henze (2000).

### *5.3.3 Miscellaneous Tests Based on Empirical Functionals of F*

In the very beginning of this section we argued that the idea behind EDF tests can be used to construct other tests based on a function $B(x)$ which must satisfy the conditions in (5.18). The EQF and ECF tests are clear classes of such tests, but sometimes statisticians have been more inventive and constructed tests based on a $B$ which is not directly expressed in terms of the EDF, EQF, or ECF. In the next few paragraphs we give some examples.

Henze and Meintanis (2002) proposed a test for exponentiality which is based on the ECF, but their test statistic is not of the form of $C_n$ as in (5.24). They only use the CF as a starting point. For the exponential distribution with rate parameter $\gamma$, $\Phi(t) = (i\gamma)/(t + i\gamma)$. As every complex-valued function, it may be written as $\Phi(t) \equiv u(t) + iv(t)$. Moreover, using $\exp(iz) = \cos(z) + i\sin(z)$, we find $u(t) = \mathrm{E}\{\cos(tX)\}$ and $v(t) = \mathrm{E}\{\sin(tX)\}$. Henze and Meintanis (2002) proved that for all $t$, $v(t) - \gamma tu(t) = 0$. The choice $B(t) = v(t) - \gamma tu(t)$ therefore makes also sense. Its empirical counterpart is obtained by replacing $u$ and $v$ by their empirical versions, $u_n(t) = (1/n)\sum_{i=1}^n \cos(tY_i)$ and $v_n(t) = (1/n)\sum_{i=1}^n \sin(tY_i)$. Because the exponential distribution is scale-invariant, the test is usually applied to the standardised observations $Y_i = X_i/\bar{X}$ $(i = 1, \dots, n)$. The test statistic becomes

$$T_n = n\int_0^\infty (u_n(t) - tv_n(t))^2 w(t)dt,$$

where $w$ can take two forms (see Table 5.1).

Also Henze (1993) proposed a test for exponentiality. He started with considering the Laplace transform, $\phi(t) = \mathrm{E}\{\exp(-tX)\} = \gamma/(\gamma + t)$. Thus $B(t) = \phi(t) - \gamma/(\gamma + t)$ is an appropriate functional. Let $\phi_n(t) = (1/n)\sum_{i=1}^n \exp(-tY_i)$ denote the empirical estimator of $\phi(t)$. Henze suggested

$$T_{n,a} = n\int_0^\infty B_n^2(t)w(t)dt \qquad (5.25)$$

with $w(t) = \exp(-at)$. Another test for exponentiality was proposed by Baringhaus and Henze (1991) who also started with the Laplace transform. They used the property that $\phi$ is a solution of the differential equation

$$(\gamma + t)\phi'(t) + \phi(t) = 0 \text{ for all } t.$$

Replacing $\phi$ and $\phi'$ with their empirical estimators, and using the weight function $w(t) = \exp(-at)$ gives their test statistic, say $D_{n,a}$. An interesting study to the effect of $a$ in extreme situations was done by Baringhaus et al. (2000). In particular they showed that

$$\lim_{a\to\infty} a^5 T_{n,a} = 6n(\bar{Y}^2 - 2)^2$$

$$\lim_{a\to\infty} a^3 D_{n,a} = 2n(\bar{Y}^2 - 2)^2,$$

in which we recognise the squared second-order component of a smooth test statistic based on the Laguerre polynomials; i.e., $\hat{\theta}_2^2 = n\frac{1}{4}(\bar{Y}^2 - 2)^2$.

Meintanis (2005) studied a similar approach based on differential equations for constructing a test for a symmetric stable distribution with shape parameter $\theta$ and scale parameter $\gamma$. He first remarks that for a symmetric stable distribution there is no $w(t)$ to make the ECF statistic $C_n$ have a closed form. Therefore he finds a differential equation for which the CF $\Phi$ is a solution.

## 5.4 The Sample Space Partition Tests

### 5.4.1 Another Look at the Anderson–Darling Statistic

The method that is discussed in this section is basically an EDF integral test that may be considered as an extension of the Anderson–Darling test, but it may also be looked at as a method that solves one of the problems related to the application of the Pearson $\chi^2$ test to test for goodness-of-fit for continuous distributions. In Section 1.1 we mentioned that one of the oldest methods for testing goodness-of-fit consists in grouping or categorising the data into $c$ groups, even when the data have a continuous distribution. Once the data are categorised, Pearson's $\chi^2$ test for the multinomial distribution can be applied. Despite the practical simplicity of this method, there are some difficult issues that need attention. We mention two: (1) how many groups, and (2) where to place the cell boundaries. When testing a composite null hypothesis there is the additional problem of how to estimate the nuisance parameters, but to keep the exposition simple we ignore this problem here.

The Anderson–Darling test statistic for testing uniformity, which is given by

$$T_n = \int_0^1 \frac{\mathbb{B}_n^2(x)}{x(1-x)} dx = n \int_0^1 \frac{\left(\hat{F}_n(x) - x\right)^2}{x(1-x)} dx,$$

may also be written as

$$\begin{aligned} T_n &= n \int_0^1 \left[ \frac{\left(\hat{F}_n(x) - x\right)^2}{x} + \frac{\left((1 - \hat{F}_n(x)) - (1 - x)\right)^2}{1 - x} \right] dx \\ &= \int_0^1 X^2(x) dx, \end{aligned}$$

where $X^2(x)$ is Pearson's $\chi^2$ statistic applied to the sample grouped into two groups with cell boundary placed at $x$. The AD statistic is thus essentially an average Pearson $\chi^2$ statistic, and by averaging the problem of the choice of the cell boundary is solved. The sample space partition test of Thas and Ottoy (2002, 2003) is an extension of the Anderson–Darling test to grouping into more than two groups.

### 5.4.2 The Sample Space Partition Test

The categorisation is equivalent to partitioning the sample space $[0, 1]$ into $c$ intervals; i.e.,

$$[0,1] = [0,d_1] \cup [d_1,d_2] \cup \cdots \cup [d_{c-1},1],$$

where $0 < d_1 < d_2 < \cdots < d_{c-1} < 1$. Let $D_c = \{d_1, \ldots, d_{c-1}\}$. By counting the number of observations in each interval, a $c$ cell array is obtained. The null hypothesis of uniformity now implies a null hypothesis in terms of a multinomial distribution for which Pearson's $X^2$ test is appropriate. In particular, for a given partition implied by $D_c$,

$$\begin{aligned} X^2_{c,n}(D_c) &= X^2_{c,n}(d_1,\ldots,d_{c-1}) \\ &= n\sum_{i=1}^{c} \frac{\left(F_n(d_{(i)}) - F_n(d_{(i-1)}) - (F_0(d_{(i)}) - F_0(d_{(i-1)}))\right)^2}{F_0(d_{(i)}) - F_0(d_{(i-1)})}, \end{aligned}$$

where $d_{(0)} \equiv 0$ and $d_{(c)} \equiv 1$, and where $d_{(1)}, \ldots, d_{(c-1)}$ are the order statistics of $d_1, \ldots, d_{c-1}$.

An important issue is the choice of $D_c$ and the number of cells ($c$) so that the test has good power. There is a vast literature on how partitions can be constructed. Some suggested that the cells should be equiprobable under the null hypothesis (e.g., Mann and Wald (1942)), whereas others argue that for the detection of, for instance, heavy-tailed alternatives unequal cells result in a test with larger power (Kallenberg et al. (1985)). Also on the choice of $c$ many different guidelines have been proposed (see, e.g., Moore (1986) for an overview).

The general form of the sample space partition (SSP) test statistic is given by

$$T_{c,n} = \int_0^1 \cdots \int_0^1 X^2_{c,n}(d_1,\ldots,d_{c-1})dd_1 \ldots dd_{c-1}.$$

For a given, but arbitrary SSP size $c$, the asymptotic null distribution of $T_{c,n}$ can be found using empirical process theory. Just as the integral EDF tests, the SSP test is consistent against any alternative (omnibus consistent), for whatever finite $c > 1$. The test based on $T_{c,n}$ is known as the SSPc test.

For $c = 2$ (Anderson–Darling), $c = 3$, and $c = 4$, the computational formulae are easily calculated. Let $X_{(i)}$ denote the $i$th order statistic ($i = 1, \ldots, n$) of the sample observations $X_1, \ldots, X_n$.

- $c = 2$:

$$T_{2,n} = A_n = -n - \frac{1}{n}\sum_{i=1}^{n}(2i-1)\left(\log(X_{(i)}) + \log(1 - X_{(n+1-i)})\right)$$

- $c = 3$:
  when $c = 3$, $T_{3,n}$ reduces to

$$T_{3,n} = 2A_n - 4W_n + K_n,$$

where $A_n$ and $W_n$ represent the Anderson–Darling and the Cramér–von Mises statistics, respectively, and

$$K_n = \int_0^1 \int_0^1 \frac{\left(B_n^2(x) - B_n^2(y)\right)^2}{|x-y|} dxdy$$
$$= -\frac{2}{n} \sum_{i=1}^{n} \sum_{j=1}^{n} \Big( X_{(i\vee j)} \log(X_{(i\vee j)}) + (1 - X_{(i\wedge j)}) \log(1 - X_{(i\wedge j)})$$
$$-(X_{(j)} - X_{(i)}) \log(X_{(j)} - X_{(i)}) + X_{(i)}(1 - X_{(i)}) + X_{(j)}(1 - X_{(j)})$$
$$-\frac{1}{6}\Big).$$

- $c = 4$:

$$T_{4,n} = 3A_n - 10.5W_n + 3K_n + 1.5n\left(\bar{X} - \frac{1}{2}\right)^2.$$

This methodology avoids the choice of the break points $d_i$ $(1 < i < c)$, but the SSP size $c$ has still to be determined by the user. As a solution to this problem, the authors proposed a data-driven version of the SSPc test by estimating a proper value for $c$ from the sample. This sample-based SSP size is denoted by $C_n$. In particular, $C_n$ is determined by means of a selection rule which has the general form

$$C_n = \text{ArgMax}_{c\in\Gamma}\{T_{c,n} - 2(c-1)\log a_n\},$$

where $\Gamma$ is a nonempty finite set containing all permissible SSP sizes (often $\Gamma = \{2,3,4\}$ or $\Gamma = \{2,3,4,5\}$ seem to be sufficiently rich), and $a_n$ is a penalty depending on the sample size $n$. Although the form of this selection rule resembles the Bayesian Information Criterion (BIC) $(a_n = n^{1/2})$ or the AIC (Akaike's Information Criterion of Akaike (1973, 1974); $a_n = e$), it has no sound theoretical justification, for $T_{c,n}$ is not a log-likelihood, as it is in AIC and BIC, nor a score statistic as it is in the modified BIC of Kallenberg and Ledwina (1997). Also a double logarithmic penalty term (LL), $a_n = \log\log n$, has been considered. As with the data-driven smooth tests, it has been shown that the selected SSP size $C_n$ converges in probability to its smallest possible value, which is $\min \Gamma$. Furthermore, for every choice of $\Gamma$ the data-driven SSP test is omnibus consistent. Despite the analogy between the data-driven SSP test and the data-driven smooth test there is an important conceptual difference. For the latter it is the data-driven mechanism that makes the data-driven smooth test omnibus consistent, at least when the maximal order is allowed to grow with the sample size (see Section 4.3.3). The extension from the fixed SSP size SPPc test to its data-driven version, on the other hand, is not necessary to make the SSPc test omnibus consistent. It will only give the test more power for alternatives that are not anticipated before sighting the data.

In a simulation study, powers of the SSP tests, their data-driven versions, and some traditional goodness-of-fit tests have been compared. From this study it was concluded that the choice of the SSP size is very important and that under many alternatives a substantial power gain is observed when $c > 2$, making the SSP tests often more powerful than the competitor tests. The data-driven SSP tests did select the "right" SSP size very often, so that that the powers of the data-driven SSP tests were among the highest under all alternatives studied.

A test defined as an average of $X^2$ statistics over many partitions is called a SSP test by the authors, but according to the terminology used by Einmahl and McKeague (2003), the test is based on a *localised* Pearson statistic, $X^2_{c,n}(d_1, \ldots, d_{c-1})$, localised at $(d_1, \ldots, d_{c-1})$. Einmahl and McKeague (2003), and also Zhang (2002), considered tests that have the general form

$$T_n = \int_0^1 P_n(x) dw(x),$$

where $w(x)$ is some weight function, and $P_n(x)$ is the localised statistic (localised at $x$). Zhang (2002) took $P_n(x)$ to be the Cressie-Read family of divergence statistics (Cressie and Read (1984)), which includes the Pearson $X^2$ statistic as a special case. This was also proposed independently by Thas and Ottoy (2003). Zhang also considered different choices of $w(x)$. The AD and CvM statistics are special cases. Einmahl and McKeague (2003) considered the (empirical) log-likelihood ratio statistic for $P_n(x)$. Thus, the methods of Einmahl and McKeague (2003) and Zhang (2002) are also extensions of the AD and CvM tests, but they are restricted to statistics $P_n(x)$ localised at one point $x$ (partitions of size $c = 2$).

## 5.5 Some Further Bibliographic Notes

A very good overview of the history of the use of empirical process theory in the context of goodness-of-fit tests is given by del Bario et al. (2000). Durbin (1973) studied the weak convergence of the estimated empirical process in the general case of locally asymptotically linear estimators. Many years before, Kac et al. (1955) already studied the particular case of the normal distribution.

The general form (2.2) of a goodness-of-fit test statistic is discussed in more detail in Romano (1988). A good and deep introduction to empirical processes is Shorack and Wellner (1986).

In their original paper, Anderson and Darling (1952) needed other, more stringent, conditions on the weight function $w(.)$ to get the asymptotic null distribution of the AD test. The conditions that we stated here are weaker because nowadays we can rely on weak convergence results in Hilbert spaces. This theory was not yet available in the 1950s.

In the literature there is no consistency in the names used for the AD and CvM tests. Often tests of the type given in (5.4) are referred to as tests of the type of Cramér–von Mises, whereas it was only years later that Anderson and Darling (1952) proposed this more general test.

The exact null distribution of the CvM statistic has been studied by many authors. We mention only three who made important contributions: Knot (1974) and Csörgö and Faraway (1996). The latter contain many corrections to errors that were published earlier. Because it turns out to be a very hard job to get the exact distribution, many approximations have been proposed. Pearson and Stephens (1962) suggest to compute approximate percentage points by fitting a Johnson's $S_B$ curve by matching the first four exact moments. A similar solution was given by Tiku (1965) who proposed to find constants $a$, $b$, and $p$ so that the distribution of the random variable $a + bX$, where $X$ has a $\chi^2_p$ distribution, matches the first three moments of the CvM statistic.

Readers interested in more properties of the Wasserstein distance are referred to Vallender (1973) and Bickel and Freedman (1981).

## 5.6 Some Practical Guidelines for EDF Tests

- Many simulation studies have indicated that the Anderson–Darling and Cramér–von Mises tests have overall very good power for detecting many different alternatives. From a power point of view, they are preferred over many other tests.
- The EDF integral statistics have simple computational forms. In the absence of nuisance parameters the critical points of the asymptotic null distribution may be used even for very small sample sizes (say $n \geq 10$). When nuisance parameters have to be estimated, we recommend using the bootstrap. When testing for composite normality, the null distributions of the AD and CvM tests are tabulated.
- When the AD test is used for testing the simple null hypothesis of uniformity, there is a very clear link with the smooth test based on Legendre polynomials, and thus all moment interpretations carry over to the AD test. In this particular case, we suggest to complement the analysis based on AD with an investigation of the individual components.
- When the AD or CvM tests are used to test for any distribution other than the uniform, then the data first have to be transformed by the PIT. Even if the test statistic decomposes in components, they are not easily interpretable because of the transformation.
- When testing for (composite) normality, we recommend using the EQF test based on the Wasserstein distance. This test statistic has a decomposition into smooth components based on Hermite polynomials, even when the nuisance parameters are estimated. The individual components may be looked at to suggest moment differences.

- When one is mainly interested in detecting "local" deviations from the hypothesised distribution ("local" means here that the true and the hypothesised densities are particularly different in some (small) interval), then the Watson test and the SSP test are most appropriate.
- The Kolmogorov–Smirnov test should only be used when stochastic orderings are of interest. The PP plot is a good plot to help understand the conclusion of the KS test.
- The overall good power properties of the EDF and EQF integral tests, and the nice feature of the interpretability of the components of smooth tests suggest that a statistic of the form

$$\sum_{j=1}^{m} \frac{1}{j} \hat{U}_j^2 \quad \text{or} \quad \sum_{j=1}^{m} \frac{1}{j} \left( \hat{U}_j^2 - 1 \right),$$

  where $m < n$ may be large, and $\hat{U}_j$ is the smooth component statistic based on the orthonormal polynomials corresponding to the hypothesised distribution. The $p$-values should be computed by the bootstrap.

# Part II
# Two-Sample and $K$-Sample Problems

# Chapter 6
# Introduction

In this second part of the book we discuss statistical methods for the *two-sample* and the *K-sample* problems. Whereas in the one-sample problem the objective is to compare the distribution of the sample observations with a hypothesised distribution, we are now concerned with comparing the distributions of two or more populations from which we have observations at our disposal. As both classes of problems are about comparing distributions, many of the methods developed for the former can be easily adapted to the latter. We indeed show that many names of tests come back (e.g., the Kolmogorov–Smirnov and the Anderson–Darling tests). It also further implies that many of the building blocks of Chapter 2 are useful again.

This part starts with an introductory chapter, followed in Chapter 7 by some extra building blocks that were not needed in Part I. In Chapter 8 we briefly discuss some graphical tools that may be helpful in comparing distributions. Chapters 10 and 11 extend the smooth and EDF tests of Part I to tests for the two- and the $K$-sample problems. In the last chapter we discuss two final methods, and we conclude with a brief discussion.

We start in Section 6.1 with defining the problem. It becomes clear that the term "two-sample problem" has many meanings. Understanding the problem in detail will help us later on to interpret so that an informative statistical analysis can be performed. The datasets that are used to demonstrate the statistical techniques are introduced in Section 6.2. The chapter is concluded with a discussion of some important tests that are not true two-sample or $K$-sample tests, but that are closely related. Some of these test statistics reappear later as components of smooth and EDF statistics.

We continue in the line of the main objective of the book. That is, we focus on classes of tests, we introduce the reader to the basic ideas and theory, and we illustrate how the methods may be used for providing informative statistical analysis. As a consequence not all tests are described. We particularly focus on continuous distributions.

O. Thas, *Comparing Distributions,* Springer Series in Statistics,
DOI 10.1007/978-0-387-92710-7_6, 

## 6.1 The Problem Defined

### *6.1.1 The Null Hypothesis of the General Two-Sample Problem*

In defining the two-sample and the $K$-sample problems it is important to be very precise about both the null and the alternative hypothesis. We start with the two-sample problem. Suppose we have two independent samples from two populations. Let $X_{11}, \ldots X_{1n_1}$ and $X_{21}, \ldots X_{2n_2}$ denote the $n_1$ and $n_2$ sample observations with distribution functions $F_1$ and $F_2$, respectively. Without loss of generality, we consider $F_1$ and $F_2$ to have the same support, say $\mathcal{S}$. We further assume that all observations are mutually independent. The notation $X_1$ and $X_2$ is used to denote random variables with distribution function $F_1$ and $F_2$, respectively. The notation $\mu_s$ and $\sigma_s^2$ $(s = 1, 2)$ is used to denote the corresponding means and variances. We define the two-sample problem as the problem concerned with testing the null hypothesis

$$H_0 : F_1(x) = F_2(x) \text{ for all } x \in \mathcal{S}. \tag{6.1}$$

Sometimes we write $H_0 : F_1 = F_2$ for short. The most general alternative hypothesis is $H_1$ : not $H_0$. Tests that are consistent for testing $H_0$ versus $H_1$ are referred to as *omnibus consistent tests.* We refer to it as the *general two-sample problem.* Sometimes less general alternative hypotheses are considered, leading to directional tests. Just as in the one-sample problem, most smooth tests (Chapter 10) are examples of directional tests. It may be informative to give one well-known example at this point: the two-sample $t$-test may be considered as a directional two-sample test. It is used to test the null hypothesis (6.1) against the directional alternative $H_1 : \mu_1 \neq \mu_2$.

We like to stress that the null hypothesis (6.1) is very nonparametric in the sense that the distributions $F_1$ and $F_2$ are not specified. Often some assumptions on $F_1$ and $F_2$ are required for the test statistic to have a proper distribution (e.g., finite first four moments), but we try to avoid these technicalities. Although (6.1) looks very similar to the one-sample null hypothesis, its nonparametric character will make a difference in finding the null distribution of a test statistic. In the one-sample problem, the distribution of the observations is very well defined under the null hypothesis, because this is exactly what is hypothesised in $H_0$. Even with a composite null hypothesis, the distribution is specified up to a very limited number of parameters. This strong distributional restriction implied by $H_0$ makes it possible, for example, to find the exact null distribution of test statistics under a simple null hypothesis, and to use the parametric bootstrap for $p$-value calculation for composite null hypotheses. For most tests, however, the distribution theory relies on the central limit theorem or the weak convergence of empirical processes. These asymptotic theories will again play a central role in finding the asymptotic null distribution of the two-sample test statistics. A parametric bootstrap procedure will not apply anymore as (6.1) does not specify any distribution. Despite the very nonparametric nature of (6.1) we are now even often in the position to obtain the exact null distribution of a test statistic, whatever $F = F_1 = F_2$ may be and whatever the sample size. The reason is that the null hypothesis (6.1) implies an invariance of the null distribution of the test statistic under permutations of the observations over the two samples. This allows for exact $p$-value calculations, however small the sample sizes are. More details of permutation tests are given in Section 7.1.

Many of the test statistics for the two-sample problem are very closely related to those discussed in Part I. This is very easy to understand. Consider the simple null hypothesis $F(x) = G(x)$, where $F$ and $G$ represent the true and the hypothesised distributions, respectively. Whereas the latter is completely specified, the former is completely unknown, but can be estimated consistently by the EDF $\hat{F}_n$. In Section 2.1.2 we gave a very generic form of test statistics in (2.2): $T_n = c(n)d(\hat{F}_n, G)$, where $c(n)$ is a scaling factor, and $d(.,.)$ is a distance or divergence functional. If we apply the same idea here, we now replace the two unknown distribution functions $F_1$ and $F_2$ by their respective EDFs, say $\hat{F}_{1n}$ and $\hat{F}_{2n}$. As $\min(n_1, n_2) \to \infty$, both EDFs converge to the true distribution functions (see Section 2.1.1 for more details on the modes of convergence). A general form of a two-sample test statistic may then be represented by

$$T_n = c(n)d(\hat{F}_{1n}, \hat{F}_{2n}),$$

where $c(.)$ and $d(.,.)$ are as before, and thus resulting in test statistics of the same form as for the one-sample problem. Later we come back to the choice of the function $d(.,.)$, and how this relates to the specification of the alternative hypothesis.

### 6.1.2 The Null Hypothesis of the General K-Sample Problem

In the $K$-sample problem we are concerned with testing whether $K$ ($K \geq 2$) independent samples come from the same population. It is thus a generalisation of the two-sample problem to $K$ samples. Denoting the $s$th distribution function by $F_s$ ($s = 1, \ldots, K$), and assuming that all $F_s$ have the same support, we may write the general $K$-sample null hypothesis as

$$H_0 : F_1(x) = F_2(x) = \ldots = F_K(x) \text{ for all } x \in \mathcal{S}.$$

Just as with the two-sample problem, we often consider the alternative hypothesis as the negation of $H_0$. Tests that are consistent against this general alternative are omnibus tests, otherwise they are directional.

One may ask why we treat the two- and the $K$-sample problem seperately. We could just as well have introduced only the $K$-sample problem, leaving $K = 2$ as a special case. There are several reasons for doing this. First there is the history argument. Many of the tests were introduced for the two-sample problem; extensions appeared only later in the statistical literature. Second, there are some tests that are only available for the two-sample problem. Third, although many $K$-sample test statistics reduce to the two-sample statistics, they apparently have a different form. The last argument is basically a didactic argument: we believe that many methods and concepts are just easier introduced in the two-sample setting.

## 6.2 Example Datasets

### *6.2.1 Gene Expression in Colorectal Cancer Patients*

In recent years there is an increasing interest in data analysis methods for high-throughput data. A typical example of these huge datasets arises from microarrays or DNA chips. Microarray experiments are used to measure the expression levels of often more than 20,000 genes simultaneously. For each gene, they essentially measure the concentration of gene-specific mRNA, which is a transcription product of the gene that triggers the productions of a specific protein. For more details on the statistical analyses of microarray experiments, see, e.g., Speed (2003), Gentleman et al. (2005), or Allison et al. (2006). These experiments are often performed for comparison purposes. For example, gene expression levels in a control group of healthy people and a group of cancer patients are measured with the aim of finding genes that are differentially expressed in the cancer groups. These genes may play an important role in the onset or the development of the cancer. The identification of such genes may be helpful in understanding the biology of the disease, or it may be used as a biomarker in a diagnostic assay to detect the cancer in an early stage. Because microarray experiments are quite costly, they are typically performed on small groups of people. Having 20 subjects in each of the two groups is considered to be a moderately large experiment. The datasets are thus massive by the dimensionality, but not in terms of the number of independent subjects in the sample. However, here we select only a few genes for illustrating the two-sample tests, thus ignoring the problem of multiplicity of tests completely.

Most textbooks on the statistical analysis of microarray experiments advise using the traditional parametric $t$-test, or the nonparametric Wilcoxon rank sum test. Some specifically designed tests have been suggested (e.g., the SAM method of Tuscher et al. (2001)), but most of them are simple modifications of the $t$-test.

The data that we present here, was collected at the VU–University Medical Center (VUmc), Amsterdam, The Netherlands. The objective of the study was to find out which genes are involved in the progression from adenoma to carcinoma in colorectal cancer. The microarray experiment was performed on RNA isolated from 68 snap-frozen colorectal tumour samples: 37 nonprogressed adenomas and 31 carcinomas. The microarray measured expression levels of 28,830 unique genes. More details on the study and its conclusions can be found in Carvalho et al. (2008). The paper also gives details on how the expression data were preprocessed (background correction, normalisation, and summarisation). In the next paragraph we give some biological background.

Not all adenomas progress to carcinomas; this happens in only a small subset of tumours. Initiation of genomic instability is a crucial step in this progression and occurs in two ways in colorectal cancer. First DNA mismatch repair deficiency leading to microsatellite instability has been most extensively studied, but it explains only about 15% of adenoma to carcinoma progression. In the other 85% of the cases where colorectal adenomas progress to carcinomas, genomic instability occurs at the chromosomal level giving rise to aneuploidy. Although for a long time these chromosomal aberrations were regarded as random noise, secondary to cancer development, it has now been well established that these DNA copy number changes occur in specific patterns and are associated with different clinical behaviour. Nevertheless, despite extensive efforts, neither the cause of chromosomal instability in human cancer progression nor its biological consequences have been fully established.

For illustrative purposes we have selected four genes. They have sequence references NM_152299, AK021616, AK0550915, and NM_012469, but we simply refer to them as genes 1, 2, 3, and 4, respectively. Figure 6.1 shows the kernel density estimates of the expression levels.

### *6.2.2 Travel Times*

A taxi company often brings clients from the central railway station to the airport. Because many of these passangers are in a hurry to catch their planes, it is important to guarantee a short travel time. Although there is a highway connection to the airport, there are frequently traffic jams. The owner of the taxi company sets up an experiment to compare five routes from the railway station to the airport:

1. Route 1: this is the route as suggested by the GPS installed in the car.
2. Route 2: this is the preferred route by a local taxi driver who has lived for many years in the area.
3. Route 3: this route has a preference for small roads through a residential area.

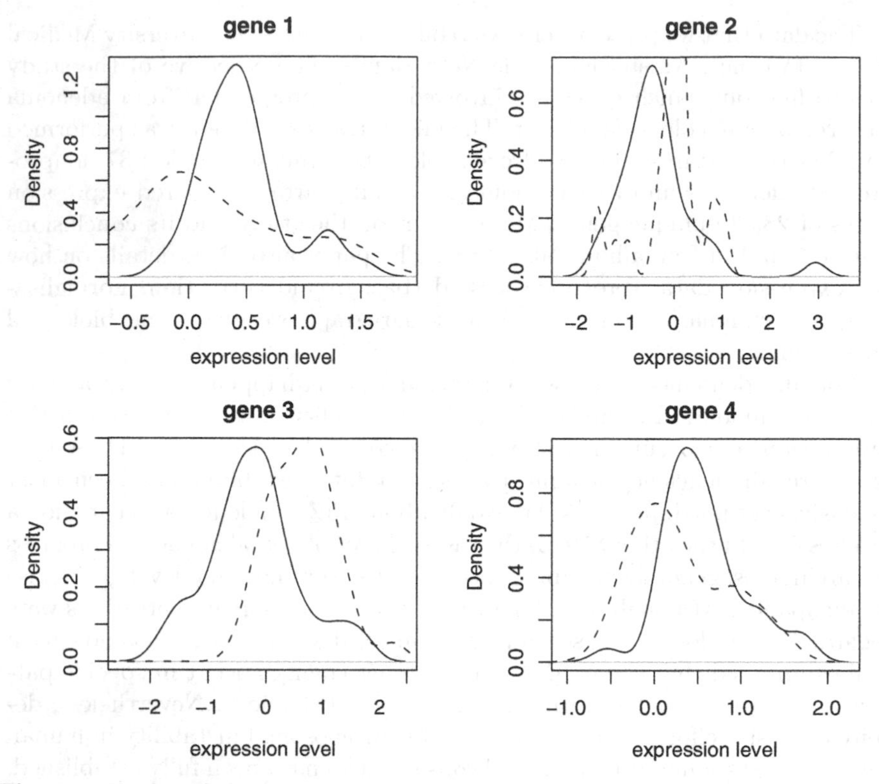


**Fig. 6.1** The kernel density estimates of the four genes. Each plot shows the density estimates of the two patient groups: the full line and the dashed line represent the non-progressed adenomas and the carcinomas, respectively

4. Route 4: this route has a preference for big roads (i.e. two lanes for each direction), but not the highway.
5. Route 5: the taxi driver first listens to the latest traffic information on the radio, and he decides to take the highway when no problems are reported; otherwise Route 1 is selected.

As the taxi drivers usually take the routes as suggested by their GPS, route 1 is considered as the reference route. In a time period of one month, 250 taxi rides from the railway station to the airport were randomly assigned to these 5 routes, resulting in a balanced design. The travel times were recorded in seconds and coverted to minutes. Boxplots of the data are shown in Figure 6.2. The dataset is referred to as the *traffic data*.

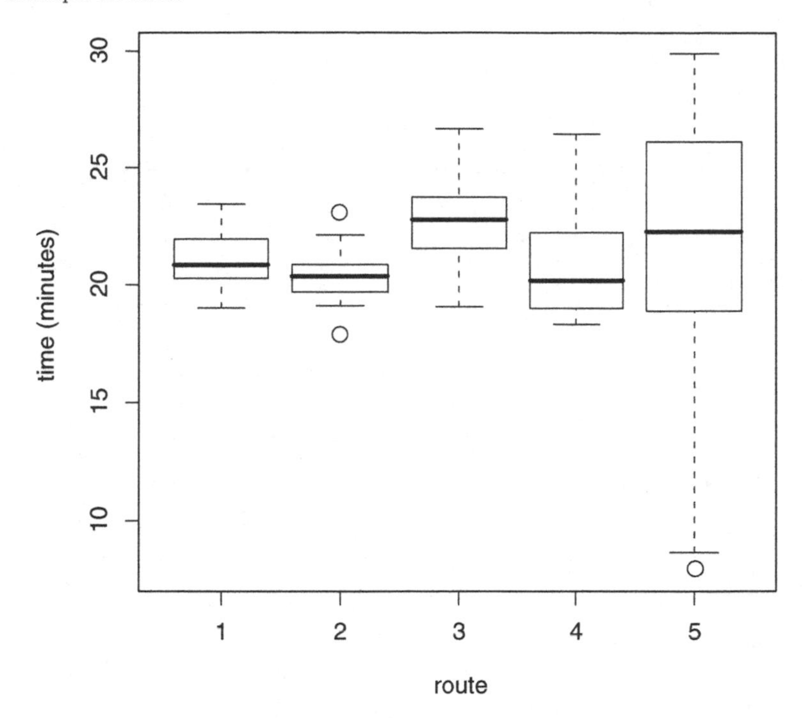


**Fig. 6.2** Boxplots of the travel times from the central railway station to the airport, with the five different routes

# Chapter 7
# Preliminaries (Building Blocks)

## 7.1 Permutation Tests

### 7.1.1 Introduction by Example

In Section 6.1.1 we argued that the $p$-values of two-sample tests cannot be based anymore on the parametric bootstrap method, because this technique presumes that the null hypothesis specifies some parameterised parametric distribution. On the other hand, most two-sample tests can be based on an asymptotic null distribution that can be derived from a central limit theorem, or from the application of the continuous mapping theorem and the weak convergence of the empirical process. Although the general two-sample null hypothesis is less parametric than the one-sample null hypothesis, we show here that this null hypothesis even allows us to obtain an *exact null distribution*. This means that the $p$-values computed from this null distribution are correct, even for very small sample sizes. Exact null distributions are often enumerated using permutations of observations. In this case we use the term *permutation null distribution* and tests based on it are referred to as *permutation tests*.

Tests based on this permutation null distribution are generally known as *permutation tests*. For a more detailed account on the principle of permutation tests we refer to the books of Good (2005) and Mielke and Berry (2001).

We first explain the concept of permutation tests in an example. Afterwards some theory is given.

*Example 7.1 (Gene expression).* Consider the gene expression data of gene 1 which was introduced in Section 6.2.1. To demonstrate the working of a permutation test we use here only the first ten observations from each of the two cancer groups. The data are presented in Table 7.1. The boxplots are shown in the left panel of Figure 7.1. At this point we do not want to go into the details on how the hypotheses and the test statistic are related.

O. Thas, *Comparing Distributions,* Springer Series in Statistics, 
DOI 10.1007/978-0-387-92710-7_7, 

**Table 7.1** The expression levels of gene 1 of the first ten patients in each disease group, and some permuted group labels. On the last line the test statistics are shown

| Patient ID | Expression level | Original group labels | 1st permuted group labels | 2nd permuted group labels | 3rd permuted group labels |
|---|---|---|---|---|---|
| 1 | 0.285 | 1 | 2 | 2 | 1 |
| 2 | 1.245 | 1 | 1 | 1 | 1 |
| 3 | 1.525 | 1 | 1 | 1 | 1 |
| 4 | 0.319 | 1 | 1 | 1 | 2 |
| 5 | -0.085 | 1 | 1 | 1 | 2 |
| 6 | 0.470 | 1 | 1 | 1 | 1 |
| 7 | 0.649 | 1 | 1 | 1 | 2 |
| 8 | 0.059 | 1 | 1 | 1 | 1 |
| 9 | 0.219 | 1 | 1 | 1 | 2 |
| 10 | 0.226 | 1 | 1 | 1 | 2 |
| 38 | 0.865 | 2 | 1 | 2 | 1 |
| 39 | 0.017 | 2 | 2 | 1 | 2 |
| 40 | -0.782 | 2 | 2 | 2 | 2 |
| 41 | 0.217 | 2 | 2 | 2 | 2 |
| 42 | -0.724 | 2 | 2 | 2 | 2 |
| 43 | 1.154 | 2 | 2 | 2 | 1 |
| 44 | 0.264 | 2 | 2 | 2 | 2 |
| 45 | 0.590 | 2 | 2 | 2 | 1 |
| 46 | 1.342 | 2 | 2 | 2 | 1 |
| 47 | -0.691 | 2 | 2 | 2 | 1 |
| $t$ | | 0.901 | 1.326 | 0.713 | 2.512 |

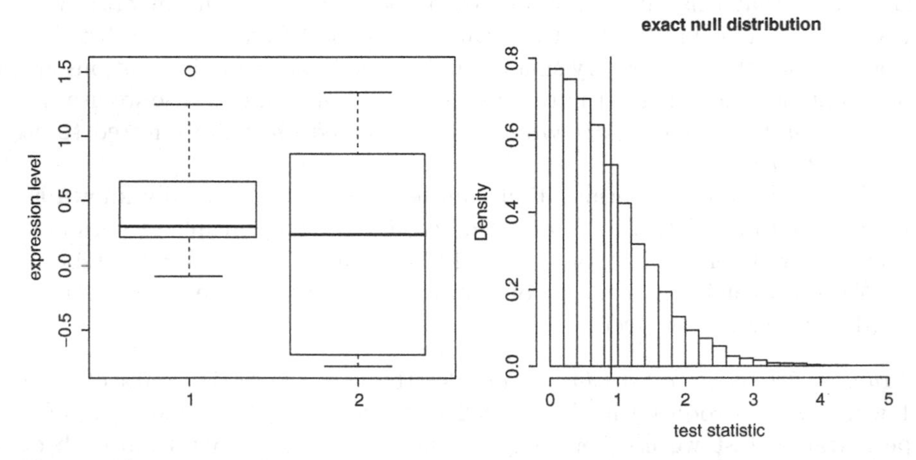


**Fig. 7.1** The boxplots (left) of the expression levels of the reduced gene 1 data, and the histogram of the permutation null distribution of the test statistic (right). The vertical reference line corresponds to the observed test statistic calculated on the original dataset ($t = 0.901$)

This is the topic of Section 9.1. Suppose the null hypothesis is the general two-sample null hypothesis,

$$H_0 : F_1(x) = F_2(x) \text{ for all } x \in \mathcal{S},$$

which has to be tested versus $H_1$ : not $H_0$, and suppose this is tested by means of the two independent samples $t$-test statistic,

$$T = \left| \frac{\bar{X}_1 - \bar{X}_2}{\sqrt{\frac{S_1^2}{n_1} + \frac{S_2^2}{n_2}}} \right|, \tag{7.1}$$

where $\bar{X}_1$ and $\bar{X}_2$ are the sample means in the two groups, and, similarly, $S_1^2$ and $S_2^2$ are the sample variances. For the reduced gene 1 data we find $t = 0.901$. The test statistic (7.1) is defined as an absolute value, because $H_1$ : not $H_0$ is suggested with both $\bar{X}_1 > \bar{X}_2$ and $\bar{X}_1 < \bar{X}_2$. The natural question in hypothesis testing is whether $t = 0.901$ is "exceptional" as compared to what is expected under the null hypothesis. In the previous sentence, the word "exceptional" only has a meaning when both the null and the alternative hypotheses are formulated. Because we expect here large values of $T$ under the alternative we should thus investigate whether $t = 0.901$ is exceptionally large. The $p$-value is used to measure this. Here

$$p = \text{Pr}_0 \left\{ T \geq t \right\}, \tag{7.2}$$

where the probability is computed under the assumption that $H_0$ holds. In a parametrical statistical setting, in which the distributional assumption of normality in the two populations is made, the null distribution of $T$ is used. In particular, this null distribution is the *sampling distribution* of the test statistic $T$ under the assumption that $H_0$ is true, and this sampling distribution gets in a frequentist statistical context an interpretation under repeated sampling from the distributions $F_1$ and $F_2$, with $F_1 = F_2$, and assuming the distributional assumption holds true. A permutation test differs from this construction in the way the null distribution is defined. In the next paragraph we illustrate the arguments that eventually result in a *permutation null distribution* which forms the basis for $p$-value calculation in permutation tests.

Suppose the null hypothesis is true; i.e., the distributions of the gene expression levels are the same for the nonprogressed adenomas and the carcinomas. If this is true, the group labels of the 20 observations in Table 7.1 are not informative, and the grouping used to compute the test statistic (7.1) is just one of the many grouping schemes that all make just as much (non)sense as the original grouping scheme; i.e., any grouping scheme would have resulted in the same responses. Consider the grouping labels in the column named "1st permuted group labels" in Table 7.1. These group labels differ from the original labels only in the first observation of group 1 and the first observation of group 2; i.e., they assign patient 1 to group 2, and patient 38 to group 1. If $H_0$ were true, the expression levels of 0.285 and 0.865 of patients 1 and 38 do not depend on their disease status, and thus it would have been just as likely to have observed these expression levels if patient 1 had a carcinoma, and patient 38 a nonprogressed adenoma. Consequently, the test statistic (7.1) calculated on the *permuted* dataset, which equals $t^\star = 1.326$, is just as likely as the test statistic calculated on the original data, $t = 0.901$, at least when $H_0$ holds. This reasoning holds for all *permutations* of the group labels over the two groups. Table 7.1 shows two more examples of permuted group labels, and the resulting values of the test statistic. There are $m = \binom{n}{n_1} = \binom{n}{n_2}$ such permutations. In the present example this gives $\binom{20}{10} = 184{,}756$ permutations. We denote the value of the test statistic computed on the $i$th permutation as $t^{(i)}$ $(i = 1, \ldots, m)$. All values of the test statistic computed for these permuted datasets are all equally likely under $H_0$. In this sense, each of the $t^{(i)}$ in the set $\{t^{(1)}, \ldots, t^{(m)}\}$ gets the same probability mass $(1/m)$ assigned, resulting in the *exact permutation null distribution* of $T$. Note that this construction is conditional on the observed expression levels. The theory presented in Section 7.1.2 shows that the permutation test will also have an unconditional interpretation. A histogram of the permutation null distribution of this example is shown in the right-hand panel of Figure 7.1. The vertical line represents the observed test statistic on the original data, i.e., using the original grouping: $t = 0.901$. The $p$-value, as defined in (7.2), can now be computed based on the exact permutation null distribution as

$$p = \text{Pr}_0\left\{T \geq t\right\} = \frac{\text{number of } t^{(i)} \geq t}{\text{total number of } t^{(i)}} = \frac{1}{m} \text{ number of } t^{(i)} \geq t.$$

In this example, $p = 0.376$. We may thus conclude that $t = 0.901$ is not sufficiently exceptional under the null hypothesis that the two distributions are the same.

Next we provide the R code for this exact test. The following R code requires the coin package, which contains many routines for exact tests. See Hothorn et al. (2006) for a good introduction to the efficient algorithmic approach taken in the coint package. The oneway_test function is the exact analogue of parametric one-way ANOVA.

```
> gene1.20<-gene1[c(1:10,38:47),]
> oneway_test(expression~group,data=gene1.20,
+ distribution="exact")

            Exact 2-Sample Permutation Test

data:  expression by group (1, 2)
Z = 0.9053, p-value = 0.3764
alternative hypothesis: true mu is not equal to 0
```

## 7.1.2 Some Permutation and Randomisation Test Theory

### 7.1.2.1 Definitions

Although we usually use the term "permutation test", it would be more correct to refer to them as "randomisation tests". The term "permutation" refers to the property that under the null hypothesis the joint distribution of the sample observations does not change when the subscripts of the observations are permuted. This property is known as the *exchangeability* of the observations. A more precise definition is given next.

**Definition 7.1 (exchangebilaty).** Let $f = f_{1 \dots n}$ denote the joint distribution of $Z_1, \dots, Z_n$, and let $\pi$ denote any permutation of the subscripts $\{1, \dots, n\}$. The random variables $Z_1, \dots, Z_n$ are said to be *exchangeable* if

$$\Pr_f \left\{ (Z_{\pi(1)}, \dots, Z_{\pi(n)}) \in B \right\} = \Pr_f \left\{ (Z_1, \dots, Z_n) \in B \right\},$$

for every Borell set $B$ of the sample space of $f$.

For the two-sample problem the observations $(Z_1, \dots, Z_n)$ that are used in the definition, are the observations of the pooled sample. In particular we adopt the convention $Z_1 = X_{11}$, $Z_2 = X_{12}$, ..., $Z_{n_1} = X_{1n_1}$, $Z_{n_1+1} = X_{21}$, ..., $Z_n = X_{2n_2}$; i.e., the first $n_1$ $Z$ observations are the observations from the first sample, and the last $n_2$ $Z$ observations are those from the second sample. As demonstrated in Example 7.1, the general two-sample null hypothesis implies the exchangeability of the pooled sample observations. More generally, for some other statistical applications the null hypothesis implies that the joint distribution of the $n$ $Z$ observations is invariant under a particular finite group of transformations of the sample space onto itself. Tests that are based on this invariance property are called "randomisation tests". Despite this minor distinction, we further use the term "permutation test".

Let $\mathcal{S}$ again denote the sample space of $Z$. Then, the sample space of the $n$ sample observations $\boldsymbol{z}^t = (z_1, \dots, z_n)$ equals $\mathcal{S}^n$. Let $T_n(\boldsymbol{Z})$ denote the test statistic. Let $G$ denote a finite group of transformations of $\mathcal{S}^n$ onto itself; i.e., for every $g \in G$ and every $\boldsymbol{z}^t = (z_1, \dots, z_n) \in \mathcal{S}^n$, $g\boldsymbol{z} = g(\boldsymbol{z}) \in \mathcal{S}^n$. With this notation we define the *randomisation hypothesis*.

**Definition 7.2 (Randomisation hypothesis).** Under the null hypothesis the distribution of $\boldsymbol{Z}$ is invariant under the transformations in $G$; i.e., for every $g \in G$, $\boldsymbol{Z}$ and $g\boldsymbol{Z}$ have the same distribution.

For the general two-sample null hypothesis, the randomisation hypothesis applies to the group of transformations that exchanges one or more of the first $n_1$ elements of $\boldsymbol{Z}$ with elements of the last $n_2$ elements of $\boldsymbol{Z}$. The null hypothesis indeed says that all $Z_i$ in the pooled sample have the same distribution, and thus the order of the elements in $\boldsymbol{Z}$ does not affect the joint distribution of the sample observations. The group $G$ of all these transformations has $m = \binom{n_1}{n} = \binom{n_2}{n}$ elements.

Before the construction of the permutation test can be described, we need a general framework in which a statistical test can be described. We start with defining a *test function* $\phi$ which maps the sample space ($\mathcal{S}^n$) of the sample onto $[0, 1]$. For a given sample, it gives the probability with which the null hypothesis should be rejected at the $\alpha$ level of significance. Most test functions work via the test statistic, which is usually real-valued. In particular, $\phi(\boldsymbol{Z})$ first maps $\mathcal{S}^n$ onto $\mathbb{R}$, and then further onto $[0, 1]$. For most samples, the test function results in 0 or 1, leaving one no doubt about rejecting or accepting the null hypothesis. However, conditional on the observed sample, if the test function returns $\gamma \in (0, 1)$, then one should reject the null hypothesis with probability $\gamma$. Fortunately, this situation does not happen often. The only reason for allowing for this is that by choosing an appropriate $\gamma$ the size of the test can be made exactly equal to the nominal size $\alpha$, which is generally not possible when using the discrete permutation null distribution. The permutation test described in the next section clarifies this concept.

#### 7.1.2.2 Construction of the Permutation Test

For a given sample, denote by

$$T_n^{(1)}(\boldsymbol{z}) \leq T_n^{(2)}(\boldsymbol{z}) \leq \cdots \leq T_n^{(m)}(\boldsymbol{z}) \tag{7.3}$$

the $m$ ordered values of $T_n(g\boldsymbol{z})$ as $g$ varies in $G$ (and $\#G = m$). For a fixed nominal level $\alpha \in (0, 1)$, define

$$k = m - \lfloor m\alpha \rfloor,$$

where $\lfloor m\alpha \rfloor$ denotes the largest integer less than or equal to $m\alpha$. Furthermore, let $m^+(\boldsymbol{z})$ and $m^0(\boldsymbol{z})$ denote the number of values $T_n^{(j)}(\boldsymbol{z})$ ($j = 1, \ldots, m$) greater than $T_n^{(k)}(\boldsymbol{z})$ and equal to $T_n^{(k)}(\boldsymbol{z})$, respectively. Let

$$a(\boldsymbol{z}) = \frac{m\alpha - m^+(\boldsymbol{z})}{m^0(\boldsymbol{z})}.$$

Define the test function $\phi$ as

$$\phi(\boldsymbol{z}) = \begin{cases} 0 & \text{if } T_n(\boldsymbol{z}) > T_n^{(k)}(\boldsymbol{z}) \\ a(\boldsymbol{z}) & \text{if } T_n(\boldsymbol{z}) = T_n^{(k)}(\boldsymbol{z}) \\ 1 & \text{if } T_n(\boldsymbol{z}) < T_n^{(k)}(\boldsymbol{z}) \end{cases}. \tag{7.4}$$

Note that this test function is indeed a formalisation of the permutation test that was introduced in a more intuitive fashion in Section 7.1.1, except that now the case $T_n(\boldsymbol{z}) = T_n^{(k)}(\boldsymbol{z})$ is explicitly included and results in a random decision of rejecting the null hypothesis with probability $a(\boldsymbol{z})$. The *permutation null distribution* enters via the ordering (7.3) of the values of the test statistic over all transformations $g \in G$. Note that this distribution is conditional on the observed sample $\boldsymbol{z}$, implying that the resulting permutation test is a *conditional test.*

With this construction of the test function, we find, for every $\boldsymbol{z} \in \mathcal{S}^n$,

$$\sum_{g \in G} \phi(g\boldsymbol{z}) = m^+(\boldsymbol{z}) + a(x)m^0(\boldsymbol{z}) = m\alpha.$$

This equality immediately gives the validity of permutation tests in the sense that they actually attain the nominal size $\alpha$. Although it may be seen easily that this property holds conditionally on the observed sample $\boldsymbol{z}$, the next theorem presents the unconditional property.

**Theorem 7.1.** *Let $\boldsymbol{Z}$ denote the pooled sample of size $n$, and let $\phi$ denote the test function (7.4) defined in terms of the test statistic $T_n(\boldsymbol{Z})$. Suppose the null hypothesis implies the randomisation hypothesis so that the distribution of $\boldsymbol{Z}$ is invariant under the finite group of transformations $G$ under this null hypothesis. Then,*

$$Pr_0\{\text{reject } H_0\} = \mathrm{E}_0\{\phi(\boldsymbol{Z})\} = \alpha.$$

*Proof.* Because $m\alpha$ is constant for fixed sample sizes $n_1$ and $n_2$, we have $m\alpha = \mathrm{E}_0\{m\alpha\}$. Hence,

$$\begin{aligned} m\alpha = \mathrm{E}_0\{m\alpha\} &= \mathrm{E}_0\left\{\sum_{g\in G}\phi(g\boldsymbol{Z})\right\} \\ &= \sum_{g\in G}\mathrm{E}_0\{\phi(g\boldsymbol{Z})\} \\ &= \sum_{g\in G}\mathrm{E}_0\{\phi(\boldsymbol{Z})\} \\ &= m\,\mathrm{E}_0\{\phi(\boldsymbol{Z})\}, \end{aligned}$$

where the last step is a consequence of the randomisation hypothesis. It follows now that $m\alpha = m\,\mathrm{E}_0\{\phi(\boldsymbol{Z})\}$, or $\alpha = \mathrm{E}_0\{\phi(\boldsymbol{Z})\}$. □

#### 7.1.2.3 Monte Carlo Approximation to the Exact Permutation Null Distribution

In the previous section we have seen that the number of permutations increases rapidly with the number of observations. To illustrate this we give a

**Table 7.2** The number of permutations ($m$) required for the exact permutation null distribution of two-sample tests with sample sizes $n_1 = n_2$

| $n_1 = n_1$ | $m = \binom{n_1}{n}$ |
|---|---|
| 5 | 252 |
| 10 | 184756 |
| 15 | 155117520 |
| 20 | 137846528820 |
| 25 | $1.264106 \times 10^{14}$ |

few examples of balanced two-sample designs in Table 7.2. The computation time thus also increases rapidly with the sample size. Fortunately the exact permutation null distribution can be very well approximated by means of Monte Carlo simulations. Instead of enumerating all $m$ permutations, a Monte Carlo approximation consists in sampling at random a large number of permutations from $G$, with each permutation having the same chance of being selected. Say we perform $B$ such permutations. This procedure is repeated $B$ times, and for each repetition the test statistic is computed. Similarly as for the construction of the exact null distribution, let $T_n^{(1)} \leq T_n^{(2)} \leq \cdots \leq T_n^{(B)}$ denote the ordered test statistics. From here on, the computations are as before, but now with $m$ replaced by $B$, which is usually much smaller than $m$.

The Monte Carlo approach is to be considered as an approximation to the exact permutation testing procedure. For $p$-value calculations based on $B$ random permutations, the asymptotic normal approximation to the binomial distribution may be applied for the calculation of standard deviations on the estimated $p$-value. This gives a standard deviation of $\sqrt{p(1-p)/B}$. As the $p$-value is unknown prior to the experiment, the size ($B$) of the Monte Carlo procedure can be designed by considering the most pessimistic situation of $p = 0.5$. For $B = 1000$, $B = 10{,}000$ and $B = 100{,}000$, this gives 0.0158, 0.005, and 0.0016, respectively. Thus with 100,000 simulation runs, the most pessimistic 95% confidence interval on the $p$-value has width $0.0031 = 0.31\%$.

*Example 7.2 (Gene expression).* In Example 7.1 we have seen the results of the exact two-sample $t$-test. Here we provide the R code of a Monte Carlo approximation. First we illustrate the concept by showing an R program that implements the algorithm explicitly.

```
> N<-10000
> t.obs<-t.test(expression~group,data=gene1.20,
+ var.equal=T)$stat
> t.star<-rep(NA,N)
> group.labels<-gene1.20$group
> permuted.data<-gene1.20
```

```
> for(i in 1:N) {
+    permuted.labels<-sample(group.labels,replace=F)
+    permuted.data[,2]<-permuted.labels
+    t.star[i]<-t.test(expression~group,data=permuted.data,
+    var.equal=T)$stat
+ }
> mean(abs(t.star)>abs(t.obs))
[1] 0.3768
```

This simple algorithm gives an approximated $p$-value of 0.3768. We now do the same analysis, but with the efficient algorithms in the coin package, which also gives a 99% confidence interval on the $p$-value.

```
> gene1.extest<-oneway_test(expression~group,data=gene1.20,
+ distribution=approximate(B=10000))
> pvalue(gene1.extest)
[1] 0.3727
99 percent confidence interval:
 0.3602585 0.3852633
```

## 7.2 Linear Rank Tests

Most of the content of this section is based on the book of Hájek et al. (1999) which gives an excellent and detailed self-contained exposition of the theory of rank statistics. Just as in the introduction to permutation tests in Section 7.1 we start defining rank tests for a single sample of i.i.d. observations, and later show how this setting includes, e.g., the two-sample situation.

In Section 7.2.1 the class of linear rank statistics is defined, and it is shown how these statistics can be constructed starting from a score generating function. This section also contains some asymptotic distribution theory, but still outside the context of hypothesis testing. Hypothesis testing and optimality properties of rank tests are discussed in Section 7.2.2.

### *7.2.1 Simple Linear Rank Statistics*

#### 7.2.1.1 Ranks and Order Statistics

Let $X_1, \ldots, X_n$ denote a sample of i.i.d. random variates with distribution function $F$. At this point we assume that the probability of having two exactly equal $X$ observations is zero, so that the ranks of the observations are uniquely defined as

$$\begin{aligned} R_i = R(X_i) &= \text{number of observations in the sample } \leq X_i \\ &= \sum_{j=1}^{n} \mathrm{I}(X_j \leq X_i) \\ &= n\hat{F}_n(X_i), \end{aligned}$$

$i = 1, \ldots, n$. Thus, the smallest observation gets rank 1 assigned, the second smallest gets rank 2, ..., and the largest observation gets rank $n$ assigned. Closely related to the ranks are the order statistics of the sample observations. These are basically the ordered sample observations, denoted as

$$X_{(1)} < X_{(2)} < \cdots < X_{(n)}.$$

Later we extend the definition so that *ties* are allowed, i.e., coinciding sample observations.

We next state some distributional properties of ranks and order statistics that may turn out useful later.

**Lemma 7.1.** *Let $X_1, \ldots, X_n$ i.i.d. $F$, and let $\boldsymbol{R}^t = (R_1, \ldots, R_n)$ and $\boldsymbol{X}^t_{(.)} = (X_{(1)}, \ldots, X_{(n)})$ denote the vector of ranks and order statistics of the sample, respectively. Then $\boldsymbol{R}$ is independent of $\boldsymbol{X}_{(.)}$, and the distribution of the rank vector is given by*

$$Pr_f\left\{\boldsymbol{R} = \boldsymbol{r}\right\} = \frac{1}{n!}$$

*for all $\boldsymbol{r}$ that consist of the integers $1, \ldots, n$ in any order. And the distribution function of $X_{(i)}$ is given by*

$$Pr_f\left\{X_{(i)} \leq x\right\} = \sum_{k=1}^{n} \binom{n}{k} (F(x))^k (1 - F(x))^{n-k}.$$

The next corollary is a consequence of the previous lemma for order statistics of a random sample of $n$ i.i.d. uniform $[0, 1]$ variates.

**Corollary 7.1.** *Let $U_{(1)} < \ldots < U_{(n)}$ denote the $n$-order statistics of a sample of $n$ i.i.d. uniform $[0, 1]$ variates. Then*

$$\mathrm{E}\left\{U_{(i)}\right\} = \frac{i}{n+1} \quad \textit{and} \quad \mathrm{Var}\left\{U_{(i)}\right\} = \frac{i(n-i+1)}{(n+1)^2(n+2)},$$

$i = 1, \ldots, n$.

The next lemma is similar to Lemma 7.1, but gives the distribution of the ranks when the $X_i$ are not identically distributed (see Theorem 3.1.2.1. in Hájek et al. (1999) or Problem 6.42 in Lehmann and Romano (2005)).

**Lemma 7.2.** *Let $X_i$ be independent with density function $f_i$ $(i = 1, \ldots, n)$, and let $\boldsymbol{R}^t = (R_1, \ldots, R_n)$ and $\boldsymbol{X}^t_{(.)} = (X_{(1)}, \ldots, X_{(n)})$ denote the vector of ranks and order statistics of the sample, respectively. For any $\boldsymbol{r}$ that consists of the integers $1, \ldots, n$ in any order, the distribution of the rank vector is then given by*

$$Pr_{f_1 \ldots f_n} \{\boldsymbol{R} = \boldsymbol{r}\} = \frac{1}{n!} \mathrm{E} \left\{ \prod_{i=1}^{n} \frac{f_i(Z_{(r_i)})}{f(Z_{(r_i)})} \right\},$$

*where $Z_{(1)} < \cdots < Z_{(n)}$ are order statistics of the random variates $Z_1, \ldots, Z_n$ that are i.i.d. with density function $f$, provided $f$ is positive whenever at least one of the $f_i$ is positive.*

The $n$-dimensional vector $\boldsymbol{R}$ has a multivariate distribution. For later purposes it is important to know its variance–covariance matrix. This is stated in the following lemma.

**Lemma 7.3.** *Consider the vector of ranks $\boldsymbol{R}$ as defined in Lemma 7.2, and assume that the ranks are computed from a sample of $n$ i.i.d. observations. Then,*

$$\mathrm{Var}\{\boldsymbol{R}\} = \frac{n+1}{12}\left(n\boldsymbol{I} - \boldsymbol{J}\boldsymbol{J}^t\right),$$

*where $\boldsymbol{I}$ is the $n \times n$ identity matrix and $\boldsymbol{J}$ is a vector with all $n$ entries equal to 1.*

We now come back to the more general case where *ties* are allowed. Ties are coinciding sample observations. As a consequence the unique ordering $X_{(1)} < \cdots < X_{(n)}$ no longer exists. First note that theoretically ties do not happen with probability one when the observations come from a continuous distribution. In practice, however, observations are only observed up to a finite precision, so that ties happen very frequently in real data situations. Although there are many ways of defining ranks in the presence of ties, we focus here on *midranks*. We adopt the formulation of Akritas and Brunner (1997).

First a slightly more general definition of the distribution function (CDF) is given. Let $F^+(x) = \Pr\{X \leq x\}$ denote the traditional right continuous CDF, and $F^-(x) = \Pr\{X < x\}$ the left continuous version. The CDF is then defined as

$$F(x) = \frac{1}{2}F^+(x) + \frac{1}{2}F^-(x),$$

which coincides with the conventional definition when $X = x$ happens with probability zero. In a similar fashion the empirical versions of $F^+$ and $F^-$ may be constructed. In particular,

$$\hat{F}_n^+(x) = \frac{1}{n}\sum_{i=1}^{n} \mathrm{I}\,(X_i \leq x)$$

$$\hat{F}_n^-(x) = \frac{1}{n}\sum_{i=1}^{n} \mathrm{I}\,(X_i < x)\,.$$

Finally,

$$\hat{F}_n^{\pm}(x) = \frac{1}{2}\hat{F}^+(x) + \frac{1}{2}\hat{F}^-(x).$$

It may also be written as

$$\begin{aligned} n\hat{F}_n^{\pm}(x) &= \text{number of observations in sample } < x \\ &+ \frac{1}{2}\text{number of observations in sample } = x. \end{aligned}$$

Whereas in the no-ties case, ranks were defined as $R(X_i) = n\hat{F}_n(X_i)$, ranks in the presence of ties are given by $R(X_i) = \frac{1}{2} + n\hat{F}_n^{\pm}(X_i)$ $(i = 1, \ldots, n)$. The two definitions coincide when there are no ties. The construction of the ranks as presented here, result in *midranks.*

The linear rank statistics (see next section) are defined in terms of ranks, but when in the case of ties the ranks may often be simply replaced by the midranks.

#### 7.2.1.2 Simple Linear Rank Statistics

Before defining rank statistics, we need to introduce *scores* and a set of real valued *regression constants.* We start with the latter. Let $c_1, \ldots, c_n$ denote these *regression constants*, and let $\bar{c} = (1/n)\sum_{i=1}^{n} c_i$. We consider then the sets

$$C_n = \left\{ \{c_1, \ldots, c_n\} : c_i \in \mathbb{R}, i = 1, \ldots, n; \sum_{i=1}^{n}(c_i - \bar{c})^2 > 0 \right\},$$

and

$$C_\infty = \left\{ \{c_1, \ldots\} : c_i \in \mathbb{R}, i = 1, \ldots;\ \lim_{n\to\infty} \frac{\sum_{i=1}^{n}(c_i - \bar{c})^2}{\max_{1\leq i\leq n}(c_i - \bar{c})^2} = +\infty \right\}.$$

The restriction of the $c_i$ implied in $C_\infty$ is that the sum $\sum_{i=1}^{n}(c_i - \bar{c})^2$ may not be dominated by one single regression constant. Although it is sufficient to have considered $C_n$ for studying rank statistics based on finite sample sizes, throughout we require $C_\infty$, because we also study the asymptotic distributional properties of the rank statistics.

Scores are defined as real-valued functions that work on $i = 1, \ldots, n$. We write $a_n(i)$, $i = 1, \ldots, n$, for a collection of scores. When studying asymptotic properties, we further impose the existence of a nonconstant square integrable real-valued function $\varphi$, defined over $[0, 1]$. Let $\Psi$ denote the set of such functions. This $\varphi$ function and the scores $a_n$ must satisfy

$$\lim_{n\to\infty} \int_0^1 \left(a_n(1 + \lfloor un \rfloor) - \varphi(u)\right)^2 du = 0, \tag{7.5}$$

where $\lfloor un \rfloor$ denotes the integer part (floor) of $un$. In Section 7.2.1.3 we give more details on how scores $a_n(i)$ may be generally constructed so that (7.5) is satisfied.

**Definition 7.3 (simple linear rank statistic).** A *simple linear rank statistic* is defined as

$$T_n = \sum_{i=1}^{n} c_i a_n(R_i), \tag{7.6}$$

in which $\{c_i\} \in C_\infty$ is a set of regression constants, and the scores $a_n(i)$ satisfy (7.5).

A simple linear rank statistic thus only depends on the data through their ranks, a function $a_n$ that determines the scores through the ranks, and a set of regression constants that are independent of the observations. The mean and the variance of $T_n$ may be easily expressed in terms of the scores and the regression constants. We assume that all $n$ observations are i.i.d., and on applying Lemma 7.1 find

$$\mu_n = \mathrm{E}\{T_n\} = \bar{a} \sum_{i=1}^{n} c_i, \tag{7.7}$$

where $\bar{a} = (1/n) \sum_{i=1}^{n} a_n(i)$ is the average score. Also

$$\sigma_n^2 = \mathrm{Var}\{T_n\} = s_a^2 \sum_{i=1}^{n} (c_i - \bar{c})^2, \tag{7.8}$$

where $s_a^2 = \frac{1}{n-1} \sum_{i=1}^{n} (a_n(i) - \bar{a})^2$. When not all observations arise from the same distribution, the expressions of $\mu_n$ and $\sigma_n^2$ may be found using Lemma 7.2.

The next theorem gives the asymptotic distribution of $T_n$.

**Theorem 7.2.** *Let $X_1, \ldots, X_n$ i.i.d. $F$, and let $T_n$ denote a simple linear rank statistic as defined in Definition 7.3. Then, as $n \to \infty$,*

$$\frac{T_n - \mu_n}{\sigma_n} \xrightarrow{d} N(0, 1).$$

In the previous theorem the finite mean and variance of $T_n$ were used. It is often more convenient to use the limiting mean and variance, and to reformulate the theorem in terms of the linear rank statistic defined as an average rather than a sum. We write $S_n = T_n/n$. We further define

$$\mu_c = \lim_{n\to\infty} \frac{1}{n}\sum_{i=1}^{n} c_i \text{ and } \sigma_c^2 = \lim_{n\to\infty} \frac{1}{n}\sum_{i=1}^{n}(c_i - \mu_c)^2,$$

and

$$\mu_\varphi = \int_0^1 \varphi(u)du \text{ and } \sigma_\varphi^2 = \int_0^1 (\varphi(u) - \mu_\varphi)^2 du.$$

With this notation Theorem 7.2 may be restated.

**Theorem 7.3.** *Let $X_1, \ldots, X_n$ i.i.d. $F$, and let $S_n = T_n/n$ with $T_n$ a simple linear rank statistic as defined in Definition 7.3. Then, as $n \to \infty$,*

$$\sqrt{n}\frac{S_n - \mu_S}{\sigma_S} \xrightarrow{d} N(0,1),$$

*where $\mu_S = \mu_c\mu_\varphi$ and $\sigma_S^2 = \sigma_c^2\sigma_\varphi^2$.*

The proofs of these theorems may be found in Chapter 6 of Hájek et al. (1999).

#### 7.2.1.3 Score Generating Functions

At first sight it may be hard to find a $\varphi$ and an $a_n$ function that satisfy condition (7.5), but usually we construct the scores $a_n(i)$ starting from some $\varphi \in \Psi$. We give two particular constructions. The scores $(i = 1, \ldots, n)$

$$a_n(i) = \varphi\left(\frac{i}{n+1}\right) \text{ and } a_n(i) = n\int_{(i-1)/n}^{i/n} \varphi(u)du \tag{7.9}$$

both satisfy (7.5). The importance of the $\varphi$ function becomes clear in Section 7.2.2 where we show that for particular testing problems the function $\varphi$ may be chosen so as to give the resulting rank test an optimality property. The scores are there naturally defined in terms of expectations. For example,

$$a_n(i) = \mathrm{E}_f\left\{\varphi(F(X_{(i)}))\right\}, \tag{7.10}$$

where $F$ is the CDF of the $X_i$, and $X_{(i)}$ is the $i$th order statistic. It is easy to demonstrate that the scores in (7.9) may be considered as approximations of (7.10). To see this for the first score in (7.9), write (7.10) as $a_n(i) = \mathrm{E}\left\{\varphi(U_{(i)})\right\}$, where $U_{(i)} = F(X_{(i)})$, or, equivalently, $U_{(i)}$ is the $i$th order statistic of a sample of $n$ i.i.d. uniform random variates. Then, if $\varphi$ is a sufficiently smooth function, we have approximately $a_n(i) \approx \varphi(\mathrm{E}\left\{U_{(i)}\right\})$.

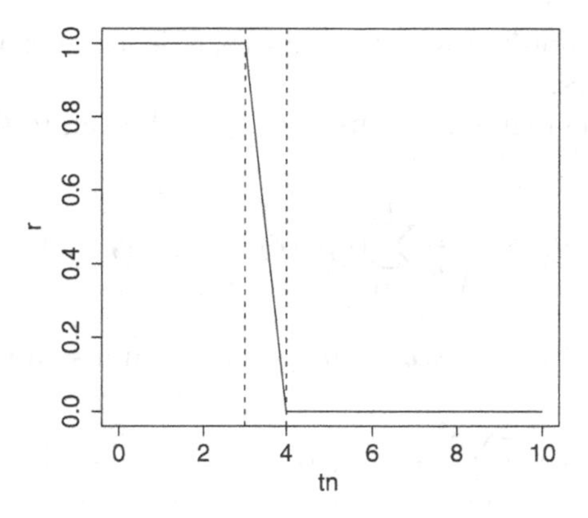


**Fig. 7.2** An example of a rank generating function, with $n = 10$, and $i = 4$

Finally, because $\mathrm{E}\left\{U_{(i)}\right\} = i/(n+1)$ (see Corollary 7.1), we get that $\mathrm{E}_f\left\{\varphi(F(X_{(i)}))\right\}$ equals approximately $\varphi\left(i/(n+1)\right)$.

There are many other similar ways of defining the scores $a_n(i)$ as an expectation of order statistics, but we do not need them any further in this book.

The scores may also be constructed starting from the *rank generating function* which is given by

$$r_n(i,t) = \begin{cases} 0 & \text{if } i \leq tn \\ i - tn & \text{if } tn \leq i \leq tn+1 \\ 1 & \text{if } tn+1 \leq i \ . \end{cases} \tag{7.11}$$

Figure 7.2 shows an example of a rank generating function. It also illustrates that this function can be seen as a continuous version of the indicator function $\mathrm{I}\left(tn < i\right)$. Scores may be now defined as

$$a_n(i) = \int_0^1 r_n(i,t)d\varphi(t)$$

or

$$a_n(i) = -\int_{(i-1)/n}^{i/n} \varphi(t)dr_n(i,t) = n\int_{(i-1)/n}^{i/n} \varphi(t)dt.$$

Note that the last expression is exactly the second approximate score of (7.9).

#### 7.2.1.4 The Rank Score Process

In this subsection we show that the rank generating function (7.11) is the core of a stochastic process, and that the linear rank statistics can be expressed in terms of this process. This provides us with another technique for finding the asymptotic distributions of rank statistics, but also of other two- and

$K$-sample goodness-of-fit rank statistics that do not belong to the class of linear rank statistics.

Suppose that the regression constants $c_1, \ldots$ belong to $C_\infty$, and define

$$d_{ni} = \left( \frac{1}{n} \sum_{j=1}^{n} (c_j - \bar{c})^2 \right)^{-1/2} (c_i - \bar{c}).$$

We now introduce the *rank score process* based on a sample $X_1, \ldots, X_n$,

$$\mathbb{S}_n(t) = \sqrt{n} \sum_{i=1}^{n} d_{ni} \left( r_n(R_i, t) - (1-t) \right), \quad t \in [0,1].$$

The centering term $(1-t)$ comes from (large $n$)

$$\begin{aligned} \mathrm{E}\{r_n(R_i,t)\} &= 0 \times \Pr\{R_i \leq tn\} \\ &\quad + \mathrm{E}\{R_i - tn | tn \leq R_i \leq tn+1\} \Pr\{tn \leq R_i \leq tn+1\} \\ &\quad + 1 \times \Pr\{tn+1 \leq R_i\} \\ &\approx \mathrm{E}\{R_i - tn | tn \leq R_i \leq tn+1\} \times 0 + (1-t) \\ &\approx 1 - t. \end{aligned} \tag{7.12}$$

The weak convergence of $\mathbb{S}_n$ is established in the next theorem.

**Theorem 7.4.** *Let $X_1, \ldots, X_n$ be i.i.d. $F$, and assume the regression constants $c_1, \ldots$ belong to $C_\infty$; then, as $n \to \infty$,*

$$\mathbb{S}_n \xrightarrow{w} \mathbb{B},$$

*where $\mathbb{B}$ is the Brownian bridge.*

Theorem 7.4, in cooperation with the continuous mapping theorem (Theorem 2.1), may be very helpful in finding the asymptotic distributions of test statistics that may be expressed as functionals of the rank score process $\mathbb{S}_n$. We illustrate this principle here on linear rank statistics. Write

$$\begin{aligned} \int_0^1 \mathbb{S}_n(t) d\varphi(t) &= \sum_{i=1}^{n} d_{ni} \int_0^1 \left( r_n(R_i,t) - (1-t) \right) d\varphi(t) \\ &= \sum_{i=1}^{n} d_{ni} \left( a_n(R_i) - \int_0^1 (1-t) d\varphi(t) \right). \end{aligned}$$

Using (7.12), we further see that

$$\begin{aligned} \int_0^1 (1-t) d\varphi(t) &= \int_0^1 \mathrm{E}\{r_n(R_i,t)\} d\varphi(t) = \mathrm{E}\left\{ \int_0^1 r_n(R_i,t) d\varphi(t) \right\} \\ &= \mathrm{E}\{a_n(R_i)\}. \end{aligned}$$

We find thus that $\int_0^1 \mathbb{S}_n(t)d\varphi(t)$ coincides with a centered simple linear rank statistic. Using the continuous mapping theorem and Theorem 2.3, we find again that the limit distribution of the linear rank statistic is a zero mean normal distribution. The variance of this normal distribution may be found using (2.10), and equals $\sigma_\varphi^2$, which agrees with Theorem 7.3, except that the factor $\sigma_c^2$ does not appear here, because the $d_{ni}$ have $\sigma_c$ in the denominator.

## 7.2.2 Locally Most Powerful Linear Rank Tests

### 7.2.2.1 Locally Most Powerful Linear Rank Tests for General Alternatives

We demonstrate the concept of a *locally most powerful linear rank test* (LMPRT) here on a general one-parameter alternative to the general $K$-sample null hypothesis: $H_0 : f_1 = \cdots = f_K$, here expressed in terms of the $K$ density functions. Let $f$ denote the common density under $H_0$. Suppose that the true densities $f_s$ $(s = 1, \ldots, K)$ can be indexed by $K$ parameters, say $\theta_s$, so that $f(.;\theta_1) = \cdots = f(.;\theta_K) = f(.)$ if and only if all $\theta_s = 0$.

More generally, as an alternative we now consider for observation $x_{si}$ $(s = 1, \ldots, K; i = 1, \ldots, n_s)$ the density function

$$f(x_i; \Delta c_{si}) \tag{7.13}$$

for some $0 < \Delta < \varepsilon$, $\varepsilon > 0$, and constants $c_{si}$. By ordering the constants we $c_1 = c_{11}$, $c_2 = c_{12}, \ldots, c_{n_1} = c_{1n_1}$, $c_{n_1+1} = c_{21}, \ldots, c_n = c_{Kn_K}$. To reduce the notational burden, we use the notations $c_i$ and $c_{si}$ interchangeably. For similar reasons, the notation $f_i$ with $i = 1, \ldots, n$ is used to denote the density function of observation $i = 1, \ldots, n$ in the pooled sample. The $\theta$ parameters are thus replaced by one parameter $\Delta$ and a set of constants $c_1, \ldots, c_n$, reducing to $\theta_s = \Delta c_{si}$ $(s = 1, \ldots, K; i = 1, \ldots, n)$ for the one-parameter alternative to the general $K$-sample null hypothesis. This setting includes the location shift model, among others. Let $\mathcal{F}_{1,\Delta}$ denote this class of densities, and write $f(.;0)$ for the common density $f$ under the null hypothesis.

A level-$\alpha$ locally most powerful linear rank test is a linear rank test that is *uniformly most powerful* (UMP) for this testing problem for some $\varepsilon > 0$ (see Section 2.9.2.2). The next theorem gives the regression constants and the scores for this LMPRT. Because of the importance of this result, we also give a taste of the proof.

**Theorem 7.5.** *Let $X_{si}$ $(i = 1, \ldots, n_s; s = 1, \ldots, K)$ denote a sample of $n = \sum_{s=1}^K n_s$ independent observations. Let $Z_{(1)} < \ldots < Z_{(n)}$ denote the order statistics of the pooled sample observations, and let $R_{si}$ denote the rank of observation $X_{si}$ in the pooled sample. The LMPRT for testing the general $K$-sample null hypothesis, $H_0 : f_1 = \cdots = f_K$, against the general one-parameter alternative $\mathcal{F}_{1\Delta}$ is then specified as follows. Let $f(.;\theta_i) \in \mathcal{F}_{1,\Delta}$, with $\theta_i = \Delta c_i$,* *denote the one-parameter alternative density function of observation* $i$, *and write* $f(.) = f(.;0)$ *for the common density function of* $X_{si}$ *under the null hypothesis, which is supposed to be completely specified. Write* $\boldsymbol{X}$ *for the* $n$-*vector of sample observations and write the linear rank test statistic (7.6) as*

$$T_n = T_n(\boldsymbol{X}) = \sum_{s=1}^{K}\sum_{i=1}^{n_s} c_{si}a_n(R_{si}), \tag{7.14}$$

*where the scores are defined as*

$$a_n(i) = \mathrm{E}_f\left\{\frac{\frac{\partial}{\partial\theta}f(Z_{(i)};\theta)\big|_{\theta=0}}{f(Z_{(i)})}\right\}.$$

*The test function of the level* $\alpha$ *LMPRT is given by*

$$\phi(\boldsymbol{X}) = \begin{cases} 0 & \text{if } T_n(\boldsymbol{X}) < c_\alpha \\ \gamma & \text{if } T_n(\boldsymbol{X}) = c_\alpha \\ 1 & \text{if } T_n(\boldsymbol{X}) > c_\alpha \end{cases},$$

*where* $\gamma$ *and* $c_\alpha$ *are chosen so that the test has size* $\alpha$.

*Proof.* We only give a sketch of the proof here, stressing the relation with the fundamental Neyman–Pearson lemma (Lemma 2.1).

For a given $\Delta > 0$, the alternative represented by $\mathcal{F}_{1,\Delta}$ is basically a simple alternative. Because $f$ is assumed known, the null hypothesis is also simple. In this setting, we may apply the Neyman–Pearson lemma which says that the test statistic should be the likelihood ratio statistic, or, equivalently, the log-likelihood ratio statistic. Because the test function of a rank test is in terms of the ranks of the observations, the joint densities used in the test function of the Neyman–Pearson should be replaced by the joint probabilities of the ranks. In particular, the joint densities under the null and the alternative hypothesis now become $\mathrm{Pr}_0\{\boldsymbol{R} = \boldsymbol{r}\}$ and $\mathrm{Pr}_\Delta\{\boldsymbol{R} = \boldsymbol{r}\}$, respectively. See, for example, Theorem 3.2.3.1 in Hájek et al. (1999) for a formal reformulation of the Neyman–Pearson lemma for rank tests. The log-likelihood ratio in the rank test function thus has general form

$$LLR(\Delta) = \log \mathrm{Pr}_\Delta\{\boldsymbol{R}=\boldsymbol{r}\} - \log \mathrm{Pr}_0\{\boldsymbol{R} = \boldsymbol{r}\} = \log \mathrm{Pr}_\Delta\{\boldsymbol{R} = \boldsymbol{r}\} + \sum_{i=1}^{n}\log(i),$$

upon using $\mathrm{Pr}_0\{\boldsymbol{R} = \boldsymbol{r}\} = 1/(n!)$ (Lemma 7.1). The last term only depends on the sample size, therefore we may further ignore it.

For $\varepsilon > 0$ small we now approximate $LLR(\Delta)$ using a Taylor series expansion about $\Delta = 0$,

$$LLR(\Delta) = LLR(0) + \Delta\frac{\partial}{\partial\Delta}LLR(\Delta)\bigg|_{\Delta=0} + o(\Delta). \tag{7.15}$$

The partial derivative becomes

$$\left.\frac{\partial}{\partial \Delta} LLR(\Delta)\right|_{\Delta=0} = \left.\frac{\frac{\partial}{\partial \Delta}\mathrm{Pr}_{\Delta}\{\boldsymbol{R}=\boldsymbol{r}\}}{\mathrm{Pr}_{\Delta}\{\boldsymbol{R}=\boldsymbol{r}\}}\right|_{\Delta=0}$$

$$= n! \left.\frac{\partial}{\partial \Delta}\mathrm{Pr}_{\Delta}\{\boldsymbol{R}=\boldsymbol{r}\}\right|_{\Delta=0},$$

in which, using Lemma 7.2,

$$\frac{\partial}{\partial \Delta}\mathrm{Pr}_{\Delta}\{\boldsymbol{R}=\boldsymbol{r}\} = \frac{1}{n!}\frac{\partial}{\partial \Delta}\mathrm{E}\left\{\prod_{i=1}^{n}\frac{f(Z_{(r_i)};\theta_i)}{f(Z_{(r_i)})}\right\}$$

$$= \frac{1}{n!}\mathrm{E}\left\{\sum_{j=1}^{n}\frac{\partial}{\partial \theta_j}\frac{\partial \theta_j}{\partial \Delta}\prod_{i=1}^{n}\frac{f(Z_{(r_i)};\theta_i)}{f(Z_{(r_i)})}\right\}$$

$$= \frac{1}{n!}\mathrm{E}\left\{\sum_{i=1}^{n}c_i\frac{\frac{\partial}{\partial \theta_i}f(Z_{(r_i)};\theta_i)}{f(Z_{(r_i)})}\right\}.$$

Hence,

$$\left.\frac{\partial}{\partial \Delta} LLR(\Delta)\right|_{\Delta=0} = \sum_{i=1}^{n}c_i\,\mathrm{E}\left\{\frac{\left.\frac{\partial}{\partial \theta_i}f(Z_{(r_i)};\theta_i)\right|_{\theta_i=0}}{f(Z_{(r_i)})}\right\}$$

$$= \sum_{i=1}^{n}c_i a_n(R_i).$$

Equation (7.15) now becomes

$$LLR(\Delta) = \Delta\sum_{i=1}^{n}c_i a_n(R_i) + o(\Delta),$$

and thus, for every $\Delta < \varepsilon$ the MPRT according to Neyman–Pearson is the one proposed in the theorem statement. □

It is important to realise here that it has been assumed in Theorem 7.5 that the common density $f(.) = f(.;0)$ is known. If this is not the case, the scores cannot be calculated. The theory presented here is thus a parametric theory for rank tests; i.e., optimality can only be guaranteed for an a priori known family of alternatives. Later in Chapter 9 we show how some well-known rank tests are LMPRT under certain parametric assumptions.

### *7.2.3 Adaptive Linear Rank Tests*

In the previous sections we have discussed the construction of rank tests that possess an optimality criterion. In particular, the LMPRT was defined. Although rank tests are often considered to be *nonparametric tests*, the construction of the LMPRT demonstrated that this type of test is actually constructed for a very parametric testing setting. Theorem 7.5 shows how the optimal scores can be obtained, but it requires knowledge of the density function $f$ of the observations, and the null hypothesis must be formulated in terms of parameters indexing these density functions. An *adaptive linear rank test* is a linear rank test based on scores that are chosen by a data-based selection rule, so that the resulting scores are in some sense optimal for the testing problem. An important consequence of choosing the scores in a data-driven manner is that it affects the null distribution of the test statistic. Later, in Chapters 9 and 10 some more details are given.

## 7.3 The Pooled Empirical Distribution Function

Under the $K$-sample null hypothesis all $K$ distribution functions coincide and can be represented by one CDF, say $H(x)$. This *pooled CDF* is defined as

$$H(x) = \sum_{s=1}^{K} \frac{n_s}{n} F_s(x). \tag{7.16}$$

Note that under the $K$-sample null hypothesis, the CDF $H(x)$ is equal to all $K$ individual CDFs. More generally, when the fractions $n_s/n$ are fixed by design, or if they are replaced by the probabilities that an observation is sampled from population $s$, $H(x)$ is simply the marginal distribution of $X$.

A natural estimator of $H(x)$ arises from substituting all $F_s$ by their EDFs; i.e.,

$$\hat{H}_n(x) = \sum_{s=1}^{K} \frac{n_s}{n} \hat{F}_{sn_s}(x), \tag{7.17}$$

and thus $\hat{H}_n$ can be seen as a weighted average of the $K$ EDFs. By substituting each $\hat{F}_{sn_s}(x)$ by $(1/n_s)\sum_{i=1}^{n_s} \mathrm{I}(X_{si} \leq x)$, we get

$$\hat{H}_n(x) = \frac{1}{n} \sum_{s=1}^{K} \sum_{i=1}^{n_s} \mathrm{I}(X_{si} \leq x).$$

Using the notation of Section 7.1.2 of the pooled sample observations $Z_i$, the last expression immediately reduces to

$$\hat{H}_n(x) = \frac{1}{n}\sum_{i=1}^{n} \mathrm{I}(Z_i \leq x). \tag{7.18}$$

Thus $\hat{H}_n(x)$ is simply the EDF using the $n$ pooled sample observations. This indeed makes sense under the $K$-sample null hypothesis, which says that all $K$ distributions or populations are equal, and thus no distinction should be made between them, nor between the observations from the $K$ samples.

Finally we rewrite (7.18) in terms of ranks. In particular, when $R_i$ denotes the rank of $Z_i$ in the pooled sample, then $\hat{H}_n$ evaluated in some $Z_i$ becomes

$$\hat{H}_n(Z_i) = \frac{R_i}{n} \text{ or } \hat{H}_n(Z_i) = \frac{R_i - 0.5}{n}, \tag{7.19}$$

where a continuity correction is applied in the latter.

## 7.4 The Comparison Distribution

Just as for the one-sample problem we can also define here a comparison distribution. Recall that in Section 2.4 the comparison distribution function was defined as the CDF of the random variable $U = G(X)$ with $X$ having CDF $F$. The random variable $U$ is said to possess information about the relative rank of $X \sim F$ compared to $X_0 \sim G$. When $F = G$, $U$ has a uniform distribution. The interpretation is best seen from the density function of $U = F(X)$, given by $r(u) = f(G^{-1}(u))/g(G^{-1}(u))$; an extensive discussion was presented in Section 3.3. In the one-sample problem only two distributions are involved, therefore it is obvious that the hypothesised $G$ plays the role of a reference distribution. In the $K$-sample problem, however, many more distributions are involved so that it is not always obvious which distribution has to serve as a reference distribution.

When $k = 2$, then again only two distributions are involved, and $U = F_1(X)$ with $X \sim F_2$, or $U = F_2(X)$ with $X \sim F_1$ can be chosen. In the more general setting of $k \geq 2$, many more such constructions may be chosen, each resulting in a different interpretation, and none containing all information against $H_0$. Instead of choosing one of the $K$ CDFs as a reference distribution, it is often preferred to use the pooled CDF (7.16) so that all $K$ comparison distributions are computed relative to this common reference distribution. Moreover, under the null hypothesis the common CDF coicides with all individual $F_s$. In particular, let $U_s = H(X_s)$ with $X_s \sim F_s$. The CDF and df of $U_s$ are then given by ($i = s, \ldots, K$)

$$R_s(u) = F_s(H^{-1}(u)) \text{ and } r_s(u) = \frac{f_s(H^{-1}(u))}{h(H^{-1}(u))}. \tag{7.20}$$

In contrast to the one-sample problem, where only the distribution $F$ of the observations was unknown, now none of the potential reference distributions is known. Thus the reference CDF will also have to be replaced by an estimator. Later, in Section 7.5.2 we give more details on the distributional properties of $\hat{R}_{nij} = \hat{F}_{in_i}(\hat{F}_{jn_j}^{-1}(u))$ and $\hat{R}_{sn} = \hat{F}_{sn_s}(\hat{H}_n^{-1}(u))$, and in Section 8.2 we illustrate how the comparison densities $r_s(u)$ can be estimated directly and used as a graphical diagnostic tool for comparing the $K$ distributions.

## 7.5 The Quantile Process

In Part I of the book we argued that all the information in the $n$ sample observations against the (simple) null hypothesis $H_0 : F = G$ is contained in the empirical process $\mathbb{B}_n(x) = \sqrt{n}(\hat{F}_n(x) - G(x))$. The core of the argument is that $\hat{F}_n$ is a very good estimator of the true CDF $F$, in the sense of the Glivenko–Cantelli theorem (Section 2.1.1). In the $K$-sample case we have $K$ independent samples, and within each sample the estimator $\hat{F}_{sn_s}$ is such a consistent estimator of $F_s$ $(s = 1, \ldots, K)$. Under the $K$-sample null hypothesis, all $K$ EDFs are estimators of the common CDF $H = F_1 = \cdots = F_K$. In the simplest case of $K = 2$ the previous reasoning brings us to considering a process that is proportional to

$$\hat{F}_{1n_1} - \hat{F}_{2n_2}, \tag{7.21}$$

but when $K > 2$, there are many more informative processes that can be built. In Section 7.5.1 we give a brief discussion on how informative processes can be built; we call them *contrast processes.* Finally, in Section 7.5.2 we construct slightly more complicated processes, which form the basis of many goodness-of-fit tests.

### 7.5.1 Contrast Processes

When $K = 2$ we argued that all the information against the null hypothesis is contained in (7.21), which we call a *contrast process* as it contrasts the two sample distributions. To study its asymptotic behavior it is more convenient to consider a properly scaled process. In particular, we define the contrast processes as $(i \neq j = 1, \ldots, K)$

$$\mathbb{C}_{nij}(x) = \sqrt{\frac{n_i n_j}{n}} \left( \hat{F}_{in_i}(x) - \hat{F}_{jn_j}(x) \right). \tag{7.22}$$

For arbitrary $K$, many systems of contrast processes may be constructed. For example, suppose $j$ is fixed by the user, say $j = 1$ (e.g., referring to a control group), then the contrast empirical processes $\{\mathbb{C}_{ni1}, i = 2, \ldots, K\}$ possess information on the difference between $F_i$ and $F_1$, where the latter now serves as a reference distribution. Another set of contrast empirical processes can be constructed as $\{\mathbb{C}_{nii+1}, i = 1, \ldots, K-1\}$, and many other constructions are possible. These contrasts can best be compared with the contrasts used in ANOVA in terms of means. When $K > 2$, it may also be interesting to compare each group with the "overall mean", or the pooled CDF $H$, i.e., a process based on $\hat{F}_{ni}(x) - \hat{H}_n(x)$.

Although we later work more often with other, though related processes, it is still educative to look for their limiting behaviour under the null hypothesis. Consider

$$\begin{aligned}
&\sqrt{\frac{n_1 n_2}{n}} \left( \hat{F}_{1n_1}(x) - \hat{F}_{2n_2}(x) \right) \\
&= \sqrt{\frac{n_1 n_2}{n}} \left\{ \hat{F}_{1n_1}(x) - H(x) - \left( \hat{F}_{2n_2}(x) - H(x) \right) \right\} \qquad (7.23) \\
&= \sqrt{\frac{n_2}{n}} \mathbb{B}_{1n_1}(x) - \sqrt{\frac{n_1}{n}} \mathbb{B}_{2n_2}(x),
\end{aligned}$$

where $\mathbb{B}_{1n_1}$ and $\mathbb{B}_{2n_2}$ are the empirical processes of samples 1 and 2, respectively. Under the null hypothesis these processes converge weakly to two independent $H$-Brownian bridges, say $\mathbb{B}_1$ and $\mathbb{B}_2$ (see Section 2.2.2 for details on the weak convergence). These convergences hold for $\min(n_1, n_2) \to \infty$. We further assume that $\lim_{n\to\infty} n_1/n = \lambda_1$ exists, and that $0 < \lambda_1 < 1$. Hence, under $H_0$,

$$\sqrt{\frac{n_1 n_2}{n}} \left( \hat{F}_{1n_1}(x) - \hat{F}_{2n_2}(x) \right) \xrightarrow{w} \sqrt{1-\lambda_1}\mathbb{B}_1(x) - \sqrt{\lambda_1}\mathbb{B}_2(x). \qquad (7.24)$$

Note that if we would have used $F_1$ or $F_2$ instead of $H$ in (7.23), the limiting processes $\mathbb{B}_1$ and $\mathbb{B}_2$ would be $F_1$- or $F_2$-Brownian bridges, but they coincide of course under $H_0$.

Consider now a process based on $\hat{F}_{in_i}(x) - \hat{H}_n(x)$. Although this looks similar to the contrast process (7.21), it is slightly more complicated because the two EDFs are not independent as they have the $n_i$ observations of the $i$th sample in common. To circumvent this problem we write $(s = 1, \ldots, K)$

$$\hat{H}_n(x) = \left( \sum_{j \neq s} \frac{n_j}{n} \hat{F}_{jn_j}(x) \right) + \frac{n_s}{n} \hat{F}_{sn_s}(x), \qquad (7.25)$$

and introduce the estimator $\hat{H}_{-sn}(x) = \sum_{j\neq s}(n_j/(n-n_s))\hat{F}_{jn_j}(x)$, which is the EDF of the pooled sample excluding all observations in the $s$th sample. Equation (7.25) now becomes

$$\hat{H}_n(x) = \frac{n-n_s}{n}\hat{H}_{-sn}(x) + \frac{n_s}{n}\hat{F}_{sn_s}(x).$$

With this in mind, it is straightforward to arrive at the following weak convergence,

$$\sqrt{\frac{n_s(n-n_s)}{n}}\left(\hat{F}_{sn_s}(x)-\hat{H}_n(x)\right)$$
$$\xrightarrow{w} (1-\lambda_s)\left(\sqrt{1-\lambda_s}\mathbb{B}_1(x)-\sqrt{\lambda_s}\mathbb{B}_2(x)\right),$$

where $\mathbb{B}_1$ and $\mathbb{B}_2$ are again two independent $H$-Brownian bridges.

## *7.5.2 Comparison Distribution Processes*

### 7.5.2.1 Construction

Despite the rather simple theory presented in the previous section, these contrast processes are usually not often used directly in practice. Many goodness-of-fit test statistics can though be written as a functional of a closely related process, of which the asymptotic theory is slightly more complex, but of which the limiting process has an important advantage over those of the previous section. In particular, the limiting processes $\mathbb{B}_1$ and $\mathbb{B}_2$ in Section 7.5.1 are $H$-Brownian bridges, and the pooled CDF $H$ is generally unknown! This is similar to the one-sample problem, where we have seen that $\sqrt{n}(\hat{F}_n(x)-G(x))$ converges to a $G$-Brownian bridge under the null hypothesis, but in the simple one-sample context the CDF $G$ was completely specified in $H_0$. However, for practical reasons the data were often transformed by the PIT, i.e., $p=G(x)$, so that transformed process $\sqrt{n}(\hat{F}_n(G^{-1}(p))-p)$ has a limiting uniform Brownian bridge, independent of the distribution $G$. If $H$ were known, a similar transformation would also result in a "contrast" process $\sqrt{n_s}\left(\hat{F}_{sn_s}(H^{-1}(p))-H(H^{-1}(p))\right)=\sqrt{n_s}\left(\hat{F}_{sn_s}(H^{-1}(p))-p\right)$ that converges to a uniform Brownian bridge under $H_0$ (note that now $\hat{H}_n$ is replaced by $H$, which was assumed known). Under $H_0$, the CDF $H$ can just as well be replaced by any other $F_j$ ($j\neq s$) if they are assumed known. However, because $H$ is never known, it seems natural to replace it with its estimator $\hat{H}_n$, resulting in the process (properly rescaled)

$$\mathbb{C}_{ns}(p)=\sqrt{n}\left(\hat{F}_{sn_s}(\hat{H}_n^{-1}(p))-p\right), \tag{7.26}$$

or, when $K=2$, it may be more convenient to avoid the pooled CDF as reference distribution, and use one of the two CDFs as a reference. For example,

$$\mathbb{C}_{n12}(p)=\sqrt{n_1}\left(\hat{F}_{1n_1}(\hat{F}_{2n_2}^{-1}(p))-p\right). \tag{7.27}$$

Note that we recycled the $\mathbb{C}$ notation for notational simplicity. It should be clear from the context. In later chapters we encounter many goodness-of-fit test statistics that can be expressed as functionals of these two processes. The asymptotic null distributions of these test statistics can usually be easily found by applying a continous mapping theorem when the limit processes of $\mathbb{C}_{ns}$ and $\mathbb{C}_{n12}$ are known. In the next section we give these limit processes.

#### 7.5.2.2 Weak Convergence

The following theorem gives the limiting process of (7.26) and (7.27). Further details, related and stronger properties (e.g., strong approximations) of these convergence statements can be found in Pyke and Shorack (1968), Csörgö and Révész (1978), Hsieh and Turnbull (1992), Hsieh (1995), Hsieh and Turnbull (1996), and Parzen (1997).

**Theorem 7.6.** *Assume that all $F_s$ have positive continuous derivatives $f_s$, and $\lim_{n\to\infty}(n_s/n) = \lambda_s > 0$ for all $s = 1, \ldots, K$. Let $\mathbb{B}_1$ and $\mathbb{B}_2$ denote two independent uniform Brownian bridges.*
*(1) Suppose further that $f_s(F_j^{-1}(p))/f_j(F_j^{-1}(p))$ is bounded on any $(a, b) \subset (0, 1)$. Then, as $n \to \infty$,*

$$\sqrt{n_s}\left(\hat{F}_{sn_s}(\hat{F}_{jn_j}^{-1}(p)) - F_s(F_j^{-1}(p))\right)$$
$$\xrightarrow{w} \mathbb{B}_1(F_s(F_j^{-1}(p))) + \sqrt{\frac{\lambda_s}{\lambda_j}\frac{f_s(F_j^{-1}(p))}{f_j(F_j^{-1}(p))}}\mathbb{B}_2(p).$$

*(2) Suppose $K = 2$, and let $\lambda = \lambda_1$. Let $h$ denote the density function corresponding to CDF $H(x) = \lambda F_1(x) + (1-\lambda)F_2(x)$. Suppose that*

$$\frac{f_1(H^{-1}(p))}{h(H^{-1}(p))} \quad \text{and} \quad \frac{f_2(H^{-1}(p))}{h(H^{-1}(p))}$$

*are bounded on any $(a, b) \subset (0, 1)$. Then, as $n \to \infty$,*

$$\sqrt{n}\left(\hat{F}_{1n_1}(\hat{H}_n^{-1}(p)) - F_1(H^{-1}(p))\right)$$
$$\xrightarrow{w} (1-\lambda)\left\{\frac{1}{\sqrt{\lambda}}\frac{f_2(H^{-1}(p))}{h(H^{-1}(p))}\mathbb{B}_1(F_1(H^{-1}(p)))\right.$$
$$\left. - \frac{1}{\sqrt{1-\lambda}}\frac{f_1(H^{-1}(p))}{h(H^{-1}(p))}\mathbb{B}_2(F_2(H^{-1}(p)))\right\}.$$

The limiting process of $\mathbb{C}_{ns}$ and $\mathbb{C}_{n12}$ that are stated in Theorem 7.6 are denoted by $\mathbb{C}_s$ and $\mathbb{C}_{12}$, respectively.

## 7.6 Stochastic Ordering and Related Properties

In the next chapters differences between distributions are not always expressed in terms of moments as we did in Part I. Sometimes *stochastic orderings* or related orderings are more relevant. Let $X_1 \sim F_1$ and $X_2 \sim F_2$. Further assume that $X_1$ and $X_2$ are independent.

**Definition 7.4 (stochastic ordering).** $X_1$ (or $F_1$) is *stochastically smaller* than $X_2$ (or $F_2$) iff

$$F_1(x) \geq F_2(x) \text{ for all } x \in \mathcal{S},$$

with strict inequality for at least one $x \in \mathcal{S}$.
$X_1$ (or $F_1$) is *stochastically larger* than $X_2$ (or $F_2$) iff

$$F_1(x) \leq F_2(x) \text{ for all } x \in \mathcal{S},$$

with strict inequality for at least one $x \in \mathcal{S}$.

These concepts are illustrated in the left panel of Figure 7.3: distribution $F_1$ (dashed line) is stochastically smaller than distribution $F_2$ (full line). The following lemma is important later.

**Lemma 7.4.** *If $X_1$ is stochastically smaller than $X_2$, then*

$$Pr\{X_1 \leq X_2\} > \frac{1}{2}.$$

*If $X_1$ is stochastically larger than $X_2$, then*

$$Pr\{X_1 \leq X_2\} < \frac{1}{2}.$$

*When $F_1 = F_2$, then $Pr\{X_1 \leq X_2\} = \frac{1}{2}$.*

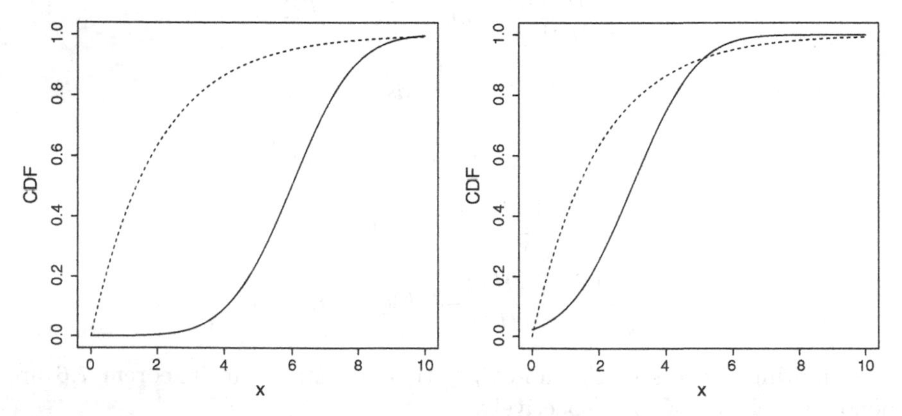


**Fig. 7.3** Distributions $F_1$ (dashed line) and distribution $F_2$ (full line). The left panel shows a situation where $F_1$ is stochastically smaller than $F_2$

An important consequence of Lemma 7.4 is that the implications hold generally only in one direction. For example, when $\Pr\{X_1 \leq X_2\} > \frac{1}{2}$ it is not necessarily true that $X_1$ is stochastically smaller than $X_2$. An example is presented in the right panel of Figure 7.3, where now the two CDFs cross so that no stochastic ordering holds. For this situation, however, we find $\Pr\{X_1 \leq X_2\} = 0.71$.

For the opposite implication to hold, distributional restrictions on $F_1$ and $F_2$ must be imposed. In Section 8.1.1.2 location-shift models are introduced. Under such a model, the opposite implication holds too.

We later work frequently with $\Pr\{X_1 \leq X_2\}$ for comparing distributions. To our knowledge there is no unambiguous terminology used in the literature for expressing the relation or ordering of $X_1$ relative to $X_2$ based on this probability. It is related to $X_1 - X_2$ being *stochastically positive* or stochastically negative (see Lehmann (1998), p. 195), but we have the feeling that this terminology does not cover its meaning. Acion et al. (2006), who advocate the use of this probability as a relevant and informative effect size, refer to it as the *probabilistic index*, and other have called it the *relative effect*. The term *stochastic improvement* was coined by Lehmann (1998), but this may be linguistically confusing when a small $X$ is contextually better than a large $X$. We suggest the following definition.

**Definition 7.5 (likely ordering (first-order)).** (1) $X_1$ (or $F_1$) is *likely larger* than $X_2$ (or $F_2$) iff

$$\Pr\{X_1 \leq X_2\} < \frac{1}{2} \text{ or, equivalently, } \Pr\{X_1 \geq X_2\} > \frac{1}{2}.$$

(2) $X_1$ (or $F_1$) and $X_2$ (or $F_2$) are *unlikely ordered* when

$$\Pr\{X_1 \leq X_2\} = \frac{1}{2}.$$

Lumley (2009) demonstrated that likely ordering is not transitive. We illustrate this by means of a simple example taken from Gillen and Emerson (2007).

*Example 7.3 (Nontransitivity of likely ordering).* Let $X_1$, $X_2$, and $X_3$ be three mutually independent random variables with multinomial distributions. In particular, $X_i$ has a discrete distribution with outcome space $\{1, 2, \ldots, 13\}$ and corresponding probabilities as shown in Figure 7.4. With these distributions we obtain

$$\Pr\{X_3 \geq X_2\} = 0.55 \quad \Pr\{X_2 \geq X_1\} = 0.5625 \text{ and } \Pr\{X_1 \geq X_3\} = 0.57.$$

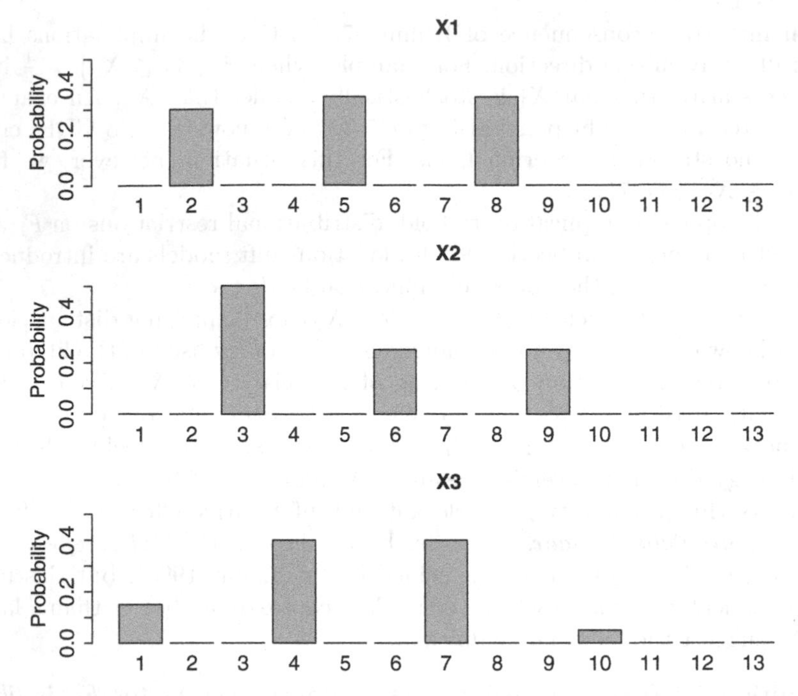


**Fig. 7.4** The distributions of $X_1$ (upper panel), $X_2$ (middle panel) and $X_3$ (lower panel)

Thus, using the terminology of Definition 7.5, we could say: $X_3$ is likely larger than $X_2$, which is likely larger than $X_1$. If the likely ordering were transitive, then we should now conclude that $X_3$ is likely larger than $X_1$. However, from $\Pr\{X_1 \geq X_3\} = 0.57 > 0.5$ we must conclude that $X_1$ is likely larger than $X_3$. This demonstrates the nontransitivity.

Higher-order likely orderings are also defined. We first illustrate the interpretation of a second-order likely ordering.

Let $X_{11}$ and $X_{12}$ be two independent observations from $F_1$, and $X_2 \sim F_2$. For example, $F_1$ corresponds to the yield of maize grown with fertilizer A, and $F_2$ is the yield when fertilizer B is used. Suppose that

$$\Pr\{\max\{X_{11}, X_{12}\} \leq X_2\} = \Pr\{X_{11} \leq X_2 \text{ and } X_{12} \leq X_2\} = 0.80.$$

This means that it is very likely that the yield of maize grown with fertilizer B is even larger than the highest yield of maize that is obtained from two fields on which fertilizer A is used. We say that the yield with fertilizer B is a *double likely larger* than the yield with fertilizer A. This brings us to the following more general definition.

**Definition 7.6 (likely ordering (higher-order)).** Let $k$ be an integer not smaller than two. Let $X_{11}, X_{12}, \ldots, X_{1k}$ i.i.d. $F_1$, and $X_2 \sim F_2$. $X_2$ (or $F_2$) is *$k$-tuple likely larger* than $X_1$ (or $F_1$) if

$$\Pr\{\max\{X_{11}, X_{12}, \ldots, X_{1k}\} \leq X_2\} > \frac{1}{2}.$$

Similarly, a *$k$-tuple unlikely ordering* of $X_1$ and $X_2$ is defined as

$$\Pr\{\max\{X_{11}, X_{12}, \ldots, X_{1k}\} \leq X_2\} = \frac{1}{2}.$$

# Chapter 8
# Graphical Tools

Most of the graphical tools that have been discussed in Chapter 3 for the one-sample problem can be adapted to the two- and the $K$-sample problems in a very straightforward way. We focus in this chapter on the QQ and PP plots and on the comparison distribution plots. Just as in Part I we start with defining the population versions of these plots, as these are easier vehicles for explaining how differences between distributions can be interpreted and understood. These graphs are again closely related to the tests discussed in the next chapters.

## 8.1 PP and QQ Plots

### *8.1.1 Population Plots*

#### 8.1.1.1 Population QQ Plot

QQ and PP plots have been described in detail in Section 3.2.2 for the one-sample goodness-of-fit problem. The functional forms of the population QQ and PP plots were there defined as $(G^{-1}(p), F^{-1}(p))$ and $(p, F(G^{-1}(p)))$ for $p \in [0,1]$, respectively (see (3.4) and (3.5)). In the present setting, particularly in the $K > 2$ case, several other constructions are possible. In general we now define the QQ plot of distribution $s$ as

$$Q_s : [0,1] \mapsto \mathcal{S}^2 : p \to (G_r^{-1}(p), F_s^{-1}(p)), \tag{8.1}$$

where $F_s$ is the CDF of population $s$ ($s = 1, \ldots, K$), and where $G_r$ serves as a reference distribution. In particular, $G_r$ can be chosen as the CDF of one of the other populations; i.e., $G_r = F_j$ ($j \neq s$). This may be a good choice when any one of the groups is a control or placebo group, for example. Similarly, $G_r$ may be set to the pooled CDF $H$ as defined in (7.16).

O. Thas, *Comparing Distributions*, Springer Series in Statistics,
DOI 10.1007/978-0-387-92710-7_8, © Springer Science+Business Media, LLC 2010

In Section 3.2.2 we argued that QQ plots may be particularly useful when the true and the hypothesised distributions $F$ and $G$ belong to the same location-scale family, so that $F^{-1}(p) = \mu + \sigma G^{-1}(p)$ and the QQ plot therefore shows a straight line. The same advantage pertains here too, but now with $G$ replaced by a reference distribution $G_r$. In many studies it is particularly the mean shift $\mu$ which is of importance. For example, when $F_1$ and $F_2$ represent the CDFs of the response in a treatment and a placebo group, respectively, then $\mu$ is the shift in mean response which is here interpretable as the treatment effect. We further ignore the scale parameter $\sigma$ (i.e., we set $\sigma = 1$) and focus on the shift parameter $\mu$ which we further denote as $\Delta$. The location-shift model for the two-sample case may be now be written as

$$F_1^{-1}(p) = \Delta + F_2^{-1}(p) \text{ or } F_2(x) = F_1(x + \Delta), \tag{8.2}$$

which is equivalent to saying that $X_1$ has the same distribution as $X_2 + \Delta$. This is illustrated in Figure 8.1. Yet another way of formulating is that the distributions of $X_1$ and $X_2$ have the same shape except that $F_1$ is translated over a distance $\Delta$. Later, in Chapter 9, we show that some well-known two-sample tests (e.g., the Wilcoxon rank sum test) may be used for testing the null hypothesis $\Delta = 0$, but only when the assumption of equal shapes holds true. The latter assumption is quite restrictive, and, particularly in small samples, hard to assess. Because $\Delta$ is basically a difference between two location parameters of distributions of the same shape, it may be estimated as the difference between the two sample means or the two sample medians. It is, however, good statistical practice to use an estimator that is directly related to the statistical test used. In particular, when the Wilcoxon rank sum test is employed, one usually estimates $\Delta$ with the *Hodges–Lehmann estimator*. See Section 9.2.7 for more details.

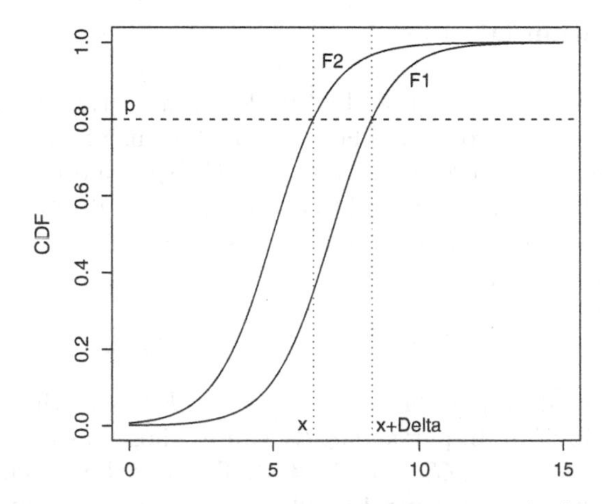


**Fig. 8.1** CDFs $F_1$ and $F_2$ representing the location-shift model (8.2)

Doksum (1974) extended the location shift model to

$$F_1^{-1}(p) = \Delta_p(p) + F_2^{-1}(p) \text{ or } F_2(x) = F_1(x + \Delta_x(x)),$$

which expresses that $X_1$ has the same distribution as $X_2 + \Delta_x(X_2)$, where $\Delta_x(x) = \Delta_p(F_2(x))$ are now *shift functions*,

$$\Delta_x(x) = F_1^{-1}(F_2(x)) - x \text{ or } \Delta_x(F_2^{-1}(p)) = \Delta_p(p) = F_1^{-1}(p) - F_2^{-1}(p). \quad (8.3)$$

The plot of $\Delta_p(p)$ versus $F_2^{-1}(p)$ (or $\Delta_x(x)$ versus $x$) thus contains the same information as the QQ plot, and may be interpreted directly in terms of the shift function or the *treatment function*, as it was referred to by Switzer (1976). Because $\Delta_p$ and $\Delta_x$ may be used interchangeably, the subscript is often omitted.

#### 8.1.1.2 Population PP Plot

The population PP plot of the $s$th distribution is now defined as

$$P_s : [0,1] \mapsto [0,1]^2 : p \to (p, F_s(G_r^{-1}(p))), \quad (8.4)$$

where $G_r$ is a reference CDF as before. Just as in the one-sample situation, the PP plot may also be recognised as a plot of the comparison distribution (7.20). This immediately gives a way of interpreting the plot: it is the CDF of $X_s$ on a scale in which the reference distribution is the uniform distribution. The PP plot of $X_s$ has thus to be compared to the CDF of a uniform, which is the 45 degree line. We briefly discuss two important classes of shapes of PP plots.

- The left-hand panel of Figure 8.2 shows four PP plots of distributions $F_s$ ($s = 1, \ldots, 4$) that are *stochastically smaller* than the reference distribution. By *stochastically smaller* we mean that $F_s(x) \geq G_r(x)$ for all $x \in \mathcal{S}$ with strict inequality for at least one $x \in \mathcal{S}$. The plots show clearly that the probability mass of $F_s$ is shifted to the smaller percentiles of the reference distribution; i.e., it is much more likely observing a small outcome from $F_s$ than from $G_r$. When $X_s$ is stochastically smaller than $Y \sim G_r$, $\Pr\{X_s \leq Y\} > \frac{1}{2}$ (note that $\frac{1}{2}$ is the probability that we expect when $F_s = G_r$) and $Y$ is thus likely larger than $X_s$.
- The opposite behavior is observed in the right-hand panel of Figure 8.2. Here the probability mass of $F_s$ is shifted to the larger percentiles of the reference distribution. Because the the PP plots do not cross the 45 degree line, we have $F_s(x) \leq G_r(x)$ for all $x \in \mathcal{S}$ with strict inequality for at least one $x \in \mathcal{S}$, and we say that $F_s$ (or $X_s$) is *stochastically larger* than the reference distribution. It is thus more likely that a realisation of $F_s$ is larger than a realisation of $G_r$. This implies that $\Pr\{X_s \geq Y\} > \frac{1}{2}$; i.e., $X_s$ is likely larger than $Y \sim G_r$.

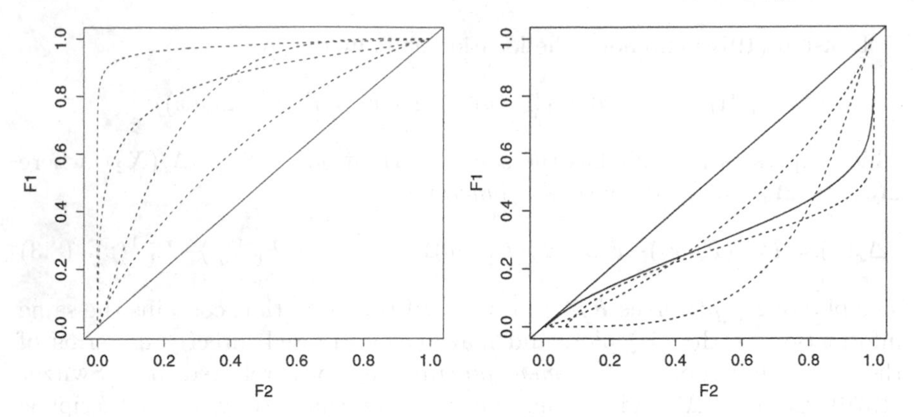


**Fig. 8.2** PP plots of stochastically smaller (left panel) and stochastically larger (right panel) distributions $F_1$ as compared to the reference distribution $F_2$

See Section 7.6 for an introduction to stochastic ordering and likely ordering. These two examples show extreme situations in the sense that none of the PP plots crosses the 45 degree line (i.e., $F_s(x) > G_r(x)$ or $F_s(x) < G_r(x)$ for all $x \in \mathcal{S}$). Consequently, the probability statements $\Pr\{X_s \leq Y\} > \frac{1}{2}$ or $\Pr\{X_s \geq Y\} > \frac{1}{2}$ also hold in any interval $[l, u]$ contained in $\mathcal{S}$. More formally, $F_s(x) > G_r(x)$, or $F_s(x) < G_r(x)$ for all $x \in \mathcal{S}$ implies $\Pr\{X_s \leq Y|(X_s, Y) \in [l, u]\} > \frac{1}{2}$ or $\Pr\{X_s \geq Y|(X_s, Y) \in [l, u]\} > \frac{1}{2}$ for all $[l, u] \subseteq \mathcal{S}$. On the other hand, when a PP plot crosses the 45 degree line, we do not have the strict order relations $F_s(x) > G_r(x)$ or $F_s(x) < G_r(x)$ anymore, but we may still have $\Pr\{X_s \leq Y\} > \frac{1}{2}$ or $\Pr\{X_s \geq Y\} > \frac{1}{2}$. However, these probability statements do not not hold anymore in all intervals $[l, u]$. This is illustrated in Figure 8.3. Suppose that $F_1$ and $F_2$ are the distributions of the length of two plant species, say S1 and S2. The PP plot now says that among the smaller plants, the S1 plants are generally smaller than the S2 plants, but among the larger plants it is just the other way around: the largest plants are more likely to belong to the S2 species. If we do not want to make this distinction between the "larger" and the "smaller" plants, we can still calculate the probability $\Pr\{X_1 \leq X_2\}$:

$$
\begin{aligned}
\Pr\{X_1 \leq X_2\} &= \int_{\mathcal{S}} \Pr\{X_1 \leq X_2|X_2 = x\} f_2(x)dx \\
&= \int_{\mathcal{S}} F_1(x)dF_2(x) \\
&= \int_0^1 F_1(F_2^{-1}(p))dp, \qquad (8.5)
\end{aligned}
$$

which summarises the likely ordering information over the whole support $\mathcal{S}$, and which contains highly relevant information on the difference between $F_1$ and $F_2$ in terms of likely orderings. Hollander and Wolfe (1999), among others, also advocated the use of $\Pr\{X_1 \leq X_2\}$ over a mean difference.

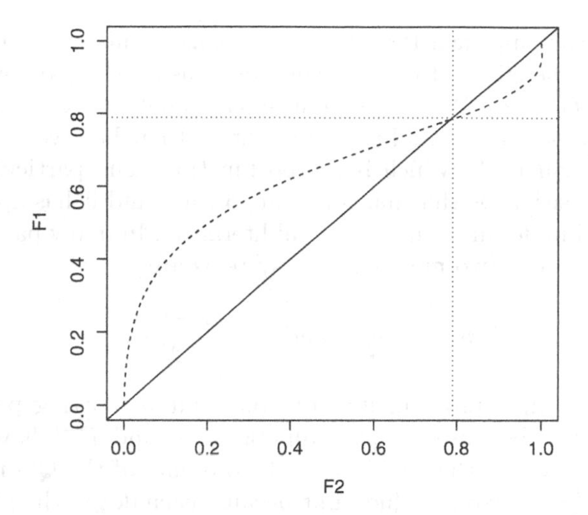


**Fig. 8.3** An example of a PP plot that crosses the 45 degree line. The horizontal and vertical reference lines correspond to the 79th percentiles

The expression (8.5) is the area under the PP plot. For the example of Figure 8.3, we find that $X_2$ is likely larger than $X_1$, with $\Pr\{X_1 \leq X_2\} = 0.55 > \frac{1}{2}$.

## *8.1.2 Empirical PP and QQ Plots*

### 8.1.2.1 Construction

In practice all unknown CDFs have to be replaced by estimators. Whereas in the one-sample case the hypothesised CDF $G$ is completely specified (or at least up to a small number of nuisance parameters), so that only the CDF of the sampled distribution has to be replaced by its EDF, we replace all CDFs $F_s$ by their estimators $\hat{F}_s$. The QQ plot thus becomes a graph of

$$(\hat{G}_{rm}^{-1}(p_i), \hat{F}_{sn_s}^{-1}(p_i)) \tag{8.6}$$

for the *plotting positions* $p_i$, $i = 1, \ldots, N$, where $N$ is user defined, and $m$ denotes the number of observations used in the estimator of $G_r$. More details on $N$ are given in the next paragraph. Similarly, the PP plot is the graph of

$$(p_i, \hat{F}_{sn_s}(\hat{G}_{rm}^{-1}(p_i))), \tag{8.7}$$

$i = 1, \ldots, N$. The PP and QQ plots for the one-sample problem have been discussed in detail in Section 3.2.2. Much attention went there to the choice of the plotting positions, because they have an important effect on the statistical properties of the empirical PP and QQ plots as estimators of their population versions. The effect of the plotting positions on the plots goes for both plots through $G^{-1}(p_i)$. When $G$ is continuous, small differences in the $p_i$ will have an effect on the plots. In the present setting, however, the reference distribution is an EDF, which is a step function. The particular choice of the plotting positions is thus much less important, and it has apparently not been discussed in detail in the statistical literature. In many papers it is even left unspecified. The two most popular systems are

$$p_i = \frac{i}{N+1} \text{ and } p_i = \frac{i-0.5}{N}.$$

In the one-sample problem it is obvious that $n$ plotting positions have to be considered, because the step functions $\hat{F}_n$ and $\hat{F}_n^{-1}$ have at most $n$ jumps. Here, however, the abcissa and the ordinate of the QQ plot can take at most $m$ and $n_s$ different values, but because each dot in the plot needs the specification of both coordinates, it seems natural to choose $N = \max(n_s, m)$ for the construction of the plotting positions. If the difference between $n_j$ and $m$ is large, the plot may look quite discontinuous, so that some prefer to interpolate between quantiles so as to get a smoother looking graph.

A PP plot, where $\hat{F}_{sn_s}(\hat{G}_{rm}^{-1}(p_i))$ is plotted versus the plotting position $p_i$, can thus have at most $\min(n_s, m)$ points at which the plot shows a jump in the abcissa of a multiple of $1/n_s$. Also here $N = \max(n_s, m)$ seems an appropriate choice. Again, when the difference in sample sizes is large, the plot may look discontinuous. Remember, though, that a PP plot can be interpreted as an EDF of the comparison distribution, and in this sense we expect it to look like a step function. Some statisticians prefer to avoid this discrete nature, and suggested to plot smoothed versions instead (see, e.g., Handcock and Morris (1999)).

#### 8.1.2.2 Sample Size Issues

Although the replacement of $G_r$ by its EDF is basically the only adaptation, it has some consequences. Maybe the most immediate consequence is the introduction of additional sampling variability. Moreover, because two estimators are involved, this variability now depends on two sample sizes. To keep the exposition simple, we discuss the variance issue for two-sample PP and QQ plots in which $F_1$ is compared with $F_2$ as a reference distribution. As before, let $\lambda_1 = \lim_{n\to\infty}(n_1/n)$ be bounded away from 0 and 1.

We start the discussion with the QQ plot (8.6). First note that both the abcissas and the ordinates are random variables, whereas in the case of a simple one-sample problem the absissa was determined by the constants $G^{-1}(p_i)$. We must thus now look at the *horizontal* and *vertical* variances of the plotted dots. Because a QQ plot is basically a plot of the quantile function $\hat{F}_{2n_2}^{-1}(p_i)$

versus $\hat{F}_{1n_1}^{-1}(p_i)$, its stochastic properties may be derived from the Bahadur–Kiefer theorem or Theorem 2.4, which have been discussed in Section 2.3.2. They give, for a fixed $p_i$ and for large sample sizes,

$$\text{Var}\left\{\hat{F}_{sn_s}^{-1}(p_i)\right\} \approx \frac{p_i(1-p_i)}{n_s f_s^2(F_s^{-1}(p_i))}.$$

Thus the horizontal and vertical variance of the $i$th plotted dot depend on the sample sizes $n_1$ and $n_2$, respectively. The numerator of the variance is the same as in the one-sample case, indicating that for many distributions the variance is larger in the tails of the distribution (often small density in the tails).

For the PP plot (8.7) the abcissas are fixed by the plotting positions, and the ordinates are determined by the empirical comparison distribution. Its stochastic properties have been studied in Section 7.5.2. In particular, Theorem 7.6 implies that for large sample sizes,

$$\begin{aligned}\text{Var}\left\{\hat{F}_{1n_1}(\hat{F}_{2n_2}^{-1}(p_i))\right\} &\approx \frac{q_i(1-q_i)}{n_1} + \frac{\lambda_1}{1-\lambda_1}\frac{p_i(1-p_i)}{n_1}r^2(p_i) \\ &\approx \frac{1}{n}\left\{\frac{q_i(1-q_i)}{\lambda_1} + \frac{p_i(1-p_i)}{1-\lambda_1}r^2(p_i)\right\}, \qquad (8.8)\end{aligned}$$

where

$$q_i = F_1(F_2^{-1}(p_i)) \text{ and } r(p_i) = \frac{f_1(F_2^{-1}(p_i))}{f_2(F_2^{-1}(p_i))}$$

are the comparison distribution and density function, respectively. The dependence on $r(.)$ makes the interpretation of this variance function slightly more complex than before. First note that the variance now has two terms, one which is proportional with $p_i(1-p_i)$ and the other term proportional with $q_i(1-q_i)$. The variance will generally again be larger near the middle of the horizontal axis. The influence of the sample sizes $n_1$ and $n_2$ through the (asymptotic) ratio $\lambda_1$ is illustrated in Figure 8.4. When the comparison density equals one, which happens under the two-sample null hypothesis, the graph of the variance function shows that the smallest variance is obtained with $\lambda_1 = 0.5$; i.e., $n_1 = n_2$. When, for a plotting position $p_i$, $r(p_i) = 1/10$ (small comparison density), which occurs at subsets of the sample space $\mathcal{S}$ where $f_1$ has low density as compared to $f_2$, the variance can be minimised with $n_1 >>> n_2$. This sample size condition basically assures that there is still a fair chance to observe at least a few observations in the first sample at places where the theoretical density is low. Finally, for the other extreme when $r(p_i) = 10$, Figure 8.4 shows that $n_1 <<< n_2$ will result in the smallest variance. It is further important to note that the variance functions for $r(p_i) = 1/10$ and $r(p_i) = 10$ are not symmetric in terms of the sample sizes. In the latter situation, which happens at intervals of $\mathcal{S}$ where we expect far less observations in the sample from the reference distribution, the variance

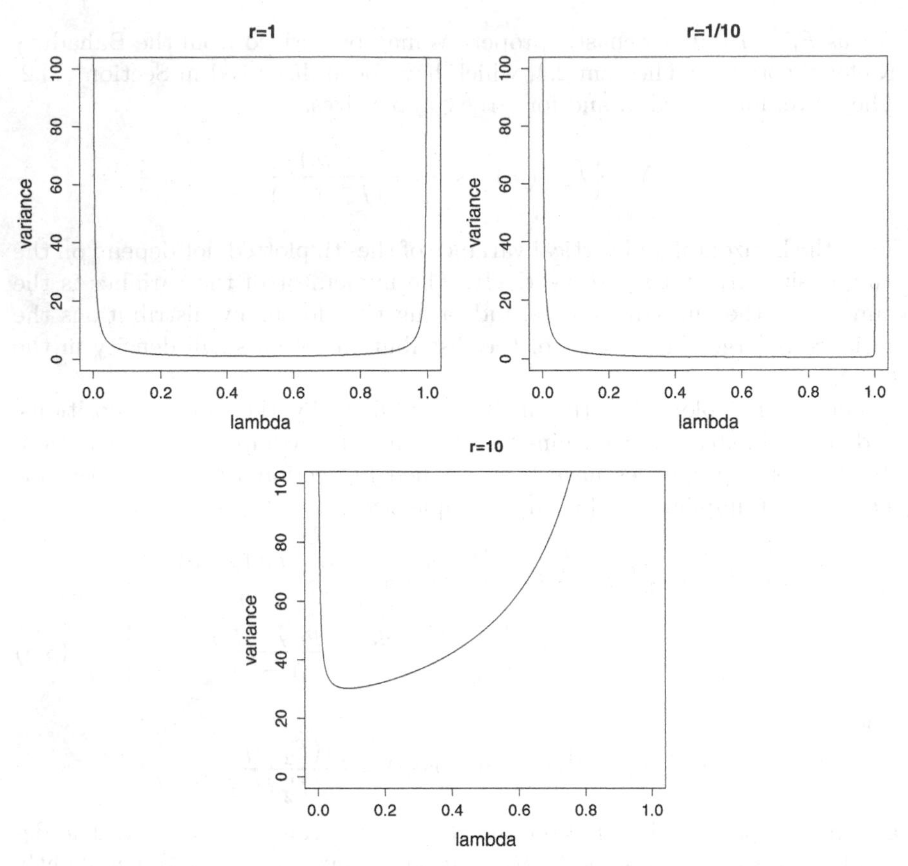


**Fig. 8.4** Variance function (8.8) (up to proportionality factor) for $r(p_i) = 1$ (top left), $r(p_i) = 1/10$ (top right), and $r(p_i) = 10$ (bottom)

is generally larger. This stresses the importance of deciding which distribution to consider as the reference. More particularly, it shows that the PP plot does not give much information in intervals of the sample space where the observations from the reference distribution are sparse. This was also demonstrated at the end of the previous section, where we argued that the choice of the number of plotting position depends on both $n_1$ and $n_2$.

Concluding this section on the consequences of the sample sizes on the quality of the plot, it seems important to have at least as many observations in the reference sample as in the other sample. Although we simplified the discussion to the situation where $F_2$ serves as the reference, our arguments suggest that it may be appropriate to work with the pooled EDF (7.17), $\hat{H}_n$, which uses all $n$ observations (the theory would only be slightly more complicated, and can be based on part (b) of Theorem 7.6).

#### 8.1.2.3 When to Use Which Plot

Before giving some specific arguments as to when to use which plot, we want to stress that it is probably the best strategy to make both plots in an exploratory phase of the statistical analysis! Most important is to know how to interpret both graphs correctly. In the next paragraph we list a few advantages and disadvantages of PP and QQ plots. Although there are not many papers with practical guidelines for two-sample PP and QQ plots, Holmgren (1995) gives a nice overview of how PP plots may be used in this setting.

A major difference between PP and QQ plots is the scale of the horizontal and vertical axes. In a QQ plot the points are shown on the same measurement scale of the observations. This is particularly informative when the two-sample problem is further analysed in a traditional parametric way in which the focus is on differences in means and/or variances. In Section 8.1.1 we have seen that pure location-scale differences between $F_1$ and $G_r$ can be expressed as $F_1^{-1}(p) = \mu + \sigma G_r^{-1}(p)$, which suggests that the empirical QQ plot will show points that are randomly scattered around a straight line. The slope of this line and the vertical shift from the 45 degree line provide information on the parameters $\mu$ and $\sigma$ in the location-scale model. Hsieh (1995) and Hsieh and Turnbull (1996) give more details on how an empirical QQ plot can be used to estimate these parameters. More generally, a QQ plot may be used to get an impression of the shift or treatment function $\Delta(x)$. The plot may also be used to estimate $\Delta(x)$; the asymptotic theory can be found in Doksum (1974). A disadvantage of a plot on the original measurement scales of $F_1$ and $G_r$ is that it makes the plot very sensitive to outliers. With the default settings of most QQ plots in statistical software, all observations are plotted, including outliers. The presence of an outlier will therefore move the vast majority of data points to a corner of the plotting area. The same argument, however, makes the QQ plot a good tool for detecting outliers. Both axes of a PP plot, on the other hand, represent probabilities. This makes a PP plot invariant to monotone increasing transformations of the measurement scales. Although this makes the PP plot not well suited for extracting information on mean differences, a PP plot still contains all information about the stochastic ordering of $F_1$ and $G_r$, as well as about the summarising probability $\Pr\{X_1 \leq Y\}$ (see Section 8.1.1 for more details on the interpretation in terms of the population PP plot). Because of the importance of the latter probability, we conclude this section with some more details on how the PP plot may be used to estimate it.

For simplicity we take $G_r = F_2$. As before, let $X_1 \sim F_1$ and $X_2 \sim F_2$. In terms of the population PP plot, we found (see (8.5))

$$\Pr\{X_1 \leq X_2\} = \int_0^1 F_1(F_2^{-1}(p))dp = \int_{\mathcal{S}} F_1(x)dF_2(x).$$

A plug-in estimator is obtained by replacing $F_1$ and $F_2$ by their respective EDFs,

$$\begin{aligned}
&\int_{\mathcal{S}} \hat{F}_1(x) d\hat{F}_2(x) \\
&= \int_{\mathcal{S}} \left( \frac{1}{n_1} \sum_{i=1}^{n_1} \mathrm{I}(X_{1i} \leq x) \right) d \left( \frac{1}{n_2} \sum_{j=1}^{n_2} \mathrm{I}(X_{2j} \leq x) \right) \\
&= \frac{1}{n_1 n_2} \sum_{i=1}^{n_1} \sum_{j=1}^{n_2} \int_{\mathcal{S}} \mathrm{I}(X_{1i} \leq x) \, d\mathrm{I}(X_{2j} \leq x) \\
&= \frac{1}{n_1 n_2} \sum_{i=1}^{n_1} \sum_{j=1}^{n_2} \mathrm{I}(X_{1i} \leq X_{2i}),
\end{aligned}$$

where in the last step we have made use of the property that the function $d\mathrm{I}(X_{2j} \leq x)$ is zero for all $x \in \mathcal{S} \setminus X_{2i}$, and it is one for $x = X_{2i}$.

*Example 8.1 (The traffic data).* Figures 8.5 and 8.6 show the PP and QQ plots of the traffic dataset. For each of routes 2, 3, 4, and 5 a PP and a QQ plot are presented. The sample of travel times of route 1 is used as reference distribution. The R-code is provided below.

```
PPplot(time~route, data=traffic, groups=c(1,2),
+ xlab="route 1",ylab="route 2",main="PP plot")
QQplot(time~route, data=traffic, groups=c(1,2),
+ xlab="route 1",ylab="route 2",main="QQ plot")
PPplot(time~route, data=traffic, groups=c(1,3),
+ xlab="route 1",ylab="route 3",main="PP plot")
QQplot(time~route, data=traffic, groups=c(1,3),
+ xlab="route 1",ylab="route 3",main="QQ plot")

PPplot(time~route, data=traffic, groups=c(1,4),
+ xlab="route 1",ylab="route 4",main="PP plot")
QQplot(time~route, data=traffic, groups=c(1,4),
+ xlab="route 1",ylab="route 4",main="QQ plot")
PPplot(time~route, data=traffic, groups=c(1,5),
+ xlab="route 1",ylab="route 5",main="PP plot")
QQplot(time~route, data=traffic, groups=c(1,5),
+ xlab="route 1",ylab="route 5",main="QQ plot")
```

We discuss each of the route comparisons separately.

- Route 2 compared with route 1 (Figure 8.5).
  The QQ plot shows points that seem to be scattered quite closely around a straight line. This indicates that the distributions of the travel times with

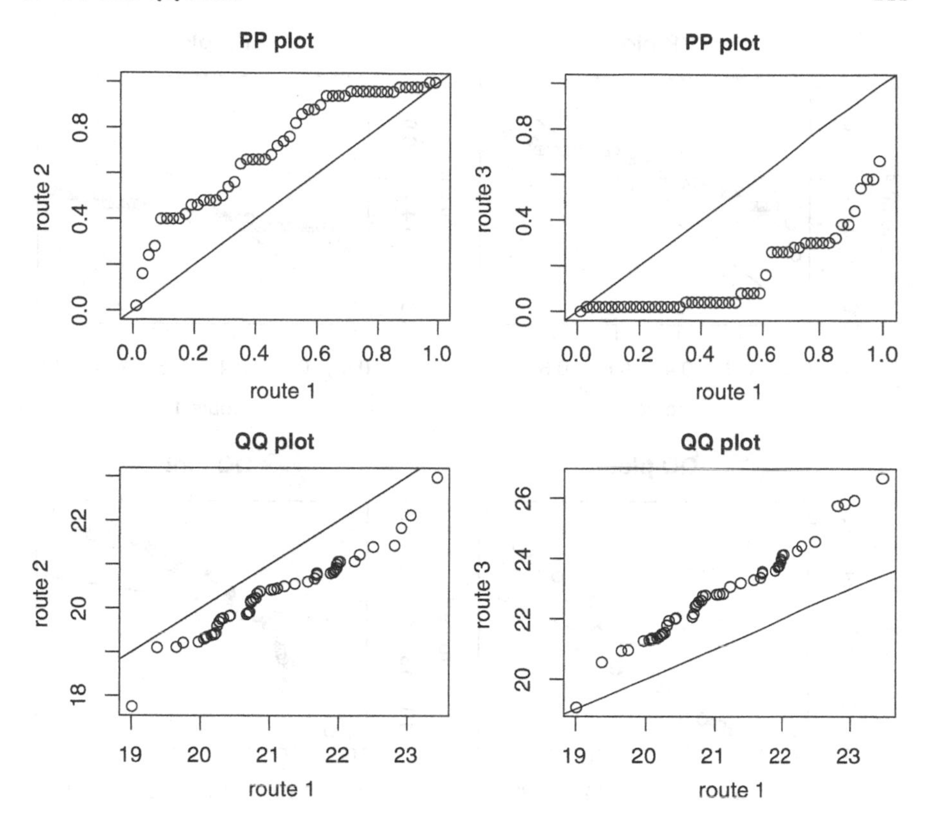


**Fig. 8.5** PP and QQ plots of the traffic dataset (routes 2 and 3 are compared with the reference route 1)

routes 2 and 1 probably only differ in terms of means and variances. This line is almost parallel to the 45 degree line, therefore we do not expect much difference in scale. The mean travel time with route 2 thus seems to be about 1 minute smaller than with route 1.

All points in the PP plot are located above the diagonal, indicating that the distributions of the travel times with route 2 are stochastically smaller than with route 1. The area under the empirical PP plot is thus larger than $\frac{1}{2}$, implying that $\Pr\{X_2 \leq X_1\} > \frac{1}{2}$, which says that it is more likely to have a smaller travel time with route 2 than with route 1.

- Route 3 compared with route 1 (Figure 8.5).

  These PP and QQ plots give almost exactly the same expression as the plots for route 2, except that now the QQ plot clearly shows points that are scattered around a line that is not parallel to the 45 degree line. This indicates that the variance of the travel times with group 3 is the largest. As the points now all lie above the 45 degree line, the mean of the travel times of route 3 seems to be larger than that of route 1.

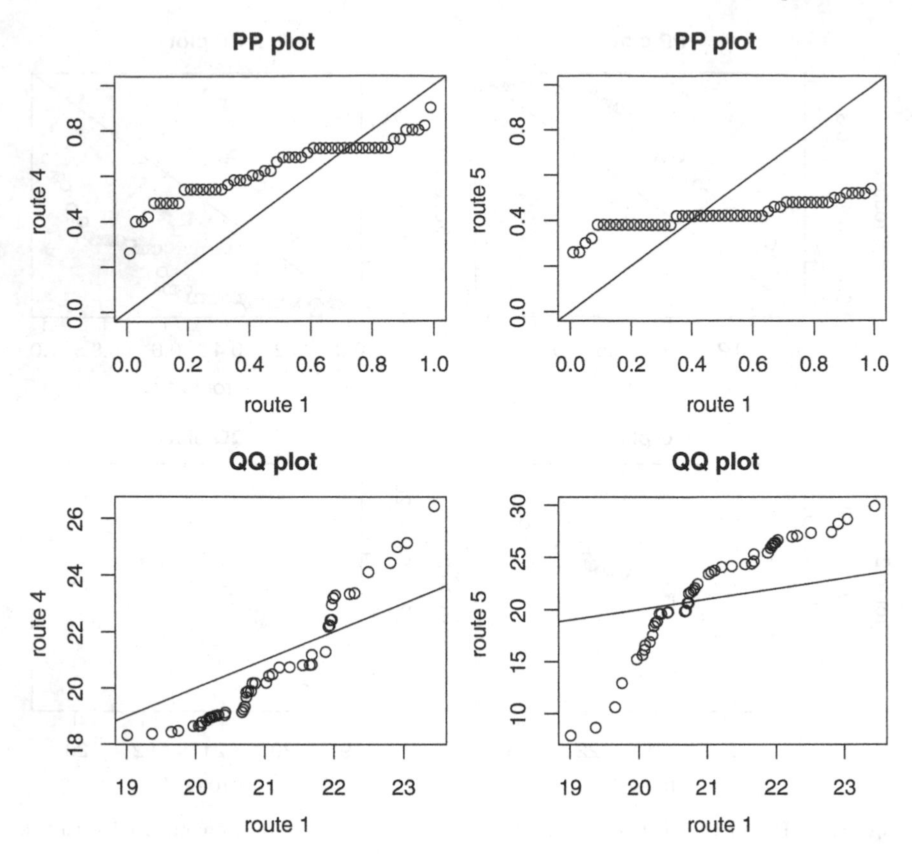


**Fig. 8.6** PP and QQ plots of the traffic dataset (routes 4 and 5 are compared with the reference route 1)

Also the PP plot is similar to the plot of route 2, except that now all points lie below the diagonal. We thus conclude that the travel times with route 3 are likely to be stochastically larger than with route 1. We may estimate that $\Pr\{X_3 \leq X_1\} < \frac{1}{2}$. We therefore say that it is more likely that route 1 is faster than route 3.

- Route 4 compared with route 1 (Figure 8.6).
  The points in the QQ plot do not show a linear relationship, so that $F_1$ and $F_2$ are probably not within the same location-scale family. The curve crosses the 45 degree line at about 22 minutes. The plot thus suggests that, given that the travel time is smaller than 22 minutes, route 4 is faster on average. For travel times larger than 22 minutes, the opposite is observed. A similar conclusion can be deduced from the PP plot, except that now the crossing of the diagonal is expressed in terms of percentiles. Here, the crossing is at the 70th percentile. Furthermore, because there is clearly a crossing, no (overall) stochastic ordering conclusion is suggested, but

conclusions in terms of likely orderings are still possible. The estimated area under the PP plot curve is now slightly larger than 0.5, which provides only weak evidence that driving along route 4 is more likely to be faster than route 1. One could, for example, summarise the conclusion as follows. When in the whole city the traffic is flowing, route 4 will probably be faster than route 1, but when dense traffic is expected throughout the whole city, route 1 may be the best option.

- Route 5 compared with route 1 (Figure 8.6).
  The PP and QQ plots are similar to those of route 4. The PP plot is now quite flat and the area under the observed curve is now close to 0.5, so that in terms of probability there seems to be no preference among routes 1 and 5. Because the crossing with the 45 degree line is now earlier than the 50th percentile, and the QQ plot shows a steep ascent before the crossing, a left-skewed distributional shape (relative to route 1) is suggested.

## 8.2 Comparisons Distributions

### *8.2.1 The Population Comparison Distribution*

Just as with the probability plots, a good way of gaining insight into the interpretation of comparison distributions is first to study the population version. The population comparison density function has been introduced in Section 7.4, and can here be written as

$$r_s(u) = \frac{f_s(G_r^{-1}(u))}{g_r(G_r^{-1}(u))},$$

where $G_r$ can represent any reference distribution function, and $g_r$ is the corresponding density. Usually the reference distribution is one of the distributions under study or it is the marginal distribution (7.16). Because the interpretation of this comparison density is completely similar to the one-sample case, which has been discussed in detail in Section 3.3.1, we continue immediately with the estimation of the comparison density.

### *8.2.2 The Empirical Comparison Distribution*

Because the comparison density is basically the density function of $G_r(X_s)$ with $X_s \sim F_s$, it can be estimated using nonparametric density estimation techniques applied to the transformed data $G_r(X_{si})$ $(i = 1, \ldots, n_s)$. For the simple one-sample problem the reference CDF is completely specified under the null hypothesis, and the transformed data are referred to as the *relative data*. Here, however, the reference distribution is unknown too and has in turn to be replaced by its EDF, resulting in the *quasi relative data*,

$$\hat{G}_{rm}(X_{si})$$

($i = 1, \ldots, n_s$), where $m$ is the size of the sample used to estimate $G_r$. The properties of quasi-relative data have been studied by Lehmann (1953) and Lin and Sukhatme (1993). Although basically all nonparametric density estimation methods can now be applied to the transformed dataset, the theoretical properties will be different because of the additional estimation of $G_r$. Appropriate theory for the histogram estimator, a kernel density estimator, and an orthogonal series density estimator were provided by Parzen (1983), Eubank et al. (1987), Alexander (1989), Cwik and Mielniczuk (1993), Mielniczuk (1992), Li et al. (1996), and Parzen (1999), but for the regression-based estimators the theory has still to be developed. We do not give theoretical details, but the interested reader may find a extensive summary in Handcock and Morris (1999).

Density estimators based on Poisson local-quadratic regression (see Section 2.8.4) are implemented in the reldist R package. We next illustrate the use of the comparison density plots on two examples.

*Example 8.2 (Traffic data: Route 2 versus route 1).* We consider the data of routes 1 and 2, and we use route 1 as the reference. In Section 8.1.2 we concluded from the QQ plot in Figure 8.5 that the distributions of the travel times with routes 1 and 2 only show a shift in means. To check this statement we now plot a comparison density, as well as the components in a decomposition of the comparison density (see Section 3.3.1 for a discussion on the decomposition in the one-sample setting). In particular, we consider

$$r(u) = \frac{f_2(F_1^{-1}(u))}{f_1(F_1^{-1}(u))} = \frac{f_{1L}(F_1^{-1}(u))}{f_1(F_1^{-1}(u))} \times \frac{f_2(F_1^{-1}(u))}{f_{1L}(F_1^{-1}(u))}, \tag{8.9}$$

where $f_{1L}$ is the density of the reference distribution shifted to have the same mean as the distribution $F_2$ under study. It is thus the density of $X_1 - \mu_1 + \mu_2$, where $\mu_1$ and $\mu_2$ are the means of $F_1$ and $F_2$, respectively. Whereas $r(u)$ contain the full information on all differences between $F_1$ and $F_2$, the first factor in (8.9) contains only information about a difference in the means, and the second factor contains the *residual* information. Only a mean correction has been applied, thus the residual comparison density contains information about all types of shape differences between the distributions $F_1$ and $F_2$ (all aspects of shape, except the mean).

Figure 8.7 shows the three plots. We first discuss the (overall) comparison density, shown in the top panel of Figure 8.7. The graph shows clearly a large comparison density before the 20th percentile of the reference distribution. This indicates that there are relatively more smaller travel times with route 2 as there are with route 1. At the larger percentiles (>80%) the density of

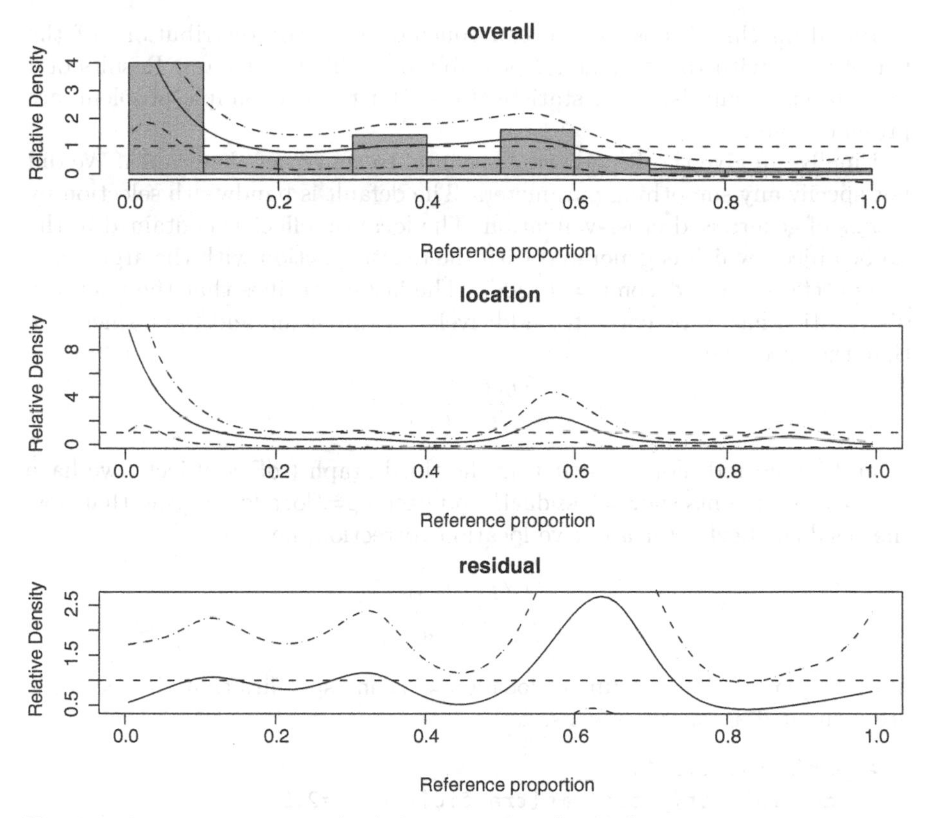


**Fig. 8.7** Comparison density (top), location effect (middle), and residual shape effect (bottom) of the route 2 traffic data, with route 1 as reference distribution. The dashed lines represent the 95% confidence intervals. In the top panel, the histogram of the quasi-relative data is also shown

travel times with route 2 is smaller than with route 1. These observations are indeed consistent with the mean-shift conclusion based on the QQ plot. We now look at the middle panel of Figure 8.7, which shows the comparison density of the mean-corrected reference distribution versus the original reference distribution. This plot thus only contains information about a potential difference in means. If the plot showed a constant density at 1, we would conclude that there is no mean shift. However, the graph clearly shows an increased comparison density at small percentiles. Note that the shape looks closely like the shape of the comparison density in Figure 3.9, which is a population comparison density of a pure location shift situation. Finally, we explore the residual comparison density in the lower panel of Figure 8.7. This plot suggests that no residual shape difference exists between $F_1$ and $F_2$, after elimination of the mean effect (note that the range of the vertical axis is much smaller than for the other two graphs).

Based on this discussion we can conclude that the distributions of the travel times with routes 1 and 2 probably only differ in means. Results of a more formal analysis, using statistical tests for the two-sample problem, are presented later.

Finally, we give the R code we have used to construct the graphs. We did not specify any smoothing parameters. The default is bandwidth selection by means of generalised cross-validation. The location effect is contained in the rdLoc object, which is generated with the reldist function with the arguments show="effect", and decomp="locadd". The latter specifies that the reference distribution has to be corrected additively for the mean, and the former says that the effect, i.e.

$$\frac{f_{1L}(F_1^{-1}(u))}{f_1(F_1^{-1}(u))},$$

should be plotted. For constructing the third graph (rdRes object), we have used the arguments show="residual" and decomp="locadd", saying that now the residual effect after additive location correction, i.e.

$$\frac{f_2(F_1^{-1}(u))}{f_{1L}(F_1^{-1}(u))},$$

must be plotted. The argument location="mean" specifies that the locations are estimated as sample means.

```
> par(mfrow=c(3,1))
> rd0<-reldist(y=traffic[traffic$route==2,2],
+ yo=traffic[traffic$route==1,2],ci=T,main="overall",
+ bar="yes")
> lines(rd0$x,rd0$ci$l,lty=2)
> lines(rd0$x,rd0$ci$u,lty=2)
> rdLoc<-reldist(y=traffic[traffic$route==2,2],
+ yo=traffic[traffic$route==1,2],main="location",
+ show="effect",decomp="locadd",ci=T,location="mean")
> lines(rdLoc$x,rdLoc$ci$l,lty=2)
> lines(rdLoc$x,rdLoc$ci$u,lty=2)
> rdRes<-reldist(y=traffic[traffic$route==2,2],
+ yo=traffic[traffic$route==1,2],main="residual",
+ show="residual",
+ decomp="locadd",ci=T,location="mean")
> lines(rdRes$x,rdRes$ci$l,lty=2)
> lines(rdRes$x,rdRes$ci$u,lty=2)
```

*Example 8.3 (Traffic data: Route 3 versus route 1).* We now do a similar exploratory analysis, but now we compare route 3 with the reference route 1. In Section 8.1.2 we have presented the QQ plot in Figure 8.5, from which we concluded that both distributions may differ in mean and in scale. To study mean and scale differences, we estimate here the comparison density and decompose it into three factors. In particular,

$$r(u) = \frac{f_2(F_1^{-1}(u))}{f_1(F_1^{-1}(u))} = \frac{f_{1L}(F_1^{-1}(u))}{f_1(F_1^{-1}(u))} \times \frac{f_{1LS}(F_1^{-1}(u))}{f_{1L}(F_1^{-1}(u))} \times \frac{f_2(F_1^{-1}(u))}{f_{1LS}(F_1^{-1}(u))}, \quad (8.10)$$

where now the first and the second factor correspond to the location and the scale effect, respectively, and the last factor again contains the information on the residual shape differences between $F_1$ and $F_2$, where now the shape refers to all distributional characteristics, expect location and scale. In this example, the locations and scales are estimated by the sample means and the sample standard deviations.

The comparison densities are shown in Figure 8.8. The overall comparison density indicates a larger density of long travel times with route 3 as com-

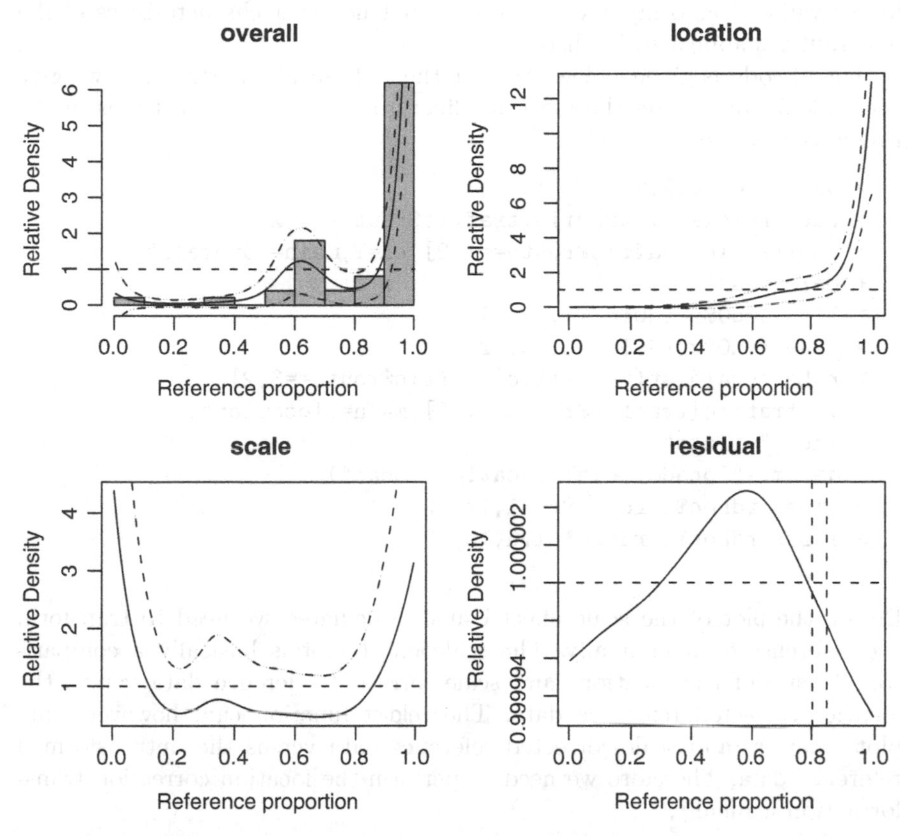


**Fig. 8.8** Comparison density (top left), location effect (top right), scale effect (bottom left), and residual shape effect (bottom right) of the route 3 traffic data, with route 1 as reference distribution. The dashed lines represent the 95% confidence intervals. In the top left panel, the histogram of the quasi-relative data is also shown

pared to route 1, and the opposite is observed for the 80% smallest travel times. There seems to be a local increase in comparison density around the 65th percentile of the reference distribution, but the confidence intervals suggest this is not significant. In the top right panel the pure location effect is visualised. Again this looks similar to the prototype shown in Figure 3.9, so that we may conclude that there is evidence that route 3 has longer travel times on average. The scale differences between the two routes can be read from the bottom left panel of Figure 8.8. This plot looks very much like the population comparison density in Figure 3.10, which results from a pure scale effect. We may thus conclude that the variance of the travel times with route 3 is the largest. After correcting for mean and scale differences, the comparison density in the bottom-right panel shows the residual shape differences. There seems to be almost no residual difference left (note the scale on the vertical axis), so that we may conclude from this exploratory analysis that the distribution of the travel times with route 3 has a larger mean and a larger variance as compared to group 1, but no other characteristics of the distributions appear to be distinct.

The R code is given below. We list the code in two parts. First we give the code for producing the location effect graph, which is exactly as in the previous example.

```
> par(mfrow=c(2,2))
> rd0<-reldist(y=traffic[traffic$route==3,2],
+ yo=traffic[traffic$route==1,2],ci=T,main="overall",
+ bar="yes")
> lines(rd0$x,rd0$ci$l,lty=2)
> lines(rd0$x,rd0$ci$u,lty=2)
> rdLoc<-reldist(y=traffic[traffic$route==3,2],
+ yo=traffic[traffic$route==1,2],main="location",
+ show="effect",
+ decomp="locadd",ci=T,location="mean")
> lines(rdLoc$x,rdLoc$ci$l,lty=2)
> lines(rdLoc$x,rdLoc$ci$u,lty=2)
```

Before the plot of the scale effect can be calculated we need to transform the reference data manually. The scale effect plot is basically a comparison density of the location- and scale-corrected reference data versus the location-corrected reference data. The reldist function can, however, only plot location- and scale-corrected reference data versus the untransformed reference data. Therefore we need to perform the location correction transformation manually.

```
> r1<-traffic[traffic$route==1,2]
> r3<-traffic[traffic$route==3,2]
> yoL<-r1-mean(r1)+mean(r3)
```

With these transformed data we produce the scale effect graph. The arguments decomp="lsadd" and show="effect" say that the reference data have to be corrected for both location and scale, and this corrected dataset has to be compared with the reference data, which are now the manually location corrected data in yoL. In the R code for the last graph we read show="residual",decomp="lsadd" to construct the residual comparison density after location and scale correction.

```
> rdScale<-reldist(y=traffic[traffic$route==3,2],yo=yoL,
+ main="scale",show="effect",decomp="lsadd",ci=T,
+ location="mean",scale="standev")
> lines(rdScale$x,rdScale$ci$l,lty=2)
> lines(rdScale$x,rdScale$ci$u,lty=2)
> rd<-reldist(y=traffic[traffic$route==3,2],
+ yo=traffic[traffic$route==1,2],main="residual",
+ show="residual",decomp="lsadd",ci=T,
+ location="mean",scale="standev")
> lines(rdRes$x,rdRes$ci$l,lty=2)
> lines(rdRes$x,rdRes$ci$u,lty=2)
```

# Chapter 9
# Some Important Two-Sample Tests

We start this chapter with some general guidelines for setting null and alternative hypotheses, while stressing their relation with the choice of a test statistic and the interplay between the null hypothesis and the distributional assumptions one is willing to make. In Section 9.2 this is illustrated for the two-sample problem in the discussion of the well-known Wilcoxon rank sum test. We study the Wilcoxon test from several points of view. From this discussion it becomes clear that its interpretation is not always as clear-cut as one would hope. For example, we demonstrate that the test may not always be used for detecting differences in means. This brings us back to the *diagnostic property* that was also important in Part I. We further elaborate on this in Section 9.3, in which we again consider the Wilcoxon test as an example. The same reasoning is applied in Section 9.5, where we discuss some of the nonparametric tests for detecting differences in scale. Section 9.6 focusses on the Kruskal–Wallis test for the $K$-sample problem, and we conclude this chapter with an introduction to adaptive tests.

The objective of this chapter is twofold. First, some of the popular and important rank tests are described. These rank tests appear again in the next chapters as components of smooth and EDF tests. The second objective is to show that the interpretation of many tests depends on the interplay between the distributional assumptions and the hypotheses. This is important for understanding how correct and informative conclusions may be obtained when using hypotheses tests in a statistical analysis.

O. Thas, *Comparing Distributions*, Springer Series in Statistics,
DOI 10.1007/978-0-387-92710-7_9,  

## 9.1 The Relation Between Statistical Tests and Hypotheses

### 9.1.1 Introduction

In the ideal situation the following procedure towards hypothesis testing could be considered scientifically optimal. First the research question should be translated into a null hypothesis and an alternative hypothesis, and only then should we look for an appropriate test. With "appropriate" we mean that the test must be consistent for testing specifically for the selected null and alternative hypotheses, and the test must be unbiased under a set of assumptions that may be reasonably expected to hold in practice, or the assumptions should be at least verifiable. As "consistency" is an asymptotic property, it may in practice be replaced by "powerful". In general a test consists of three components: the test statistic, say $T_n$, a null distribution, and a descision rule. We do not provide more details on the latter as we assume that the decision rule is implicitely included in the procedure for $p$-value calculation which relies on the null distribution, together with a specification of a significance level, say $\alpha$. We illustrate this general outline by means of the traditional two-sample $t$-test.

*Example 9.1 (The two-sample t-test).* Among marine biologists it is well known that PCBs easily accumulate in the fat tissue of shrimp. In this experiment two groups of 18 shrimp (actually 18 samples of shrimp, because the PCB concentration is measured by extracting the fat of 100 g shrimp) were grown under different conditions: the "classical" method, and a more expensive new method. The shrimp were randomised over the two groups. The research question is to assess whether the *mean* PCB concentration is affected by the growing conditions. It is very relevant here to phrase the question in terms of means, because this mean can be used to estimate the total intake of PCBs for a person that eats, for example, 1 kg of shrimp each year. It is particularly this accumulated amount of PCB that is likely to be associated with health risks.

From the description of the problem we may deduce that the null and alternative hypotheses have to be formulated in terms of means. Let $\mu_1$ and $\mu_2$ denote the means of PCB concentrations in shrimp grown with the classical and the new method, respectively. The null hypothesis is thus $H_0 : \mu_1 = \mu_2$. Because the new growing technique is more expensive than the old one, we are not interested in detecting $\mu_1 < \mu_2$, as this would only result in the conclusion to stick with the old method. Therefore the alternative hypothesis is set to $H_1 : \mu_1 > \mu_2$.

Now that the hypotheses are specified we start searching for an appropriate test. In this phase we may take into account what distributional assumptions we are prepared to assume. Suppose that we know from previous experiments that the PCB concentrations are quite well normally distributed. In this case we may opt for the two-sample Student's $t$-test or the Welch test. The former may only be used when the two population variances are equal. The test statistic is given by

$$T_n = \frac{\bar{X}_1 - \bar{X}_2}{\sqrt{S_p^2\left(\frac{1}{n_1}+\frac{1}{n_2}\right)}}, \tag{9.1}$$

where $S_p^2$ is the pooled variance estimator, and $\bar{X}_1$ and $\bar{X}_2$ are the two sample means. Under the normality condition the null distribution of $T_n$ is a $t$-distribution with $n_1 + n_2 - 2$ degrees of freedom, which can be used for the calculation of the one-sided $p$-value. This is an exact result under the assumed conditions.

We conclude this example with one more comment. Suppose we were not in the position to make the distributional assumption of normality. Consequently, the null distribution of $T_n$ is no longer a $t$-distribution. However, if the test were to be used on large samples, the central limit theorem guarantees that $T_n$ has asymptotically a standard normal null distribution, and thus $T_n$ can still be used as a test statistic. The assumption of equality of variance can still not be discarded when the pooled variance estimator is used.

This example illustrates that the test statistic measures the information in the sample against the null hypothesis in favor of the alternative hypothesis. Also note that the $t$-test statistic in (9.1) is basically a scaled estimator of $\mu_1 - \mu_2$, and that the hypotheses are also formulated in terms of $\mu_1 - \mu_2$. The example also suggests that good statistical practice should keep the following sequence in mind.

1. Formulate the null and alternative hypotheses.
2. What distributional assumptions can be made (they should be verifiable).
3. Select an appropriate test statistic for which
   (a) the null distribution is known, or can be enumerated or approximated (taking 1 and 2 into account)
   (b) the test is consistent (or powerful) for testing $H_0$ against $H_1$.
4. Compute the $p$-value.

Usually steps 2 and 3(a) are difficult, and they go hand in hand. This is illustrated next for the two-sample $t$-test.

*Example 9.2 (The two-sample t-test).* In Example 9.1 the null hypothesis was set at $H_0 : \mu_1 = \mu_2$, and normality of $F_1$ and $F_2$ was assumed. Suppose now that the researcher knows that this assumption is unlikely to hold, whereas he or she has reason to believe that equality of variances is to be expected. A solution may exists then in extending the null hypothesis to the general two-sample null hypothesis, $H_0 : F_1 = F_2$, which implies the *randomisation hypothesis* (see Definition 7.2 in Section 7.1.2). This allows the formulation of the test as a permutation test. When the same test statistic (9.1) is used the test is still consistent for detecting differences in means, as specified in $H_1$. This consistency property is a consequence of the form of the $t$-test statistic, which is a scaled estimator of $\mu_1 - \mu_2$.

In the remainder of this section we give some more examples of tests. We state the test statistics, null and alternative hypotheses, distributional assumptions, and null distributions. In particular, we look for the *least restrictive null hypotheses* and the *most general alternative hypothesis.* With the least restrictive null hypothesis, we mean that the null distribution of the test statistic is valid under this null hypothesis, and assuming that the distributional assumptions hold true; the null hypothesis combined with the assumptions should imply as few restrictions as possible. The most general alternative hypothesis consists of the largest set of conditions for which the test is consistent. We make this exercise for some well known tests that are closely related to the general two- and $K$-sample problems. Among others, we include the two-sample $t$-test, the Wilcoxon rank sum test (equivalent to the Mann–Whitney test), the Mood test, the $F$-test in a one-way ANOVA, and the Kruskal-Wallis rank test. All these test statistics reappear in later chapters. With this discussion we hope to provide a deeper understanding of how these tests work, and how statistical analyses based on them should be interpreted.

It is very often difficult to distinguish between the null hypothesis and the distributional assumption. To conclude this introduction we illustrate this for the two-sample $t$-test.

*Example 9.3 (The two-sample t-test).* Consider again the parametric two-sample Student $t$-test. This test is usually presented as the parametric test for testing the null hypothesis $H_0 : \mu_1 = \mu_2$ against, e.g., $H_1 : \mu_1 > \mu_2$, because it relies on the distributional assumption that the observations of both samples are normally distributed with equal variances. It is thus assumed that $F_1$ and $F_2$ are normal distributions with equal variances. The only parameters that are left unspecified are the common variances and the means, but the latter are the subject of the null hypothesis. Combining the null hypothesis with these distributional assumptions brings us back to the general two-sample null hypothesis (6.1), but then with the additional normality assumption. We can even go one step further: if we have large samples, the central limit theorem replaces the normality requirement, and, we could thus just as well have specified the null hypothesis as (6.1), i.e., $H_0 : F_1 = F_2$, with only the "mild" distributional assumption that the variances of $F_1$ and $F_2$ are the same. Or, we could still set $H_0 : \mu_1 = \mu_2$, again assuming equal variances. The latter is less restrictive, and using $H_0 : \mu_1 = \mu_2$ instead of $H_0 : F_1 = F_2$ does not affect the asymptotic null distribution. The null hypothesis in terms of the means is therefore the least restrictive null hypothesis in this example. The alternative hypothesis, however, remains $\mu_1 > \mu_2$, as the $t$-test statistic is based on an estimator of the difference between the two sample means and the test is thus only consistent for this alternative (under the assumptions).

**Table 9.1** Summary of the modes of useage of the two-sample $t$-test statistic

| $H_0$ | Assumptions | Null distribution | $H_1$ |
|---|---|---|---|
| $\mu_1 = \mu_2$ | $F1$ and $F_2$ normal<br>$\sigma_1^2 = \sigma_2^2$ | $t_{n_1+n_2-2}$ | $\mu_1 \neq$ (or $<$ or $>$) $\mu_2$ |
| $\mu_1 = \mu_2$ | $\sigma_1^2 = \sigma_2^2$<br>$n_1$ and $n_2$ large | $N(0,1)$ | $\mu_1 \neq$ (or $<$ or $>$) $\mu_2$ |
| $F_1 = F_2$ | | permutation | $\mu_1 \neq$ (or $<$ or $>$) $\mu_2$ |

The general two-sample null hypothesis $H_0 : F_1 = F_2$ still makes sense here if we agree to combine the two-sample $t$-test statistic with its *exact permutation null distribution*. This is approximately what we have done in Example 7.1. It is, however, only the least restrictive null hypothesis when the sample sizes are small without the normality assumption. The alternative hypothesis, however, again remains unchanged: $\mu_1 > \mu_2$. The reason is again that the test statistic (9.1) only measures deviations from $H_0$ in terms of differences between the means. The discussion given in this example is summarised in Table 9.1.

## 9.2 The Wilcoxon Rank Sum and the Mann–Whitney Tests

### *9.2.1 Introduction*

The Wilcoxon rank sum test and the Mann–Whitney test are among the most popular nonparametric tests. They were proposed by Wilcoxon (1945) and Mann and Whitney (1947). These tests are included in most basic statistics courses, in which they are often presented as the nonparametric analogues of the two-sample $t$-test. More specifically, it is frequently suggested that these nonparametric tests should be used when the normality assumption underlying the $t$-test does not hold. In this section we demonstrate that the Wilcoxon and the Mann–Whitney tests are actually tests for testing hypotheses that are not necessarily expressed in terms of means, so that they may not just be used as surrogates for the $t$-test. The hypotheses that they test for, however, are of interest in their own right, so that it is statistically more correct to first think about which hypotheses have to be tested, and only then select the appropriate test. The Wilcoxon and the Mann–Whitney statistics are equal up to a monotonic transformation, but from a didactical point of view we like to start with the latter.

It is only when some additional distributional assumptions on $F_1$ and $F_2$ are imposed, that the Wilcoxon or Mann–Whitney test statistic can be used for testing hypotheses about means. Under some of these assumptions, the optimality theory of Section 7.2.2 applies. This is the topic of Section 9.2.5.

### 9.2.2 The Hypotheses

The easiest way to introduce the Mann–Whitney test is by considering it as a test for testing the general two-sample null hypothesis

$$H_0 : F_1(x) = F_2(x) \text{ for all } x \in \mathcal{S}$$

against the alternative

$$H_1 : \Pr\{X_1 \leq X_2\} \neq \frac{1}{2},$$

where $X_1$ and $X_2$ have distribution functions $F_1$ and $F_2$, respectively. The probability used in the formulation of the alternative hypothesis is calculated as

$$\pi = \Pr\{X_1 \leq X_2\} = \int_{\mathcal{S}} \Pr\{X_1 \leq X_2 | X_2 = x\} f_2(x)dx = \int_{\mathcal{S}} F_1(x)dF_2(x),$$

which is the area under the PP plot and which allows formulating conclusions in terms of likely orderings. See Sections 7.6 and 8.1.1.2 for more details. Thus, if $F_1 = F_2$, we find

$$\pi = \Pr\{X_1 \leq X_2\} = \int_{\mathcal{S}} F_1(x)dF_2(x) = \int_0^1 udu = \frac{1}{2},$$

which explains the alternative hypothesis. Although $H_1$ is not directly interpretable in terms of means, it is very informative and it has a simple interpretation. For example, suppose $\Pr\{X_1 \leq X_2\} = 0.9$, in which the indices 1 and 2 refer to a placebo and a treatment group in a randomised trial, and further suppose that a large response is an indication of an improvement of the illness. This statement says that it is much more likely that the response of *a* treated patient is larger than the response of *a* nontreated patient (independently sampled patients). If this conclusion is statistically significant, it is very relevant evidence to a physician that most of his patients will be better off with the treatment. This interpretation is further illustrated in the example of Section 9.2.8. Note, however, that this is not a causal interpretation, as $X_1$ and $X_2$ refer to two independent (and different) patients. In this context, $\pi$ may be considered as an *effect size* parameter. In Section 7.6 it was related to likely ordering. In particular, when $\pi > (<) \frac{1}{2}$ we say that $X_2$ ($X_1$) is likely larger than $X_1$ ($X_2$). Stochastic ordering was also discussed in Section 8.1.1.2, where it was used to indicate inequality of distribution functions.

Note that $F_1(x) > F_2(x)$ for all $x \in \mathcal{S}$ implies that $\Pr\{X_1 \leq X_2\} > \frac{1}{2}$. Stochastic ordering implies likely ordering, but not necessarily the other way around.

### *9.2.3 The Test Statistics*

We start with the Mann–Whitney test statistic, which is based on an estimator of $\pi = \Pr\{X_1 \leq X_2\}$. For later purposes it is important to remember that $\Pr\{X_1 \leq X_2\} = \Pr\{X_1 < X_2\}$ when $X_1$ and $X_2$ are continuous random variables. A naive estimator of $\Pr\{X_1 \leq X_2\}$ is given by the empirical estimator of this probability, obtained by counting the number of events $X_{1i} \leq X_{2j}$ $(i = 1, \ldots, n_1;, j = 1, \ldots, n_2)$ in the sample, and dividing it by the total number of pairs $(X_{1i}, X_{2j})$. As the total number of pairs equals $n_1 n_2$, the estimator may be written as

$$\hat{\pi} = \frac{1}{n_1 n_2} \sum_{i=1}^{n_1} \sum_{j=1}^{n_2} \mathrm{I}(X_{1i} \leq X_{2j}) . \tag{9.2}$$

Later, in Section 9.2.6, we focus on $\hat{\pi}$ as an estimator, but here we consider $\hat{\pi}$ as a test statistic. The Mann–Whitney statistic is usually defined as $MW = \sum_{i=1}^{n_1} \sum_{j=1}^{n_2} \mathrm{I}(X_{1i} \leq X_{2j})$, so that $\hat{\pi} = MW/(n_1 n_2)$. It is more conventional to write $MW$ as a *rank statistic* based on the ranks of the observations in the pooled sample. Let $Z_i$ $(i = 1, \ldots, n)$ denote an observation of the pooled sample $\{X_{11}, X_{12}, \ldots, X_{1n_1}, X_{21}, \ldots, X_{2n_2}\}$. First, we assume that there are no ties; i.e., there occur no two equal $Z_i$ observations. Note that this happens theoretically with probability one if the observations arise from a continuous random variable. In practice, however, rounding and measuring with a finite accuracy often introduce ties in the data. As before, the rank of $Z_i$ in the pooled sample is defined as

$$R_i = R(Z_i) = (\text{number of observations in the sample } \leq Z_i) = \sum_{j=1}^{n} \mathrm{I}(Z_j \leq Z_i).$$

With this definition the $MW$ statistic becomes

$$\begin{aligned} MW &= \sum_{i=1}^{n_1} \sum_{j=1}^{n_2} \mathrm{I}(X_{1i} \leq X_{2j}) \\ &= \sum_{j=1}^{n_2} (\text{number of } X_{1i} \leq X_{2j}) \end{aligned}$$

$$= \sum_{j=1}^{n_2} ((\text{number of } X_{1i} \text{ and } X_{2i} \leq X_{2j}) - (\text{number of } X_{2i} \leq X_{2j}))$$
$$= \sum_{j=1}^{n_2} (R(X_{2j}) - \text{number of } X_{2i} \leq X_{2j})$$
$$= \sum_{j=1}^{n_2} R(X_{2j}) - \sum_{j=1}^{n_2} j$$
$$= \sum_{j=1}^{n_2} R(X_{2j}) - \frac{n_2(n_2+1)}{2}.$$

Set $U = \sum_{j=1}^{n_2} R(X_{2j})$, which is known as the Wilcoxon rank sum statistic. It is the sum of the ranks of the observations in the second sample, for which the ranking is relative to the pooled sample, and it is thus related to the Mann–Whitney statistic by

$$MW = U - \frac{n_2(n_2+1)}{2}.$$

This equality implies that the tests based on $MW$ and $U$ will be completely equivalent. We therefore prefer to rename both tests the *Wilcoxon–Mann–Whitney test* (WMW test). The $U$ statistic can be easily recognised as a simple linear rank statistic of the form $T_n = \sum_{i=1}^{n} c_i a_n(R_i)$, which was introduced in Definition 7.3. We adopt the convention to order the pooled sample observations $Z_1, \ldots, Z_n$ so that the first $n_1$ $Z_i$s are the original first sample observations. Let $c_i = 0$ for observation $i$ in the first sample and $c_i = 1$ when observation $i$ is in the second sample, and $a_n(R_i) = R(Z_i)$.

### *9.2.4 The Null Distribution*

In this section we discuss the null distributions of the WMW statistic under the general two-sample null hypothesis $H_0 : F_1 = F_2$. Many rank tests have the advantage that their null distributions only depend on the sample sizes ($n_1$ and $n_2$), and not on the distribution $F_1 = F_2$ of the observations under the null hypothesis. Such tests are said to be *distribution free*. We next give some more details on the exact null distribution of the $U$ statistic. For finite sample sizes, a rank statistic typically has a discrete distribution. Its probability function can be written as

$$\Pr\{U = u | H_0\} = \frac{1}{\binom{n}{n_2}} h(n_2, u) = \frac{n_1! n_2!}{n!} h(n_2, u), \tag{9.3}$$

where $h(n_2, u)$ is the number of size $n_2$ subsets $\{i_1, \ldots, i_{n_2}\}$ of $\{1, \ldots, n\}$ so that $\sum_{j=1}^{n_2} i_j = u$. In general there are two ways for computing this probability function:

1. The *exact null distribution* can be enumerated as a permutation distribution as outlined in Section 7.1. This requires $m = \binom{n}{n_2}$ permutations, but because the test statistic only depends on the data through the sample sizes $n_1$ and $n_2$, the enumeration of the permutation distribution must be performed only once for each set of sample size. From the test statistics $U^{(i)}$ $(i = 1, \ldots, m)$ calculated under the group of $m$ permutations, the count function $h(n_2, u)$ in (9.3) is simply the number of $U^{(i)}$ that equals $u$. When $n_1$ and $n_2$ are large, the exact permutation distribution may be approximated by Monte Carlo simulations, as explained in Section 7.1.2.3.
2. The major problem in the calculation of (9.3) is the computation of $h(n_2, u)$. Instead of computing them based on $m = \binom{n}{n_2}$ permutations (or Monte Carlo approximations), more intelligent algorithms have been proposed. These numerical techniques are beyond the scope of this book; we refer the interested reader to Cheung and Klotz (1997) and van de Wiel (2001) and the references therein.

In the examples of Section 9.2.8 we give more details on how these methods are implemented in the R software.

When both sample sizes $n_1$ and $n_2$ are large, Theorem 7.2 may be applied directly to find the limiting null distribution of the $U$ and the $MW$ statistics. In particular, under $H_0$, as $\min(n_1, n_2) \to \infty$,

$$\frac{U - \frac{n_2(n+1)}{2}}{\sqrt{n_1 n_2 (n+1)/12}} \xrightarrow{d} N(0, 1). \tag{9.4}$$

By the equivalence of the $U$ and the $MW$ statistics, a similar asymptotic result also holds for the Mann–Whitney statistic MW. Thus both the exact permutation and the asymptotic version of the WMW test are tests for the same null and alternative hypotheses. In particular, for the MW statistic, under $H_0$, as $\min(n_1, n_2) \longrightarrow \infty$,

$$\frac{MW - \frac{n_1 n_2}{2}}{\sqrt{n_1 n_2 (n+1)/12}} \xrightarrow{d} N(0, 1). \tag{9.5}$$

Finally we like to stress that the mean $\mu_n = \mathrm{E}\{MW\} = (n_1 n_2)/2$ and the variance $\sigma_n^2 = \mathrm{Var}\{MW\} = n_1 n_2 (n+1)/12$ which appear in (9.5) are computed under the null hypothesis $H_0 : F_1 = F_2$, using the formulae (7.7) and (7.8).

In view of later discussions it is important to have expressions for $\mathrm{E}\{MW\}$ and $\mathrm{Var}\{MW\}$ under arbitrary $F_1$ and $F_2$. Although the mean and variance may be calculated by means of a very general result of Chernoff and Savage

(1958), we prefer to give more direct calculations, for didactical reasons. From the definition of MW we immediately find

$$\begin{aligned}
\mathrm{E}\{MW\} &= \mathrm{E}\left\{\sum_{i=1}^{n_1}\sum_{j=1}^{n_2} \mathrm{I}\left(X_{1i} \leq X_{2j}\right)\right\} \\
&= \sum_{i=1}^{n_1}\sum_{j=1}^{n_2} \mathrm{E}\left\{\mathrm{I}\left(X_{1i} \leq X_{2j}\right)\right\} \\
&= \sum_{i=1}^{n_1}\sum_{j=1}^{n_2} \Pr\left\{X_{1i} \leq X_{2j}\right\} \\
&= n_1 n_2 \Pr\left\{X_1 \leq X_2\right\} = n_1 n_2 \pi, \qquad (9.6)
\end{aligned}$$

which reduces to $(n_1 n_2)/2$ under $H_0 : F_1 = F_2$. For the calculation of the variance the algebra is slightly more complicated as it also involves covariances. For details we refer to Lehmann (1951) or Birnbaum and Klose (1957), who arrive at the expression

$$\mathrm{Var}\{MW\} = n_1 n_2 \left\{(n_1 - 1)\phi_1^2 + (n_2 - 1)\phi_2^2 + \pi(1-\pi)\right\}, \qquad (9.7)$$

where

$$\begin{aligned}
\phi_1^2 &= \mathrm{Var}_{f_2}\left\{F_1(X_2)\right\} = \int_{\mathcal{S}} F_1^2(x) dF_2(x) - \pi^2 \\
\phi_2^2 &= \mathrm{Var}_{f_1}\left\{F_2(X_1)\right\} = \int_{\mathcal{S}} F_2^2(x) dF_1(x) - (1-\pi)^2 .
\end{aligned}$$

In Section 9.2.6 we express $\phi_1^2$ and $\phi_2^2$ in terms of covariances, but for now it sufficient to recognise that the first terms in $\phi_1^2$ and $\phi_2^2$ are again probabilities. In particular, let $X_1, X_{11}$ and $X_{12}$ denote i.i.d. random variables from $F_1$, and let $X_2, X_{21}$ and $X_{22}$ denote i.i.d. random variables from $F_2$. The the first term in $\phi_1^2$ equals $\Pr\left\{\max(X_{11}, X_{12}) \leq X_2\right\}$, and, similarly, the first term in $\phi_2^2$ equals $\Pr\left\{\max(X_{21}, X_{22}) \leq X_1\right\}$. Both probabilities reduce to $\frac{1}{3}$ under $H_0 : F_1 = F_2$, so that (9.7) becomes $n_1 n_2(n+1)/12$ under the null hypothesis.

### *9.2.5 The WMW Test as a LMPRT*

The WMW test statistic is also the LMPRT (Section 7.2.2) under the location-shift model with the additional distributional assumption that $F_1$ and $F_2$ are logistic distributions. The one-parameter logistic distribution with location parameter $\mu$ has density and cumulative distribution function

$$f(x;\mu) = \frac{\exp(-(x-\mu))}{\left(1+\exp(-(x-\mu))\right)^2} \text{ and } F(x;\mu) = \frac{1}{1+\exp(-(x-\mu))}.$$

Using the notation of Section 7.2.2, the location-shift alternative

$$f_1(x) = f_2(x - \Delta)$$

corresponds to (7.13) with $f_1(x) = f(x;0)$ and $f_2(x-\Delta) = f(x;\Delta)$ ($f$ being the standard logistic density), and with $c_{ji} = 0$ for $j = 1$ and $i = 1, \ldots, n_1$, and $c_{ji} = 1$ for $j = 2$ and $i = 1, \ldots, n_2$.

The optimal scores, as given in Theorem 7.5, are defined in terms of the expectations of

$$\frac{\frac{\partial}{\partial\mu} f(X;\mu)\Big|_{\mu=0}}{f(X;0)} = \frac{1-\exp(-X)}{1+\exp(-X)} = 2F(X) - 1.$$

Thus, with $X_{(i)}$ the order statistics of a sample of $n$ i.i.d. logistic variates,

$$a_n(i) = \mathrm{E}_f\left\{2F(X_{(i)}) - 1\right\} = 2\frac{i}{n+1} - 1,$$

incorporating a continuity correction. The LMPRT statistic (7.14) becomes

$$\begin{aligned} T_n &= \sum_{j=1}^{2}\sum_{i=1}^{n_j} c_{ji} a_n(R_{ji}) = \sum_{i=1}^{n_2}\left(2\frac{R(X_{2i})}{n+1} - 1\right) \\ &= \frac{2}{n+1}\left(\sum_{j=1}^{n_2} R(X_{2j}) - \frac{n_2(n+1)}{2}\right), \end{aligned}$$

in which Corollary 7.1 is used. Hence, the test statistic is, up to a scaling factor, equal to the Wilcoxon rank sum test statistic. In Section 9.4 several more optimal rank tests are discussed for other parametric location-shift models.

Finally we examine the *asymptotic efficiency* (see Section 2.9.2.2) of the Wilcoxon test under sequences of local alternatives and under several distributional assumptions on $F_1$ and $F_2$. In particular we consider the sequence $\Delta_n = \delta/\sqrt{n}$, with $\delta > 0$. The WMW test is compared to the two-sample $t$-test, because this is the most powerful test under the traditional normality assumption. Pitman (1948) showed that in the present context the asymptotic relative efficiency (ARE) may be calculated as

$$\mathrm{ARE}_{W,t} = 12\sigma_f^2\left(\int_{\mathcal{S}} f^2(x)dx\right)^2,$$

where $f$ is the density function of the location-shift model, and $\sigma_f^2$ is the variance of this distribution. Table 9.2 shows some AREs. From this table we learn that using the WMW test for detecting a location shift is often much better than using the two-sample $t$-test. Even in the situation of normality, for which the two-sample $t$-test is known to be the most powerful test, the

**Table 9.2** Some AREs of the Wilcoxon versus the two-sample $t$-test for several distributions $f$

| $f$ | Normal | Uniform | Logistic | Double exponential | Cauchy | Exponential |
|---|---|---|---|---|---|---|
| ARE | 0.955 | 1.000 | 1.097 | 1.500 | $\infty$ | 3.000 |

WMW test does only slightly worse: an ARE of 0.955! The WMW test is particularly powerful for distributions with heavy tails. Hodges and Lehmann (1956) showed that the ARE of the WMW test versus the two-sample $t$-test can never be smaller than 0.864. This property, together with the observation that the ARE is often much larger than one, suggests that it is generally better to choose the WMW test, particularly in situations where the data analyst does not know a priori how the observations will be distributed. Note, however, that the recommendation still relies on the conditions to make the WMW test informative for detecting a shift in means.

### *9.2.6 The MW Statistic as an Estimator of $\pi$*

Because the WMW test is basically useful for inference on the probability $\pi = \Pr\{X_1 \leq X_2\}$, it is informative to also report an estimate of this parameter. In Section 9.2.2 we have illustrated that $\pi$ may be a very usefull and good interpretable effect size parameter. We have also already argued that $\pi$ may be estimated by $\hat{\pi} = MW/(n_1 n_2) = (1/(n_1 n_2)) \sum_{i=1}^{n_1} \sum_{j=1}^{n_2} I_{ij}$ (Equation (9.2)), with $I_{ij} = \mathrm{I}(X_{1i} \leq X_{2j})$. Lehmann (1951) showed that $\hat{\pi}$ is the *uniform minimum variance unbiased estimator* within a large class of continuous distributions. From this expression the variance of $\hat{\pi}$ may be found immediately,

$$\mathrm{Var}\{\hat{\pi}\} = \frac{1}{n_1 n_2}\pi(1-\pi)\left[1 + (n_1 - 1)\rho_1 + (n_2 - 1)\rho_2\right], \tag{9.8}$$

with $\rho_1 = \mathrm{corr}(I_{ij}, I_{kj})$ $(i \neq k)$ and $\rho_2 = \mathrm{corr}(I_{ij}, I_{ik})$ $(j \neq k)$. These correlations may be rewritten as $\rho_i = (p_i - \pi^2)/(\pi - \pi^2)$ $(i = 1, 2)$ with $p_1 = \Pr\{I_{ij} I_{kj} = 1\} = \Pr\{\max(X_{1i}, X_{1k}) \leq X_{2j}\}$ $(i \neq k)$ and $p_2 = \Pr\{I_{ij} I_{ik} = 1\} = \Pr\{X_{1i} \leq \min(X_{2j}, X_{2k})\}$ $(j \neq k)$. Note the relation with (9.7) when discussing the WMW test. When we write

$$\nu = \frac{n_1 n_2}{1 + (n_1 - 1)\rho_1 + (n_2 - 1)\rho_2}, \tag{9.9}$$

the variance of $\hat{\pi}$ becomes

$$\mathrm{Var}\{\hat{\pi}\} = \frac{\pi(1-\pi)}{\nu}. \tag{9.10}$$

In the next paragraph we describe a method for confidence interval estimation.

Many authors proposed confidence intervals for $\pi$. Newcombe (2006a,b) gives a good overview of the confidence intervals that have been proposed in the statistical literature. He evaluated ten confidence intervals in a simulation study (Newcombe (2006b)). Some of the confidence intervals are actually based on parametric assumptions on $F_1$ and $F_2$, so that the variance of the estimator of $\pi = \Pr\{X_1 \leq X_2\} = \int_0^1 F_1(F_2^{-1}(p))dp$ depends on the parameters of $F_1$ and $F_2$ and has a rather simple expression. Replacing these parameters by their consistent estimators results in a confidence interval. From his simulation study he concluded that these confidence intervals usually have coverages close to the nominal level, even when the parametric assumptions are not satisfied. Here, however, we prefer using a method that does not explicitly rely on such parametric assumptions. Many of the confidence intervals that have been described in the literature differ with respect as to how $\nu$ is estimated in the denominator of (9.10). We give the confidence interval of Halperin et al. (1987) with the modification of Mee (1990). We first define some statistics. Unbiased estimators of $p_1$ and $p_2$ are given by

$$\hat{p}_1 = \frac{1}{n_1 n_2(n_1-1)} \sum_{h\neq i=1}^{n_1} \sum_{j=1}^{n_2} I_{ij} I_{hj} \text{ and } \hat{p}_2 = \frac{1}{n_1 n_2(n_2-1)} \sum_{i=1}^{n_1} \sum_{k\neq j=1}^{n_2} I_{ij} I_{ik}. \tag{9.11}$$

With these estimators $\rho_i$ may be estimated as $\hat{\rho}_i = (\hat{p}_i - \hat{\pi}^2)/(\hat{\pi} - \hat{\pi}^2)$, $i = 1, 2$. These estimators are now combined with an expression of the jacknife estimator of $\nu$ of Sen (1967), resulting in

$$\hat{\nu} = \frac{n_1 n_2}{\frac{1+(n_1-1)\hat{\rho}_1}{1-1/n_2} + \frac{1+(n_2-1)\hat{\rho}_2}{1-1/n_1}}.$$

The $1-\alpha$ confidence interval of $\pi$ is then

$$\left\{\pi : \frac{|\pi - \hat{\pi}|}{\sqrt{\pi(1-\pi)}} \leq \frac{z_{\alpha/2}}{\sqrt{\hat{\nu}}}\right\},$$

which is an interval with lower and upper bounds given by

$$\frac{1}{1+C}\left[\hat{\pi} + \frac{1}{2}C \pm \sqrt{C\left(\hat{\pi}(1-\hat{\pi}) + \frac{1}{4}C\right)}\right],$$

with $C = z_{\alpha/2}/\sqrt{\hat{\nu}}$.

Finally, Mee (1990) further suggested an improvement of the interval in the case where $\hat{\pi}$ is very close to one of the extreme probabilities 0 or 1, $\hat{\pi}$ or $1-\hat{\pi} < 1/(2\sqrt{n_1 n_2})$. He suggested estimating $\nu$ after adding a constant to the second sample observations $X_{2j}$ ($j = 1, \ldots, n_2$). The constant has to be chosen so that $\hat{\pi}$ or $1-\hat{\pi} = 1/(2\sqrt{n_1 n_2})$.

### 9.2.7 The Hodges–Lehmann Estimator

In Section 8.1.1 we have introduced the location-shift model in which the parameter $\Delta$ quantifies the shift. Although under the location-shift model assumptions the shift parameter may be well estimated as the difference in sample means or sample medians, it is good statistical practice to use an estimator which is closely related to the test statistic used for testing hypotheses about that parameter. If the WMW test is used, the *Hodges–Lehmann* estimator is the most natural choice. The rationale of this estimator is to find a shift, say $\hat{\Delta}$, so that when this $\hat{\Delta}$ is added to the observations in the second sample, the Wilcoxon rank sum statistic $U$ equals its mean under the null hypothesis, $n_2(n+1)/2$. Thus the $p$-value of the WMW test applied to this aligned dataset would be maximal. This Hodges–Lehmann estimator $\hat{\Delta}$ is usually calculated as

$$\hat{\Delta} = \text{median}\left(\{X_{1i} - X_{2j} : i = 1, \ldots, n_1; j = 1, \ldots, n_2\}\right);$$

i.e., it is the median of all $n_1 n_2$ differences between an observation from the first and an observation from the second sample. See Hodges and Lehmann (1983) for an overview of the applicability of their estimator.

The $1-\alpha$ confidence interval of $\Delta$ is also based on the Wilcoxon rank sum test (Bauer (1972)). Let $u_{\alpha/2}$ denote the $1-\alpha/2$ percentile of the exact null distribution of the Wilcoxon rank sum statistic $U$. Define

$$c_\alpha = \frac{n_2(n+1)}{2} + 1 - u_{\alpha/2}, \tag{9.12}$$

and let $D_{(1)}, \ldots, D_{(n_1 n_2)}$ denote the $n_1 n_2$ order statistics of the pairwise differences $X_{1i} - X_{2j}$ ($i = 1, \ldots, n_1; j = 1, \ldots, n_2$). The lower and upper limits are then given by $D_{(c_\alpha)}$ and $D_{(n_1 n_2 + 1 - c_\alpha)}$, respectively.

For large $n_1$ and $n_2$ a large sample approximation of the confidence interval may be considered. It is based on the normal approximation of the WMW test statistic. This only requires replacing (9.12) by

$$c_\alpha = \left\lfloor \frac{n_1 n_2}{2} - z_{\alpha/2}\sqrt{\frac{n_1 n_2(n+1)}{12}} \right\rfloor,$$

where $z_{\alpha/2}$ is the $1-\alpha/2$ percentile of the standard normal distribution, and $\lfloor x \rfloor$ denotes the largest integer not larger than $x$.

### 9.2.8 Examples

In this section we give three examples. In the first two examples the travel times of routes 2 and 3 are compared to the travel times with the reference route 1. These could be considered as didactical examples, as everything is very clear-cut. The final example in which the data of gene 3 are analysed, is a bit more elaborate. In all three examples we start off with presenting the results of the WMW tests (exact and asymptotic approximation), and by examining several distributional assumptions we formulate conclusions. Sometimes we also present results of the $t$-test. These examples are primarily meant as a didactical aid; in reality we usually do not start with presenting results of a test, but rather we first think about the problem, formulate hypotheses and assumptions, verify assumptions, and finally perform the appropriate test.

*Example 9.4 (The traffic data: Routes 1 and 2).* The next R code gives the exact and asymptotic WMW test results. For the first two WMW tests we have used the coin R package of Hothorn et al. (2006), as their methods allow for exact (or approximated exact) $p$-values for larger sample sizes. We also use the wilcox.test function in the standard R **stats** library.

```
> wilcox_test(time~route,data=traffic12,
+ distribution=approximate(B=10000),conf.int=T)

        Approximative Wilcoxon Mann-Whitney Rank Sum Test

data:  time by route (1, 2)
Z = 3.5919, p-value = 4e-04
alternative hypothesis: true mu is not equal to 0
95 percent confidence interval:
 0.38 1.15
sample estimates:
difference in location
                  0.81

> wilcox_test(time~route,data=traffic12,
+ distribution="asymptotic",conf.int=T)

        Asymptotic Wilcoxon Mann-Whitney Rank Sum Test

data:  time by route (1, 2)
Z = 3.5919, p-value = 0.0003283
alternative hypothesis: true mu is not equal to 0
95 percent confidence interval:
 0.3799401 1.1399433
sample estimates:
difference in location
             0.8099407
```

```
> wilcox.test(time~route,data=traffic12,conf.int=T)

        Wilcoxon rank sum test with continuity correction

data:  time by route
W = 1771, p-value = 0.0003327
alternative hypothesis: true location shift is not equal to 0
95 percent confidence interval:
 0.3799494 1.1399467
sample estimates:
difference in location
             0.8099852
```

At this point we still have not assessed any of the assumptions that may help fine-tuning the conclusions based on the WMW test, but according to Table 9.3 we may always interpret the $p$-value obtained from the exact (or Monte Carlo approximation) permutation WMW test, as this test does not require any assumption. Provided that the variance used in the normalisation of the WMW test statistic is appropriate, the conclusion from this test can be in terms of the probability $\Pr\{X_1 \leq X_2\}$, in which $X_1$ and $X_2$ denote random variables of the distributions of travel times with routes 1 and 2, respectively. From the first wilcox_test call, we read $p = 4 \times 10^{-4}$, by which we very convincingly reject the null hypothesis $H_0 : F_1 = F_2$ at the 5% level of significance. For assessing the correctness of the variance used in the traditional WMW statistic, we look at the output of the wmw.diagnose function.

**Table 9.3** Summary of the modes of useage of the WMW statistic. The first column (Asymp.) indicates whether the null distribution (Null distr.) is based on asymptotic theory. Details on the columns $\mathcal{F}_{NI}$ and $\mathcal{F}_{ND}$ are postponed to Section 9.3

| Asymp. | $H_0$ | $\mathcal{F}_{NI}$ | $\mathcal{F}_{ND}$ | $\hat{\sigma}^2$ | Null distr. | $H_1$ |
|---|---|---|---|---|---|---|
| no | $F_1 = F_2$ | | | $\sigma^2_{MW}$ | permutation | $\Pr\{X_1 \leq X_2\} \neq \frac{1}{2}$ |
| yes | $F_1 = F_2$ | | | $\sigma^2_{MW}$ | $N(0,1)$ | $\Pr\{X_1 \leq X_2\} \neq \frac{1}{2}$ |
| no | $\mu_1 = \mu_2(b)$ | $\mathcal{F}^{LS}_{NI}$ | | $\sigma^2_{MW}$ | permutation | $\mu_1 \neq \mu_2$ |
| yes | $\mu_1 = \mu_2(b)$ | $\mathcal{F}^{LS}_{NI}$ | | $\sigma^2_{MW}$ | $N(0,1)$ | $\mu_1 \neq \mu_2$ |
| yes | $\mu_1 = \mu_2(b)$ | $\mathcal{F}^{S}_{NI}$ | $\mathcal{F}^{SH}_{ND}$ | $\sigma^2_{MW}$ | $N(0,1)$ | $\mu_1 \neq \mu_2$ |
| yes | $\mu_1 = \mu_2(b)$ | $\mathcal{F}^{S}_{NI}$ | | $\hat{\sigma}^2_{FP}$ | $N(0,1)$ | $\mu_1 \neq \mu_2$ |
| yes | $\mu_1 = \mu_2$ | | $\mathcal{F}^{LSM}_{ND}$ | (a) | bootstrap | $\mu_1 \neq \mu_2$ |

($a$) the modified statistic of Babu and Padmanabhan (2002) has to be used.
($b$) the hypotheses may also be formulated in terms of $\Pr\{X_1 \leq X_2\}$.

```
> wmw.diagnose(time~route,data=traffic12)

Estimation of p112=Pr(max(X21,X22)<=X1) and
p112=Pr(max(X11,X12)<=X2), and Var(MW)

   p112 =  0.1467
   p221 =  0.5765
   Estimated Var(MW) =  0.002733231
   Null Var(MW) =  0.003366667
   Ratio Estimated / Null =  0.81

   WMW test may be too conservative
```

In the output we read the estimated variance of $\hat{\pi} = MW/(n_1 n_2)$ using the estimator in (9.19). The WMW statistic uses the variance $(n+1)/12n_1n_2$, which is true under the general two-sample null hypothesis, or, more generally, when $\Pr\{\max(X_{11}, X_{12}) \leq X_2\} = \Pr\{\max(X_{21}, X_{22}) \leq X_1\} = \frac{1}{3}$ and $\Pr\{X_1 \leq X_2\} = \frac{1}{2}$. The estimated variance is smaller than the null variance, thus the WMW test may be slightly conservative. However, when the null hypothesis is rejected, as it the case here, the conservativeness is not an issue anymore.

Because the WMW $p$-value is two-sided, we must still decide upon which of the two travel times is stochastically the best. This may be most conveniently done by estimating the probability $\Pr\{X_1 \leq X_2\}$. As the wilcox_test function gives the standardised Wilcoxon statistic, and the wilcox.test function gives the Mann–Whitney statistic, we use the latter to obtain the estimate

$$\widehat{\Pr}(X_1 \leq X_2) = 1 - \widehat{\Pr}(X_1 \geq X_2) = 1 - \frac{1771}{50 \times 50} = 0.2916.$$

Note that the wilcox.test function gives the Mann–Whitney statistic in terms of the count of the events $X_{1i} \geq X_{2j}$. This estimate, as well as an approximate 95% confidence interval may be computed using the following R code.

```
> pr12(time~route,data=traffic12)
Estimation of Pr(X1<=X2), and the Halperin-Mee confidence
   interval
Estimate =  0.2916
The  0.95  confidence interval:
0.2056955 0.3955165
```

From these results we may strongly conclude that it is much more likely to have a shorter travel time with route 2 than with route 1. As the $p$-value is based on the exact permutation distribution, and because the conclusion is formulated in terms of $\Pr\{X_1 \leq X_2\}$, of which the test statistic is a direct estimator, the formal correctness of the conclusion does not further depend on any assumptions. In the next paragraph we assess some of the assumptions of Table 9.3, so that perhaps we may also get to conclusions formulated in terms of other characteristics of the distributions $F_1$ and $F_2$.

From the QQ plots and the comparison distributions discussed in Examples 8.1 and 8.2, respectively, we have concluded that the distributions of the travel times with routes 1 and 2 differ only in location. $F_1$ and $F_2$ are thus of equal shape. Using this characteristic we may now also conclude that the mean travel time with route 2 is smaller than with route 1; the conclusion may also be formulated in terms of medians, or in terms of the location shift parameter $\Delta$. In the R output of the wilcox_test function we read the Hodges–Lehmann point estimate $\hat{\Delta} = 0.81$. The asymptotic 95% confidence interval is $[0.38, 1.14]$, and the exact (approximated using 10,000 Monte Carlo simulations) interval is $[0.38, 1.15]$. We may thus conclude that the mean driving time with route 2 is about 0.81 minutes ($\approx$49 seconds) faster than with route 1. More precisely, with a confidence of 95% we conclude that route 2 is about 0.38 to 1.15 minutes (23 to 69 seconds) faster than route 1.

Although Table 9.3 mentions that we may use both the exact permutation and the asymptotic null distribution, we still prefer the exact distribution, but in this example this would not have changed the conclusions.

Finally we note that the conclusion could also have been formulated in terms of a stochastic ordering. On the PP plot in Figure 8.1 we see that all points lie above the 45 degree line. At the rejection of the general two-sample null hypothesis we could thus also have concluded that $F_2(x) > F_1(x)$ for all $x$; i.e., $X_1$ is stochastically larger than $X_2$.

*Example 9.5 (The traffic data: Routes 1 and 3).* We start again by giving the results of the WMW test.

```
> traffic13<-traffic[traffic$route==1|traffic$route==3,]
> traffic13$route<-as.factor(as.numeric(traffic13$route))
> wilcox_test(time~route,data=traffic13,
+ distribution=approximate(B=10000),conf.int=T)

        Approximative Wilcoxon Mann-Whitney Rank Sum Test

data:  time by route (1, 3)
Z = -5.8254, p-value < 2.2e-16
alternative hypothesis: true mu is not equal to 0
95 percent confidence interval:
 -2.15 -1.19
sample estimates:
difference in location
               -1.69

> wilcox_test(time~route,data=traffic13,
+ distribution="asymptotic",conf.int=T)
```

```
          Asymptotic Wilcoxon Mann-Whitney Rank Sum Test

data:  time by route (1, 3)
Z = -5.8254, p-value = 5.697e-09
alternative hypothesis: true mu is not equal to 0
95 percent confidence interval:
 -2.169944 -1.170050
sample estimates:
difference in location
             -1.690062

> wilcox.test(time~route,data=traffic13,conf.int=T)

          Wilcoxon rank sum test with continuity correction

data:  time by route
W = 405, p-value = 5.816e-09
alternative hypothesis: true location shift is not equal to 0
95 percent confidence interval:
 -2.169942 -1.170048
sample estimates:
difference in location
             -1.690067

> pr12(time~route,data=traffic13,alpha=0.05)
Estimation of Pr(X1<=X2), and the Halperin-Mee confidence
   interval
Estimate =  0.838
The  0.95  confindence interval:
0.7509696 0.8987173
```

From the (approximate) exact test we again conclude a strong significant rejection of the general two-sample null hypothesis at the 5% level of significance, with an extremely small $p$-value. At this point we may only conclude that $F_1 \neq F_2$ at the 5% level of significance. We use the R function wmw.diagnose again to assess the appropriateness of the variance used in the WMW test statistic.

```
> wmw.diagnose(time~route,data=traffic13)

Estimation of p112=Pr(max(X21,X22)<=X1) and
p112=Pr(max(X11,X12)<=X2), and Var(MW)

  p112 =  0.743
  p221 =  0.0572
```

```
Estimated Var(MW) =  0.001574478
Null Var(MW) =  0.003366667
Ratio Estimated / Null =  0.47

WMW test may be too conservative
```

The estimated variance is again smaller than the null variance so that the WMW test is likely to be conservative, which, however, is no problem here because the null hypothesis is rejected. We may thus conclude that $\Pr\{X_1 \leq X_3\} \neq \frac{1}{2}$. The probability is now estimated as $1 - 405/(50 \times 50) = 0.838$ with an approximate 95% confidence interval of $[0.75, 0.90]$. This clearly says that it is much more likely to have a longer taxi ride with route 3 as compared to route 1. No assumptions were needed to state this conclusion formally.

From the discussions of QQ plots and comparison distributions in Figures 8.1 and 8.3, we have concluded that $F_1$ and $F_3$ differ probably only in location and scale. This means, using the terminology of Table 9.3 that the two distributions do not have the same shape, and that $\sigma_1^2 \neq \sigma_3^2$. At first sight, these characteristics do not correspond to any of the assumptions listed in Table 9.3. However, from the boxplots in Figure 6.2 we see that the distributions of the travel times with routes 1 and 3 are quite symmetric. According to Table 9.3, we might formulate a conclusion in terms of means if we would have used the modified test statistic in combination with its asymptotic null distribution. This is a nice example of the nonparametric Behrens–Fisher problem; see Section 9.3.3. From the output of the wmw.diagnose function, however, we have concluded that the use of the null variance is safe here. Although this argument is correct, it is not really the best statistical approach to the problem: a conservative test is safe, but the test consequently has less power than when the correct variance was used in the construction of the test statistic.

In this example we can arrive at a conclusion in terms of means quite directly by using the concept of stochastic ordering. The PP plot in Figure 8.1 shows that all points are at one side of the 45 degree line. In particular it suggests $F_3(x) < F_1(x)$ for all $x$. The WMW test thus also resulted in a rejection in favour of this alternative, which immediately also implies that $\mu_1 < \mu_3$. Note, however, that the WMW test does not formally test against stochastic ordering, but as the PP plot is very extreme (not one point at the other side), we are confident in this decision. Formal tests against stochastic ordering are usually based on test statistics closely related to the WMW test; see, e.g., Carolan and Tebbs (2005) for a discussion.

To conclude this example, we mention that it would also have been possible to apply an ordinary Welch $t$-test, because the data do not deviate strongly from normality.

*Example 9.6 (The gene expression data: Gene 3).* We first explore the data. Figure 9.1 shows the normal QQ plots of the expression values in the two groups, a two-sample QQ plot and the two boxplots. These exploratory

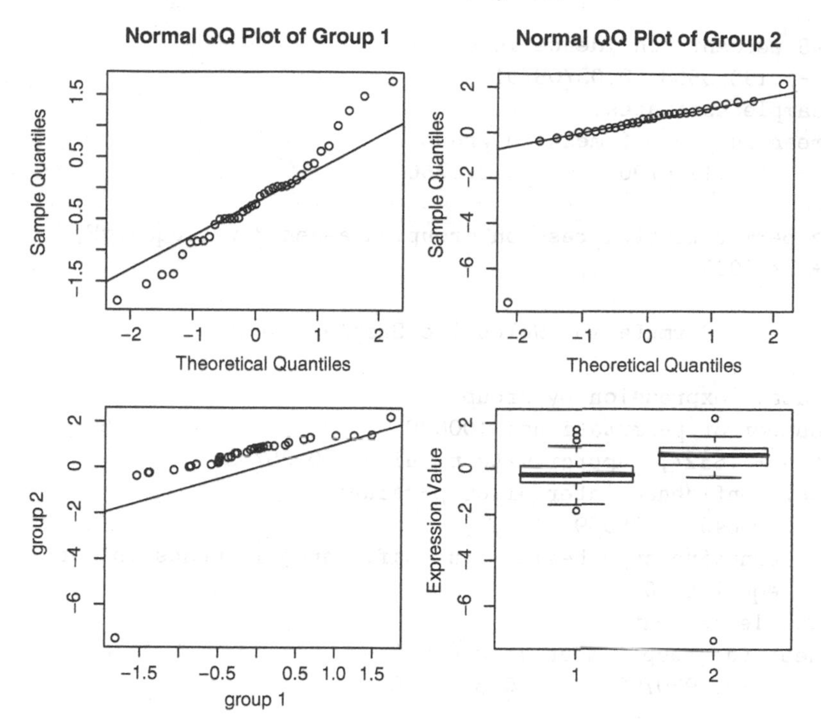


**Fig. 9.1** The normal QQ plots (top panels), two-sample QQ plot (bottom left), and the boxplots (bottom right) of the expression values in the two groups for gene 3

graphs demonstrate that the expression values in group 1 are not normally distributed, and that there is an outlier in group 2. At this point, however, there is no reason to remove the outlier from the dataset.

We now present the results of the statistical analysis, but this time we also have included the analysis with the parametric $t$-test and an exact permutation-based Welch $t$-test. The latter is also known as a *Studentised permutation test.* This test is not distribution free under the equal means null hypothesis, but Janssen and Pauls (2005), who studied the behavior of this test in a simulation study, concluded that in many situations its level is close to the nominal level. This test is implemented in the perm.t.test function in the cd package. We first look at the output of the $t$-tests.

```
> t.test(expression~group,data=gene3)

        Welch Two Sample t-test

data:  expression by group
t = -1.8327, df = 42.9, p-value = 0.07379
alternative hypothesis: true difference in means is not
   equal to 0
```

```
95 percent confidence interval:
 -1.19375868  0.05708701
sample estimates:
mean in group 1 mean in group 2
     -0.1930790       0.3752569

> perm.t.test(expression~group,data=gene3,var.equal=F,
+ B=10000)

        Permutation Welch Two Sample t-test

data:  expression by group
number of permutations: 100000
t = -1.8327,  approximate p-value = 0.05703
95% confidence interval of p-value:
  0.05846  0.05559
alternative hypothesis: true difference in means is not
   equal to 0
sample estimates:
mean in group 1 mean in group 2
     -0.1930790       0.3752569
```

From the output of the parametric Welch $t$-test, we read a $p$-value of 0.0738. As the data are not normally distributed we cannot thrust this value, particularly because it is close to the nominal significance level of 5%. The $p$-value of the permutation test version of the Welch test equals 0.057 and on using 100,000 random permutations, its 95% confidence interval does not include 5% so that we may be quite sure that the $p$-value is not smaller than the significance level. Thus, despite the rather small $p$-value, we may not formally reject the null hypothesis and conclude a difference in means. Because Figure 9.1 demonstrated the presence of an outlier, we have a strong belief that the $p$-value is influenced by this outlier. Indeed, when the outlier is removed the $p$-value becomes very much smaller (results not shown), but as there seems to be no good reason for believing that the outlier is a faulty observation, it may not be removed from the data and we have to stick to the larger $p$-values.

Next we present the results of the analysis with the WMW test. Because the WMW test is based on ranks, it is insensitive to outliers.

```
> wilcox_test(expression~group,data=gene3,
+ distribution=approximate(B=10000))

        Approximative Wilcoxon Mann-Whitney Rank Sum Test

data:  expression by group (1, 2)
Z = -4.082, p-value < 2.2e-16
```

```
alternative hypothesis: true mu is not equal to 0

> wmw.diagnose(expression~group,data=gene3)

Estimation of p112=Pr(max(X21,X22)<=X1) and
   p112=Pr(max(X11,X12)<=X2), and Var(MW)

  p112 =  0.6629
  p221 =  0.1194
  Estimated Var(MW) =  0.003322051
  Null Var(MW) =  0.005013078
  Ratio Estimated / Null =  0.66

  WMW test may be too conservative

> pr12(expression~group,data=gene3,alpha=0.05)
Estimation of Pr(X1<=X2), and the Halperin-Mee confidence
   interval
Estimate =  0.7890148
The  0.95  confindence interval:
0.6618551 0.8772263
```

The WMW $p$-value is now much smaller than 5% so that we may strongly reject the general two-sample null hypothesis. Moreover, because the estimated variance of the statistic is smaller than the null variance, we may again formulate the conclusion in terms of the probability $\Pr\{X_1 \leq X_2\}$, which is estimated as 0.789. Recalling that group 1 and group 2 correspond to the nonprogressed adenomas and the carcinomas, respectively, we conclude that it is much more likely that the carcinomas have a larger expression value for gene 3 as compared to the nonprogressed adenomas. Because the two-sample QQ plot in Figure 9.1 has all points except the outlier above the 45 degree line, it seems also quite safe to formulate the conclusion in terms of means.

## 9.3 The Diagnostic Property of Two-Sample Tests

In Part I of the book we introduced the *diagnostic property* in Section 4.2.1.3 for the simple null hypothesis, and later, in Section 4.5.6 for the composite null hypothesis. In this section we illustrate that much of the discussion we had about the interpretability of the WMW test in terms of the null and alternative hypotheses, can be clarified when the WMW test is looked at from a semiparametric perspective, similar to the approach taken in Section 4.5.

Many arguments given before lead to the conclusion that the hypotheses and the test statistic are often related so that the test statistic is a standardised estimator of the parameter involved in the hypotheses. Such parameters are often low-dimensional, and they usually summarise a few characteristics of the distributions of the observations, leaving infinitely many other parameters unspecified under the null hypothesis. The semiparametric null hypothesis is thus less restrictive than the general two-sample null hypothesis. In Section 4.5 we contrasted the semiparametric null hypothesis with the full parametric null hypothesis, which in the one-sample problem is indeed completely parametric, at most up to a $p$-dimensional nuisance parameter. In the two-sample problem, on the other hand, even the general two-sample null hypothesis, $H_0 : F_1 = F_2$, does not specify the distribution of the observations completely. It only restricts the two distribution functions to coincide. Of course, this is also a very stringent restriction on $F_1$ and $F_2$, involving infinitely many one-dimensional summarising parameters on $F_1$ and $F_2$. We demonstrate the semiparametric formulation in the next paragraphs. First we give a general formulation and next it is applied to the WMW setting.

### 9.3.1 The Semiparametric Framework

Let $\boldsymbol{X}_1^t = (X_{11}, \ldots, X_{1n_1})$ and $\boldsymbol{X}_2^t = (X_{21}, \ldots, X_{2n_1})$ denote two samples of i.i.d. observations with density functions $f_1$ and $f_2$, respectively. We also use the notation $X_1$ and $X_2$ to denote two independent random variables with density functions $f_1$ and $f_2$, respectively. Let $\{b_j\}$ $(j = 1, \ldots, k < \infty)$ represent a set of functions mapping $\mathcal{S} \times \mathcal{S}$ on $\mathbb{R}$. We further assume that $(f_1, f_2)$ is within a family $\mathcal{F} \subset L_2(\mathcal{S}, H) \times L_2(\mathcal{S}, H)$ so that the expectations and squared expectations of $b_j(X_1, X_2)$ exist and are finite (detailed conditions are omitted here). More important, the set $\mathcal{F}$ may also impose further restrictions on $f_1$ and $f_2$ (e.g., equal variances, or belonging to a location-shift model).

A semiparametric null hypothesis may be formulated in terms of the set

$$\mathcal{F}_0 = \{(f_1, f_2) \in \mathcal{F} : \mathrm{E}_{f_1 f_2} \{b_j(X_1, X_2)\} = \theta_{0j}, j = 1, \ldots, k\}, \qquad (9.13)$$

where $\theta_{01}, \ldots, \theta_{0k}$ are constants. In particular,

$$H_0 : (f_1, f_2) \in \mathcal{F}_0 \text{ versus } H_1 : (f_1, f_2) \in \mathcal{F} \setminus \mathcal{F}_0.$$

Thus $\mathcal{F}_0$ expresses both the assumptions on $f_1$ and $f_2$ and the null hypothesis. The latter is expressed here in terms of the $k$ functions in $\{b_j\}$ It is important to see that usually $\mathcal{F}_0$ is not uniquely characterised by a combination of $\mathcal{F}$ and $\{b_j\}$; i.e., $\mathcal{F}_0$ can have resulted from several combinations of sets of assumptions and null hypotheses.

Equation (9.13) suggests that the null hypothesis is naturally expressed as $H_0 : \mathrm{E}\{b_j(X_1, X_2)\} = \theta_{0j}$ $(j = 1, \ldots, k)$. This expectation often has an interpretation and can be represented by a parameter, say $\theta_j =$

$\mathrm{E}_{f_1 f_2}\{b_j(X_1, X_2)\}$, which typically has an interpretation in terms of how $f_1$ and $f_2$ compare to each other. However, when the null hypothesis is combined with the restrictions on $X_1$ and $X_2$ as specified in $\mathcal{F}$, the null hypothesis may sometimes be reformulated. For example, for the WMW test it is natural to take $b_1(x_1, x_2) = \mathrm{I}(x_1 \le x_2) - \frac{1}{2}$ so that $\mathrm{E}\{b_1(X_1, X_2)\} = 0$ gives $\Pr\{X_1 \le X_2\} = \frac{1}{2}$. However, when $\mathcal{F}$ restricts $f_1$ and $f_2$ to belong to a location-shift family, $\Pr\{X_1 \le X_2\} = \frac{1}{2}$ and $\Pr\{X_1 \le X_2\} \ne \frac{1}{2}$ are equivalent to $\mu_1 = \mu_2$ and $\mu_1 \ne \mu_2$, so that often the hypothesis is formulated using the means. We turn again to the more natural formulation of $H_0$ in terms of the $\theta_j$ parameters. Along the lines of Henze and Klar we say now that a test is *diagnostic* for testing $H_0 : \boldsymbol{\theta} = \mathbf{0}$ versus $H_1 : \boldsymbol{\theta} \ne \mathbf{0}$ when the test possesses the properties of asymptotically unbiasedness and consistency w.r.t. the semiparametric null and alternative hypotheses. Such tests may again be constructed as in Section 4.5.2. In particular, when the test statistic is constructed from an estimator of $\boldsymbol{\theta}^t = (\theta_1, \ldots, \theta_k)$, it often takes the form

$$T_n = n\left(\hat{\boldsymbol{\theta}} - \boldsymbol{\theta}_0\right)^t \hat{\boldsymbol{\Sigma}}^{-1}\left(\hat{\boldsymbol{\theta}} - \boldsymbol{\theta}_0\right), \tag{9.14}$$

where $\hat{\boldsymbol{\Sigma}}$ is an estimator of $\boldsymbol{\Sigma} = \mathrm{Var}\left\{\sqrt{n}\hat{\boldsymbol{\theta}}\right\}$. Its distribution theory uses the (asymptotic) multivariate normality of $\hat{\boldsymbol{\theta}}$. For many test statistics of this form the conditions for (asymptotic) unbiasedness and consistency often reduce to the following requirements.

1. For all $(f_1, f_2) \in \mathcal{F}_0$,

$$\lim_{n\to\infty}\left(\mathrm{E}_{f_1 f_2}\left\{\hat{\boldsymbol{\theta}} - \boldsymbol{\theta}_0\right\}\right) = \mathbf{0}$$

   and

$$\hat{\boldsymbol{\Sigma}} \text{ is a } \sqrt{n} \text{ consistent estimator of } \boldsymbol{\Sigma} \tag{9.15}$$

   (convergence in $P_{f_1 f_2}$ probablility).
2. For all $(f_1, f_2) \in \mathcal{F} \setminus \mathcal{F}_0$,

$$\lim_{n\to\infty} \inf_{(f_1,f_2)\in\mathcal{F}\setminus\mathcal{F}_0} \Pr_{f_1 f_2}\{T_n > c_\alpha\} = 1,$$

   where $c_\alpha$ denotes the asymptotic $\alpha$ level critical value of the test statistic $T_n$. For the class of tests that we consider here, this reduces to

$$\lim_{n\to\infty}\left(\mathrm{E}_{f_1 f_2}\left\{\hat{\theta}\right\} - \theta\right) \ne 0. \tag{9.16}$$

   Moreover, still for all $(f_1, f_2) \in \mathcal{F} \setminus \mathcal{F}_0$,

$$\hat{\boldsymbol{\Sigma}} \text{ is bounded in } \Pr_{f_1 f_2}\text{-probability.} \tag{9.17}$$

In the one-sample problem the functions $b_j$ (in Chapter 4 we used the notation $h_j$) depended only on one random observation variable, whereas here they are functions of $X_1$ and $X_2$. This makes the search for an appropriate variance estimator slightly more complicated.

### 9.3.2 Natural and Implied Null Hypotheses

Several times before we drew attention to the interplay among distributional assumptions, hypotheses, and null distributions. We make the discussion slightly more formal here. It is most natural to have a null hypothesis $\mathcal{F}_0$ as in (9.13) and a test statistic as in (9.14) that are both expressed in terms of the same parameter $\theta$. When this happens for a given $\mathcal{F}_0$ and $T_n$ combination we say that $\mathcal{F}_0$ is the *natural null hypothesis.*

It often occurs, however, that $\mathcal{F}_0$ together with some of the restrictions imposed by $\mathcal{F}$ allows for a reformulation of $\mathcal{F}_0$ so that it expresses restrictions on $f_1$ and $f_2$ in terms of different parameters. For example, Potthof (1963) showed that $\Pr\{X_1 \leq X_2\} = \frac{1}{2}$ is equivalent to $\mu_1 = \mu_2$ when $f_1$ and $f_2$ are symmetric. See also Hilgers (2007) for a more recent account. Thus, when these additional distributional assumptions are part of $\mathcal{F}$, the null hypothesis in $\mathcal{F}_0$ may just as well be reformulated in terms of the means $\mu_1$ and $\mu_2$, say

$$\mathcal{F}_0^{Imp} = \{(f_1, f_2) \in \mathcal{F} : \mathrm{E}_{f_1 f_2}\{X_1 - X_2\} = 0\}.$$

Such null hypotheses are referred to as *implied null hypotheses*, and the restrictions in $\mathcal{F}$ that allowed the step from the natural to the implied null hypothesis are collected in $\mathcal{F}_{NI} \subseteq \mathcal{F}$. The remaining restrictions in $\mathcal{F}$ (i.e., $\mathcal{F}_{ND} = \mathcal{F} \backslash \mathcal{F}_{NI}$) are the distributional assumptions required for a certain null distribution to be valid. For the WMW test, for example, $\mathcal{F}_{ND}$ may further specify that the variances of $f_1$ and $f_2$ coincide, so that asymptotically the WMW test statistic has a standard normal distribution. Instead of keeping the presentation general, we illustrate the semiparametric setting in the next section on the WMW test.

### 9.3.3 The WMW Test in the Semiparametric Framework

For a more systematic study of the applicability of the WMW test we proceed in two steps:

1. Which implied null hypothesis may be tested?
2. Which null distributions are appropriate?

These two steps go hand in hand with the distributional assumptions in $\mathcal{F}_{NI}$ and $\mathcal{F}_{ND}$.

#### 9.3.3.1 Implied Null Hypothesis

The Mann–Whitney form of the WMW test statistic is directly related to an estimator of $\theta = \pi = \Pr\{X_1 \leq X_2\}$; see Section 9.2.3. It is thus natural to think of the WMW test as a test for testing $\pi = 1/2(= \theta_0)$ versus $\pi \neq 1/2$. Because $\pi = \int_{\mathcal{S}} F_1(x)dF_2(x) = \int_{\mathcal{S}}\int_{\mathcal{S}} \mathrm{I}(x_1 \leq x_2)\, dF_1(x_1)dF_2(x_2)$ the null hypothesis of the semiparametric testing problem can be formulated as in (9.13) with $b_1(x_1, x_2) = \mathrm{I}(x_1 \leq x_2)$ and $\theta_0 = 0.5$. This is the natural null hypothesis, say $\mathcal{F}_0^{Nat}$, or $H_0^{Nat} : \pi = \frac{1}{2}$. Inasmuch this is a one-dimensional testing problem ($k = 1$ in $\mathcal{F}_0$) we consider a (not squared) test statistic of the form

$$T_n = \frac{\hat{\pi} - \frac{1}{2}}{\hat{\sigma}} \tag{9.18}$$

in which $\hat{\pi} = MW/(n_1 n_2)$. The traditional WMW test uses an estimator $\hat{\sigma}^2$ that is a consistent estimator of $\mathrm{Var}\{\hat{\pi}\}$ under the general two-sample null hypothesis, in which case it reduces to $(n+1)/12n_1n_2$. Thus the question may arise whether $\hat{\sigma}_n^2$ is consistent under a set of weaker assumptions. This is demonstrated in the next section.

The WMW test is still often introduced as the nonparametric test for testing $H_0 : \mu_1 = \mu_2$ versus $H_1 : \mu_1 \neq \mu_2$ when the normality assumption is violated. These hypotheses may be implied when the natural null hypothesis is combined with some distributional assumptions in $\mathcal{F}_{NI}$

- The distributions $f_1$ and $f_2$ belong to a location-shift model; i.e.,

$$\mathcal{F}_{NI}^{LS} = \{(f_1, f_2) : f_1(x) = f_2(x - \Delta) \text{ for all } x \in \mathcal{S} \text{ and } \Delta \in \mathbb{R}\}.$$

- The distributions $f_1$ and $f_2$ are symmetric. Let $m_1$ and $m_2$ denote the medians of $f_1$ and $f_2$, respectively. Then

$$\mathcal{F}_{NI}^{S} = \{(f_1, f_2) : f_1(m_1 - x) = f_1(m_1 + x) \text{ and } f_2(m_2 - x) = f_2(m_2 + x) \text{ for all } x \in \mathcal{S}\}.$$

Note that in the restrictions of the set $\mathcal{F}_{NI}^{S}$ the means can be replaced by the medians.

The general two-sample null hypothesis may also be implied. This happens, for example, with $\mathcal{F}_{NI}^{LS}$.

#### 9.3.3.2 Null Distributions

The simplest situation arises when the general two-sample null hypothesis is implied. Then Section 9.2.4 may be consulted. Both the exact permutation null distributions and the standard normal distribution (for large sample sizes) are valid with (9.18) with $\hat{\sigma}^2 = (n+1)/12n_1n_2$. Obviously the exact permutation null distribution may also be used in conjunction with the general two-sample null hypothesis. The standard normal distribution, however, is also valid for the natural and the implied null hypotheses, provided the variance estimator in (9.18) is consistent under the null hypothesis. This may ask for some additional assumptions, specified in $\mathcal{F}_{ND}$.

It is instructive to start the discussion with the general expression of $\text{Var}\{\hat{\pi}\}$, which has been given in (9.7). This expression shows that the variance $\hat{\sigma}^2 = (n(n+1))/12n_1n_2$, which is also used in (9.4) and (9.5), is also appropriate when

$$\phi_1^2 = \phi_2^2 = \frac{1}{12},$$

under the null hypothesis. This happens when $\Pr\{\max(X_{11}, X_{12}) \leq X_2\} = \Pr\{\max(X_{21}, X_{22}) \leq X_1\} = \frac{1}{3}$. For the standardised WMW statistic to have a proper distribution it must further be avoided that $\phi_1^2 = \phi_2^2 = 0$, but because this only occurs when $\text{Var}_{f_2}\{F_1(X_2)\} = \text{Var}_{f_1}\{F_2(X_1)\} = 0$, we further ignore this situation.

A sufficient condition for using this variance estimator is therefore

$$\begin{aligned}\mathcal{F}_{ND} = \Big\{(f_1, f_2) &: \Pr\{\max(X_{11}, X_{12}) \leq X_2\} \\ &= \Pr\{\max(X_{21}, X_{22}) \leq X_1\} = \frac{1}{3}\Big\}.\end{aligned}$$

However, it is often more convenient to describe other characteristics of $f_1$ and $f_2$ in $\mathcal{F}_{ND}$ that combined with the restrictions in $\mathcal{F}_0$ (or $\mathcal{F}_0^{Imp}$) guarantee that $\phi_1^2 = \phi_2^2 = \frac{1}{12}$. We give two examples:

- The distributions $f_1$ and $f_2$ belong to a location-shift model; i.e., $\mathcal{F}_{ND}^{LS} = \mathcal{F}_{NI}^{LS}$.
- The distributions $f_1$ and $f_2$ are symmetric with equal variances; i.e.,

$$\mathcal{F}_{ND}^{SH} = \left\{(f_1, f_2) \in \mathcal{F}_{NI}^{S} : \text{Var}_{f_1}\{X_1\} = \text{Var}_{f_2}\{X_2\}\right\}.$$

These two examples demonstrate clearly that the restrictions in $\mathcal{F}_{NI}$ and $\mathcal{F}_{ND}$ are not always mutually exclusive.

When we are not willing to assume the location-shift model, or symmetric distributions $F_1$ and $F_2$ with equal variances, the traditional MW test statistic of (9.4) is not appropriate. Some authors have referred to this less restrictive setting as the *nonparametric Behrens–Fisher problem*. The WMW test is in this situation, however, not valid anymore, but a Welch-type correction can be made by replacing the variance of $MW$ in the denominator of the standardised test statistic (9.5) by an estimator that is consistent under less restrictive assumptions.

Fligner and Policello (1981) proposed an estimator of $\text{Var}\{MW\}$ that is consistent under mild assumptions. This estimator is basically a plug-in estimator based on expression (9.7), obtained by replacing the unknown CDFs $F_1$ and $F_2$ by their empirical counterparts $\hat{F}_{1n_1}$ and $\hat{F}_{2n_2}$. Their variance estimator can be most conveniently expressed in terms of *placements*. The placements of the observations $X_{1i}$ are defined as $P_{1i} = n_2 \hat{F}_{2n_2}(X_{1(i)})$, and those of $X_{2i}$ are defined as $P_{2i} = n_1 \hat{F}_{1n_1}(X_{2(i)})$; i.e., $P_{1i}$ is the number of observations in the second sample that are not larger than $X_{1(i)}$, and, similarly, $P_{2i}$ is the number of observations in the first sample that are not larger than $X_{2(i)}$. Their variance estimator is

$$\frac{1}{n_1 n_2}\left(\frac{n_1-1}{n_1}\sum_{i=1}^{n_2}\left(P_{2i}-\bar{P}_2\right)^2+\frac{n_2-1}{n_1}\sum_{i=1}^{n_1}\left(P_{1i}-\bar{P}_1\right)^2+\frac{\bar{P}_1\bar{P}_2}{n_1}\right), \quad (9.19)$$

where $\bar{P}_1$ and $\bar{P}_2$ are the averages of the first and the second set of placements. Fligner and Policello (1981) proved that with this estimator the standardised $MW$ statistic has asymptotically a standard normal distribution under the natural null hypothesis, assuming that $F_1$ and $F_2$ are both symmetric, i.e., when $(f_1, f_2) \in \mathcal{F}_{ND}^S = \mathcal{F}_{NI}^S$. When the symmetry assumption is dropped, its limiting distribution holds under $H_0 : \Pr\{X_1 \leq X_2\} = \frac{1}{2}$. The test based on this appropriately standardised statistic is again consistent.

Suppose the variance estimator of Fligner and Policello (1981) is used and that the symmetry assumption does not hold, but we still want to test $H_0 : \mu_1 = \mu_2$. Then $\mu_1 = \mu_2$ does not imply $\Pr\{X_1 \leq X_2\} = \frac{1}{2}$, and $\theta_n = \frac{1}{2}$ should be replaced by an estimator of $\Pr\{X_1 \leq X_2\}$, say $\hat{\theta}_n$, that is consistent under the null hypothesis $\mu_1 = \mu_2$. This is clearly not an obvious problem, because the null hypothesis only imposes a very mild restriction on the densities $f_1$ and $f_2$ which makes the relation to the probability $\Pr\{X_1 \leq X_2\}$ not straightforward. Babu and Padmanabhan (2002) proposed a bootstrap procedure, which, on the one hand, works indeed without the symmetry assumption, but requires the location-scale assumption; i.e.,

$$\mathcal{F}_{ND}^{LSM} = \left\{(f_1, f_2) : f_1(x) = f\left(\frac{x-\mu_1}{\sigma_1}\right)\right.$$

and

$$\left. f_2(x) = f\left(\frac{x-\mu_2}{\sigma_2}\right) \text{ for all } x \in \mathcal{S}\right\},$$

with $f$ some arbitrary continuous zero-mean density function. The test of Babu and Padmanabhan (2002), however, does not really fit nicely within the framework presented here, because its test statistic is not of the form (9.14) anymore, so that the relation between the natural and the imposed hypotheses is disrupted.

The discussions on how the WMW test can be used and/or adopted to test for other null hypotheses illustrates the interplay among the hypotheses, the assumptions, and the null distribution that has to be used for $p$-value calculations. In the examples in Section 9.2.8 we also demonstrate how assumptions, such as symmetry and equal shape, can be assessed.

As a side remark we mention that the variance estimator of Fligner and Policello [1981] is not the only possibility. For example, Zaremba (1962) suggested estimating the variance by replacing the probabilities $\Pr\{\max(X_{11}, X_{12}) \leq X_2\}$ and $\Pr\{\max(X_{21}, X_{22}) \leq X_1\}$ by their direct estimators; as in (9.11). In the next section we come back to variance estimation in a more general setting. From a small sample simulation study, however, Fligner and Policello [1981] concluded that their estimator results in a better-behaving test.

Finally, we conclude the discussion on the WMW with a summary of the different situations in which the test can be used. This is presented in Table 9.3.

In this section the variance estimator of Fligner and Policello (1981) was used as the least restrictive estimator. This expression of the variance was derived directly from the simple form of $\hat{\pi}$ or MW (Equation (9.2)). It is, however, at this point of interest to know a method that is more generally applicable so that we may use it for other linear rank tests as well. Details are given in the next section.

### *9.3.4 Empirical Variance Estimators of Simple Linear Rank Statistics*

We present here two general methods for obtaining consistent estimators of the variance of simple linear rank statistics. Simple linear rank statistics were introduced in Section 7.2.1 and they have the form

$$T_n = \sum_{i=1}^{n} c_i a_n(R_i),$$

where the regression constants $\{c_i\}$ and the score function $a_n$ are as given in Definition 7.3.

The estimators that we consider here are consistent under many semiparametric null hypotheses. The first method is based on comparison distribution processes, and the second approach uses the jackknife principle. The former method results in the expression of the asymptotic variance in which all unknown population parameters are subsequently replaced by their sample estimators so as to find a computational form of a consistent variance estimator. The latter approach avoids the calculation of an explicit expression of a variance estimator.

#### 9.3.4.1 The Asymptotic Variance of a Simple Linear Rank Statistic

Many of the simple linear rank statistics for the two sample problem have regression constants defined as $c_i = 1/n_1$ when observation $i$ comes from the first sample, and $c_i = 0$ or $c_i = -1/n_2$ when observation $i$ comes from the second sample (or visa versa). The test statistic $T_n$ decomposes thus into two terms, say $(1/n_1)\sum_{i=1}^{n_1} a(R_i)$ and $(1/n_2)\sum_{i=n_1+1}^{n} a(R_i)$ in which the index $i$ refers to the pooled sample observations $Z_1, \ldots, Z_n$. In this section we focus first on one such term.

On using the comparison distribution process of Section 7.5.2 we may write

$$\frac{1}{n_1}\sum_{i=1}^{n_1} a(R_i) = \int_{\mathcal{S}} a(n\hat{H}(z))d\hat{F}_1(z).$$

We rewrite this expression so that the comparison distribution process $\mathbb{C}_{n1}$ of (7.26) can be recognised. We use the notation $a^*(p) = a(np)$. In the following calculations we have used twice integration by parts.

$$\begin{aligned}
\frac{1}{n_1}\sum_{i=1}^{n_1} a(R_i) &= \int_{\mathcal{S}} a^*(\hat{H}(z))d\hat{F}_1(z) \\
&= \int_0^1 a^*(p)d\hat{F}_1(\hat{H}^{-1}(p)) \\
&= a^*(1) - \int_0^1 \hat{F}_1(\hat{H}^{-1}(p))da^*(p) \\
&= a^*(1) - \int_0^1 \hat{F}_1(\hat{H}^{-1}(p))a'(p)dp \\
&= a^*(1) - \int_0^1 \left(\hat{F}_1(\hat{H}^{-1}(p)) - F_1(H^{-1}(p))\right) a'(p)dp \\
&\quad - \int_0^1 F_1(H^{-1}(p))a'(p)dp \\
&= \int_0^1 a^*(H(F_1^{-1}(p))dp - \frac{1}{\sqrt{n}}\int_0^1 \mathbb{C}_{n1}(p)a'(p)dp,
\end{aligned}$$

where $a'(p) = da^*(p)/dp$. From this last expression, the weak convergence of $\mathbb{C}_{n1}$ according to Theorem 7.6, and the continuous mapping theorem, the variance may be found immediately. We state the result in the following lemma.

**Lemma 9.1.** *Consider a simple linear rank statistic of the form* $S_n = (1/n_1)\sum_{i=1}^{n_1} a(R_i)$. *Then, for all* $(f_1, f_2) \in \mathcal{F}$,

$$\begin{aligned}
&\lim_{n\to\infty} \text{Var}_{f_1 f_2}\left\{\sqrt{n}S_n\right\} \\
&= \int_0^1\int_0^1 a'(s)a'(t)\,\text{Cov}_{f_1 f_2}\left\{\mathbb{C}_1(s),\mathbb{C}_1(t)\right\} dsdt \\
&= 2(1-\lambda)\left\{\frac{1-\lambda}{\lambda}\int\int_{x<y} a'(H(x))a'(H(y))F_1(x)(1-F_1(y))dF_2(x)dF_2(y)\right. \\
&\quad \left. + \int\int_{x<y} a'(H(x))a'(H(y))F_2(x)(1-F_2(y))dF_1(x)dF_1(y)\right\}.
\end{aligned}$$

Back to the simple linear rank statistic $T_n$. Because $H(z) = \lambda F_1(z) + (1-\lambda) F_2(z)$, we have the equality (integration by parts)

$$\int_{\mathcal{S}} a^*(\hat{H}(z))d\hat{F}_2(z) = -\frac{\lambda}{1-\lambda}\int_{\mathcal{S}} a^*(\hat{H}(z))d\hat{F}_1(z).$$

Thus when $c_i = 1/n_1$ and $c_i = -1/n_2$ for observation $i$ coming from the first or the second sample, respectively, the simple linear rank statistic may be written as

$$T_n = \frac{1}{1-\lambda}\int_{\mathcal{S}} a^*(\hat{H}(z))d\hat{F}_1(z), \tag{9.20}$$

and the asymptotic variance of $\sqrt{n}T_n$ follows again from Lemma 9.1.

A consistent estimator of the asymptotic variance may be obtained by replacing the double integration by an average, and using the EDFs for the unknown distribution functions. The result is summarised in the following lemma for the $S_n$ statistic.

**Lemma 9.2.** *Consider a simple linear rank statistic of the form* $S_n = (1/n_1)\sum_{i=1}^{n_1} a(R_i)$. *Let* $X_{i(1)} < X_{i(2)} < \cdots < X_{i(n_i)}$ *denote the order statistics of the i the sample observations* ($i = 1, 2$). *Then, for all* $(f_1, f_2) \in \mathcal{F}$, *the asymptotic variance of* $\sqrt{n}S_n$ *is consistently estimated by* $\hat{\sigma}^2_{\text{cd}}$ *which is given by*

$$2\frac{n_2}{n}\left\{\frac{n_2}{n_1}\frac{1}{\frac{1}{2}n_2(n_2-1)}\sum_{i=1}^{n_2-1}\sum_{j=i+1}^{n_2} a'(\hat{H}(X_{2(i)}))a'(\hat{H}(X_{2(j)}))\right.$$
$$\times\hat{F}_1(X_{2(i)})(1-\hat{F}_1(X_{2(j)})) + \frac{1}{\frac{1}{2}n_1(n_1-1)}$$
$$\left.\times\sum_{i=1}^{n_1-1}\sum_{j=i+1}^{n_1} a'(\hat{H}(X_{1(i)}))a'(\hat{H}(X_{1(j)}))\hat{F}_2(X_{1(i)})(1-\hat{F}_2(X_{1(j)}))\right\}.$$

#### 9.3.4.2 The Jackknife Estimator of the Asymptotic Variance

Shao (1988) and Shao (1993) studied jackknife variance estimators of linear rank statistics, and they showed that under mild regularity conditions on the score function this variance estimator is consistent. The estimator is computed as follows. Let $T_n(i)$ denote the simple linear rank statistic $T_n$ based on all pooled sample observations except $Z_i$ ($i = 1, \ldots, n$). The jackknife variance estimator is given by

$$\hat{\sigma}^2_j = (n-1)\sum_{i=1}^{n}\left\{T_n(i) - \frac{1}{n}\sum_{i=1}^{n} T_n(i)\right\}^2.$$

This estimator has the advantage that it is generally easily computable, because it basically only requires the formula of the test statistic. Note that although the formula of $\hat{\sigma}_j^2$ above shows only one summation, it also contains summations in the expression of $T_n(i)$. The computational burden is therefore of the same complexity as for the estimator of the previous section ($\hat{\sigma}_{cd}^2$).

## 9.4 Optimal Linear Rank Tests for Normal Location-Shift Models

In Section 9.2.5 we have demonstrated that the WMW test is the LMPRT for detecting a location shift when the distributions $F_1$ and $F_2$ are logistic distributions. In this section we give some other LMPRT for other location-shift models. In general the two-sample location-shift alternative, $f_1(x) = f_2(x - \Delta)$, corresponds to (7.13) with $f$ the assumed distribution so that we can write $f_1(x) = f(x)$ and $f_2(x - \Delta) = f(x - \Delta)$, and thus $c_{ji} = 0$ for $j = 1$ and $i = 1, \ldots, n_1$, and $c_{ji} = 1$ for $j = 2$ and $i = 1, \ldots, n_2$. The optimal scores, as given in Theorem 7.5, are defined as the expectations of

$$\frac{\frac{\partial}{\partial \Delta} f(X_{(i)} - \Delta)\big|_{\Delta=0}}{f(X_{(i)})} = \frac{\frac{\partial}{\partial z} f(z) \frac{\partial z}{\partial \Delta}\big|_{\Delta=0, z=X_{(i)}}}{f(X_{(i)})} = -\frac{\frac{\partial}{\partial x} f(x)\big|_{x=X_{(i)}}}{f(X_{(i)})},$$

with $X_{(i)}$ the order statistics of a sample of $n$ i.i.d. random variates with density function $f$. Using $f'(x) = \partial f(x)/\partial x$, we get

$$a_n(i) = \mathrm{E}_f\left\{-\frac{f'(X_{(i)})}{f(X_{(i)})}\right\}.$$

When $f$ is the standard normal density, $-f'(x)/f(x) = x$, and thus

$$a_n(i) = \mathrm{E}_f\left\{X_{(i)}\right\} = \mathrm{E}\left\{\Phi^{-1}(U_{(i)})\right\},$$

with $U_{(i)}$ the order statistics of $n$ i.i.d. uniform variates, and $\Phi^{-1}$ the quantile function of a standard normal distribution. The resulting linear rank test is known as the *Fisher–Yates–Terry–Hoeffding* or *normal scores* test. The expectations, however, do not have a simple expression, but they are tabulated in, e.g., Pearson and Hartley (1972). Instead of using the exact optimal scores, defined as expectations, van der Waerden (1952, 1953) suggested using the approximate scores (see Section 7.2.1.3),

$$a_n(i) = \Phi^{-1}\left(\frac{i}{n+1}\right),$$

**Table 9.4** Some AREs of the van der Waerden test versus the WMW for several distributions $f$

| $f$ | Normal | Uniform | Logistic | Double exponential | Cauchy | Exponential |
|---|---|---|---|---|---|---|
| ARE | 1.047 | $\infty$ | 0.955 | 0.847 | 0.708 | $\infty$ |

which are directly available in almost all software packages. The linear rank statistic is thus

$$T_n = \sum_{i=1}^{n_2} \Phi^{-1}\left(\frac{R(X_{2i})}{n+1}\right)$$

with mean zero and variance

$$\text{Var}\{T_n\} = \frac{n_1 n_2}{n(n-1)} \sum_{j=1}^{2} \sum_{i=1}^{n_j} \left(\Phi^{-1}\left(\frac{i}{n+1}\right)\right)^2.$$

The test based on these approximate scores is known as the *van der Waerden* test. Note that just as with the WMW test, this variance estimator is a consistent estimator of the true variance under the general two-sample null hypothesis.

It is interesting to investigate how the van der Waerden test compares to the more popular WMW test. Table 9.4 shows the AREs of the van der Waerden test relative to the Wilcoxon test for several location-shift models. As expected, the van der Waerden test is more powerful under a normal model, but only marginally (ARE $= 1.047$). These AREs further suggest that the van der Waerden scores result in more powerful tests for location shifts in distributions with a finite upper or lower bound in their support. For example, the uniform has support $[0, 1]$, and the exponential $[0, +\infty]$. Chernoff and Savage (1958) compared the van der Wearden test with the two-sample $t$-test and they concluded that the ARE is never smaller than 1, and it equals 1 for the normal shift model. This heavily supports the use of the van der Waerden rank test over the two-sample $t$-test!

## 9.5 Rank Tests for Scale Differences

The discussions on the Wilcoxon rank sum test of Sections 9.2 and 9.3 tried to stress that the WMW test may be used for testing several hypotheses. Most important, we have demonstrated that in its most nonparametrical nature, it tests the general two-sample null hypothesis versus the alternative that $\Pr\{X_1 \leq X_2\} \neq \frac{1}{2}$, and in its most *natural* usage it tests $H_0 : \Pr\{X_1 \leq X_2\} = \frac{1}{2}$. On the other hand, most contributions to the theory of the WMW test concern the use of the WMW test for testing against location-shift alternatives. In particular, the WMW test is the LMPRT for a location shift under the parametric assumption of logistic distributions. Similarly, the van der Waerden test (Section 9.4) is the LMPRT for the normal shift model. Despite the optimality properties of these rank tests in location-shift models, we prefer to present rank tests in their most nonparametrical nature, or in a semiparametric setting and look for null hypotheses and asumptions for which the test is (at least asymptotically) diagnostic. In the semiparametric setting of Section 9.3, which is particularly useful for studying the diagnostic property we have seen that several combinations of null hypothesis formulations and sets of distributional assumptions may lead to the same testing situation. It is therefore interesting to investigate these combinations more closely. We keep this kind of presentation in this section too. In this sense the title of this section may not be the most accurate; the tests presented here are often optimal for testing for scale differences, however, only if additional assumptions are imposed.

As rank tests for scale differences usually do not work very well unless very stringent hard-to-assess assumptions are made on $F_1$ and $F_2$, we do not give too many details. We restrict our atttention to methods and test statistics that are useful for understanding the methods presented in the next chapters. In Section 9.5.1 we first briefly introduce the scale difference model, and in Section 9.5.2 some related LMPRT tests. The more non- or semiparametric discussion follows in Section 9.5.3.

### 9.5.1 The Scale-Difference Model

The two-sample scale-difference model may be formulated using the notation introduced in Section 7.2.2 for a general one-parameter alternative to the general two-sample null hypothesis. Consider

$$f_1(x) = \exp(-\Delta) f_2(x \exp(-\Delta)) \text{ or } F_1(x) = F_2(x \exp(-\Delta)), \qquad (9.21)$$

where $f_1 = f$ is some known density function, and $\Delta > 0$. Note that the factor $\exp(-\Delta)$ in front of $f_2$ is required to make the right-hand side of the equation integrate properly to one. The parameter $\Delta$ is introduced through $\exp(-\Delta)$ so that $\Delta = 0$ corresponds to the null hypothesis. This corresponds to a linear rank statistic with $c_{1i} = 0$ for $i = 1, \ldots, n_1$, and $c_{2i} = 1$ for $i = 1, \ldots, n_2$. According to Theorem 7.5, the optimal scores are given by

$$a_n(i) = \mathrm{E}_f \left\{ \frac{\frac{\partial}{\partial \theta} f(X_{(i)}; \theta)\big|_{\theta=0}}{f(X_{(i)})} \right\} = \mathrm{E}_f \left\{ -1 - X_{(i)} \frac{f'(X_{(i)})}{f(X_{(i)})} \right\}, \qquad (9.22)$$

where $f'(x) = df(x)/dx$. With these scores the LMTRT statistic for the scale difference model with $f_1 = f$ has the form

$$T_n = \sum_{i=1}^{n_2} a_n(R(X_{2i})).$$

Note that it is assumed in (9.21) that the medians of the two populations are equal. The scale difference model (9.21) is very unambiguous concerning the interpretation of what is exactly meant by a scale difference. Moreover, the scale-difference is explicitly measured by $\Delta$. However, when this model does not hold, i.e., when $F_1$ and $F_2$ do not only differ by the factor $\exp(-\Delta)$ in their argument, we might need a more general understanding of scale differences. In Section 9.5.3 we approach this testing problem from a completely different viewpoint by defining scale differences in terms of probabilities involving stochastic orderings.

van Eeden (1964) gave two conditions that must be satisfied by a measure for scale difference.

C1 The difference in scale is invariant under the substitution of $X_i$ with $-X_i$ ($i = 1, 2$).

C2 If $X_1$ has a larger (equal or smaller) dispersion than $X_2$, then $-X_1$ has a larger (equal or smaller) dispersion than $-X_2$.

Finally we mention that the location-shift and the scale-difference models can be combined into a *location-scale-difference model*,

$$F_1\left(\frac{x-\mu_1}{\sigma_1}\right) = F_2\left(\frac{x-\mu_2}{\sigma_2}\right), \tag{9.23}$$

for all $x$, and with obvious notation.

### 9.5.2 The Capon and Klotz Tests

The Capon test is the scale-test version of the normal scores test for a location shift under a normality assumption. It is a straightforward application of the optimal scores in (9.22), resulting in

$$a_n(i) = \mathrm{E}\left\{X_{(i)}\right\}^2 = \mathrm{E}\left\{\Phi^{-1}(U_{(i)})\right\}^2,$$

where the $X_{(i)}$ and the $U_{(i)}$ are the order statistics of $n$ i.i.d. observations from a normal and uniform distribution, respectively. The Klotz test replaces these scores with the approximate scores

$$a_n(i) = \left[\Phi^{-1}\left(\frac{i}{n+1}\right)\right]^2.$$

More details may be found in Section 4.2 of Hájek et al. (1999). This test is the LMPRT in this very parametric setting.

## 9.5.3 Some Other Important Tests

### 9.5.3.1 Measures for Differences in Scale

There exists a vast literature on statistical tests for the two-sample scale problem, though not as extended as for the location-shift problem. Most of these tests are motivated from an optimality viewpoint, as, e.g., the Capon test, or they are based on a test statistic that has an intuitively appealing form for detecting differences in scale. The popular Ansari–Bradley, Siegel–Tukey, and Mood tests, among others, belong to this last group. In this section we present some of these tests in an unconventional manner. Instead of building up arguments for proving that the test statistics are appropriate for testing $\Delta = 0$ in the scale-difference model (9.21), we start with considering the test statistic as an estimator of a population parameter of which we further investigate its interpretability as a scale-difference measure. As before, we consider test statistics of the form

$$T_n = \sqrt{n}\frac{\hat{\theta} - \theta_0}{\hat{\sigma}}$$

so that the natural null hypothesis is formulated as $H_0 : \theta = \theta_0$, where $\theta = \mathrm{E}\left\{\hat{\theta}\right\}$ (or at least in the limit as $n \to \infty$). The natural alternative hypothesis is then $H_1 : \theta \neq \theta_0$ (or one-sided). When $T_n$ can be used for detecting differences in scales of $f_1$ and $f_2$, we would expect that $\theta - \theta_0$, or at least

$$\delta = \lim_{n\to\infty} \left\{\mathrm{E}_{f_1 f_2}\left\{\hat{\theta}\right\} - \theta_0\right\} \tag{9.24}$$

is related to an appropriate measure for difference in scale. When $\hat{\sigma}^2$ satisfies the conditions (9.15) and (9.17), then the test based on $T_n$ is consistent for $\delta \neq 0$.

As some of the important two-sample tests for scale differences are related to the same measure for scale differences, we discuss the interpretation of these measures first.

1. Suppose the medians of $X_1$ and $X_2$ are known and equal to $m_1$ and $m_2$, respectively. Consider

$$\pi^{(1)} = \Pr\left\{|X_1 - m_1| < |X_2 - m_2|\right\}.$$

Under the two-sample null hypothesis this probability equals $\frac{1}{2}$. When the location-scale model (9.23) holds $\pi^{(1)} = \frac{1}{2}$, $< \frac{1}{2}$ or $> \frac{1}{2}$ is equivalent to $\sigma_1 = \sigma_2$, $\sigma_1 > \sigma_2$, and $\sigma_1 < \sigma_2$. But even when the location-scale model does not hold, the probability $\pi^{(1)}$ has a straightforward interpretation as a dispersion difference measure. It basically describes a difference in dispersion as the stochastic ordering of the absolute deviations of an observation from its median. van Eeden (1964) showed that it satisfies conditions C1 and C2.

2. The following probability does not involve the medians. Let $X_{11}$ and $X_{12}$ i.i.d. $F_1$, and $X_{21}$ and $X_{22}$ i.i.d. $F_2$. Define
$$\pi^{(2)} = \Pr\{|X_{11} - X_{12}| < |X_{21} - X_{22}|\}.$$
This again measures a dispersion difference as a likely ordering, but now related to the absolute deviations between two observations from the same distribution. When $F_1 = F_1$, $\pi^{(2)} = \frac{1}{2}$. A large $\pi^{(2)}$ indicates that the observations from $F_2$ are more "dispersed" among each other than those from $F_1$. Another way to phrase this is to say that it is more likely to find two $F_1$ observations close to each other, than if the observations came from $F_2$. Conditions C1 and C2 are satisfied for $\pi^{(2)}$.
3. Suppose the medians $m_1$ and $m_2$ coincide, and let $m = m_1 = m_2$. Consider
$$\pi^{(3)} = \Pr\{X_2 \leq X_1 \leq m\} + \Pr\{m \leq X_1 \leq X_2\}, \tag{9.25}$$
which equals $\frac{1}{4}$ under the two-sample null hypothesis. When $\pi^{(3)}$ is larger, it indicates that the $X_1$ observations tend to be clustered closer around the median than the $X_2$ observations. This again quantifies an aspect of dispersion differences between $F_1$ and $F_2$. However, van Eeden (1964) showed that $\pi^{(3)}$ does not satisfy C2 when both $F_1$ and $F_2$ are asymmetric distributions. He also showed that $\pi^{(3)} - \frac{1}{4} = \pi^{(1)} - \frac{1}{2}$ when $F_1$ or $F_2$ is symmetric.
4. Let $X_1$, $X_{11}$ and $X_{12}$ i.i.d. $F_1$, and $X_2$, $X_{21}$ and $X_{22}$ i.i.d. $F_2$. Suppose that $m_1 = m_2$, and define
$$\pi^{(4)} = \Pr\{X_{11} \leq X_2 \leq X_{12}\} - \Pr\{X_{21} \leq X_1 \leq X_{22}\}.$$
When $F_1 = F_2$, $\pi^{(4)} = 0$. When $\pi^{(4)} > 0$ we may say that the probability mass of $F_2$ is more concentrated than the probability mass of $F_1$, again reflecting an aspect of a dispersion difference. The assumption that $m_1 = m_2$ is very important here. If, for example, the median of $X_2$ is much larger than the median of $X_1$, so that the two distributions are completely separated, both probabilities $\Pr\{X_{11} \leq X_2 \leq X_{12}\}$ and $\Pr\{X_{21} \leq X_1 \leq X_{22}\}$ are zero, resulting in $\pi^{(4)} = 0$ whatever the variances of $X_1$ and $X_2$. van Eeden (1964) again showed that condition C2 is generally not satisfied unless $F_1$ or $F_2$ is symmetric.
5. The next measure is closely related to $\pi^{(4)}$. First write $\pi^{(4)}$ as
$$\begin{aligned}
\pi^{(4)} &= \int_{\mathcal{S}} F_1(x)(1 - F_1(x))dF_2(x) - \int_{\mathcal{S}} F_2(x)(1 - F_2(x))dF_1(x) \\
&= \int_{\mathcal{S}} F_1(x)dF_2(x) - \int_{\mathcal{S}} F_1^2(x)dF_2(x) \\
&\quad - \int_{\mathcal{S}} F_2(x)dF_1(x) + \int_{\mathcal{S}} F_2^2(x)dF_1(x)
\end{aligned}$$

$$
\begin{aligned}
&= \Pr\{X_1 \le X_2\} - \Pr\{\max(X_{11}, X_{12}) \le X_2\} \\
&\quad - \Pr\{X_2 \le X_1\} + \Pr\{\max(X_{21}, X_{22}) \le X_1\} \\
&= (2\Pr\{X_1 \le X_2\} - 1) + (\Pr\{\max(X_{21}, X_{22}) \le X_1\} \\
&\quad - \Pr\{\max(X_{11}, X_{12}) \le X_2\}),
\end{aligned}
$$

in which we did not assume that $m_1 = m_2$. The last equation shows a decomposition of $\pi^{(4)}$ into two terms, each representing a different order of likely ordering of $X_1$ and $X_2$. In particular, the first term, $2\Pr\{X_1 \le X_2\} - 1$, measures the likely ordering of $X_1$ and $X_2$ as we have discussed in Section 9.2.2 in the context of the WMW test. The second term in the decompostion,

$$\Pr\{\max(X_{21}, X_{22}) \le X_1\} - \Pr\{\max(X_{11}, X_{12}) \le X_2\}$$

is related to what we call *double likely ordering*, and we let $\pi^{(5)}$ denote this component. To get a better understanding of $\pi^{(5)}$ we first focus on $\Pr\{\max(X_{21}, X_{22}) \le X_1\}$. Under the null hypothesis,

$$\Pr\{\max(X_{21}, X_{22}) \le X_1\} = \int_S F_2^2(x) dF_1(x) = \int_0^1 u^2 du = \frac{1}{3}.$$

An easy interpretation is to read it as a stronger form of the likely ordering used before. Whereas a large $\Pr\{X_2 \le X_1\}$ says that it is likely to have an observation from $F_2$ that is smaller than an observation from $F_1$, a large $\Pr\{\max(X_{21}, X_{22}) \le X_1\}$ says that even the largest of two observations from $F_2$ is still very likely to be smaller than an observation from $X_1$.

#### 9.5.3.2 The Ansari–Bradley Test

The Ansari–Bradley (AB) test statistic is usually given by (Ansari and Bradley (1960))

$$T_{ABn} = \frac{\sum_{i=1}^{n_1} \left| R(X_1 i) - \frac{n+1}{2} \right| - \frac{1}{2} n_1 (n+1) + \mu_n}{\sigma_n},$$

where

$$\mu_n = \begin{cases} \frac{1}{4} n_1 (n+2) & \text{when } n \text{ is even} \\ \frac{1}{4} n_1 \frac{(n+1)^2}{n} & \text{when } n \text{ is odd} \end{cases}$$

and where

$$\sigma_n^2 = \begin{cases} \frac{n_1 n_2 (n^2 - 4)}{48(n-1)} & \text{when } n \text{ is even} \\ \frac{n_1 n_2 (n+1)(n^2+3)}{48 n^2} & \text{when } n \text{ is odd.} \end{cases}$$

For simplicity, however, we consider here a test statistic that is asymptotically equivalent. Let

$$T_n = \sqrt{n}\frac{AB_n - \frac{1}{4}}{\sigma_{AB}} \quad (9.26)$$

with

$$\sigma_{AB} = \frac{1}{48}\frac{n_1}{n_2},$$

and

$$AB_n = \frac{1}{n_1 n}\sum_{i=1}^{n_1}\left|R(X_{1i}) - \frac{n+1}{2}\right|. \quad (9.27)$$

To understand the intuition behind this statistic it is instructive to assume that the medians of the two distributions coincide. In terms of the ranks the median equals approximately $(n+1)/2$. Thus the terms $|R(X_{1i}) - (n+1)/2|$ assign small scores to observations close to the median and large scores to observations farther away from the median. Under the null hypothesis of equal scale, and assuming that the medians are equal, we expect these terms, which only correspond to the contributions of the observations of the first sample, to be uniform between approximately 0 and $n/2$. A small value of $AB_n$ suggests a concentration of the ranks of the first sample observations near the median, indicating a smaller variance in the first sample.

Under the general null hypothesis $H_0 : F_1 = F_2$, the theory of linear rank tests gives that $T_n$ is asymptotically standard normal distributed. Ansari and Bradley (1960) showed that their test is consistent for testing $\Delta = 0$ in the scale-difference model (9.21). It has been shown that the AB test is equivalent to the Siegel–Tukey (Siegel and Tukey (1960)) test.

van Eeden (1964) showed that for the AB test the parameter $\delta$ of Equation (9.24) equals

$$\delta_{AB} = \frac{1}{4} - \int_S \left|\lambda F_1(x) + (1-\lambda)F_2(x) - \frac{1}{2}\right| dF_1(x), \quad (9.28)$$

where $\lambda = \lim_{n\to\infty}(n_1/n)$ is assumed to be bounded away from 0 and 1. This parameter can be related to scale under additional assumptions. In particuar, when $X_1$ and $X_2$ have common median $m$, this reduces to

$$\begin{aligned}\delta_{AB} &= (1-\lambda)\left[\Pr\{X_2 \geq X_1 \geq m\} + \Pr\{X_2 \leq X_1 \leq m\} - \frac{1}{4}\right] \\ &= (1-\lambda)\left(\pi^{(3)} - \frac{1}{4}\right), \quad (9.29)\end{aligned}$$

with $\pi^{(3)}$ as in (9.25). Using the notation of Section 9.3,

$$\mathcal{F}_{NI}^{M} = (f_1, f_2) : m_1 = m_2\}$$

with $m_i$ the median of $f_i$ ($i = 1, 2$). Thus, when $(f_1, f_2) \in \mathcal{F}_{NI}^{M}$ the AB test may perhaps be appropriate for testing the implied null hypothesis $H_0 : \pi^{(3)} = \frac{1}{4}$. However, a further requirement is that the variance estimator used in (9.26) must be consistent under the implied null hypothesis. The variance

$$\sigma^2_{AB} = \frac{1}{48}\frac{n_1}{n_2},$$

however, is only valid under very restrictive assumptions in $f_1$ and $f_2$, which are moreover very hard to interpret. We therefore do not give further details on these conditions here; the interested reader is referred to van Eeden (1964) who gave an explicit formula for $\text{Var}\{AB_n\}$. The only simple case for which $\sigma^2_{AB}$ is a valid variance is when the general two-sample null hypothesis $H_0 : F_1 = F_2$ holds true. Therefore, when $\sigma^2_{AB}$ is used, the AB test is testing the general two-sample null hypothesis, and it is consistent for $\pi^{(3)} \neq \frac{1}{4}$ when $(f_1, f_2) \in \mathcal{F}^M_{NI}$. Hence, these arguments suggest that

$$\mathcal{F}^{LSM}_{ND} = \left\{(f_1, f_2) : f_1\left(\frac{x-m}{\sigma_1}\right) = f_2\left(\frac{x-m}{\sigma_2}\right)\right\}$$

is a safe set of restrictions.

When AB is to be used for testing the implied null hypothesis $H_0 : \pi^{(3)} = \frac{1}{4}$, a consistent estimator of $\text{Var}\{AB_n\}$ under this less restrictive null hypothesis is required. See also the next section and Section 9.3.4) for such estimators.

When $F_1$ and $F_2$ have different medians, the interpretation of (9.28) is not straightforward. Moses (1963) studied the asymptotic behaviour of $AB_n$ under alternatives with unequal medians, and he showed that the AB test may sometimes be even consistent for detecting location shifts; i.e., $F_1(x) = F_2(x-\gamma)$ with $F_1$ and $F_2$ thus having equal variances. This example clearly suggests that one should be very careful with interpreting the $p$-values resulting from the AB test.

#### 9.5.3.3 The Shukatme Test

Sukhatme's test (Sukhatme (1957)) is very closely related to the AB test. The test statistic is an immediate estimator of the sum of probabilities $\pi^{(3)} = \Pr\{X_2 \geq X_1 \geq m\} + \Pr\{X_2 \leq X_1 \leq m\}$, where $m$ is taken as the common median of the two populations. The test statistic is

$$T_{Sn} = \sqrt{n}\frac{S_n - \frac{1}{4}}{\sigma_S} \tag{9.30}$$

with

$$\sigma^2_S = \frac{(n+7)n}{48n_1n_2}$$

and

$$S_n = \frac{1}{n_1 n_2} \sum_{i=1}^{n_1} \sum_{j=1}^{n_2} \psi(X_{1i}, X_{2j}),$$

where

$$\psi(x, y) = 1 \text{ if } y \geq x \geq m \text{ or } y \leq x \leq m, \text{ and } \psi(x, y) = 0 \text{ otherwise.}$$

Obviously $\delta_S = \pi^{(3)} - \frac{1}{4}$, demonstrating that the natural null hypothesis is $H_0 : \pi^{(3)} = \frac{1}{4}$, but the test only makes sense when the medians of $f_1$ and $f_2$ coincide; i.e., $(f_1, f_2) \in \mathcal{F}_{NI}^M$. The variance used in (9.30) is again computed under the general two-sample null hypothesis. Conditions for

$$\sigma_S^2 = \frac{(n+7)n}{48 n_1 n_2}$$

being also valid for less restrictive null hypotheses may be derived from a general expression for Var $\{S_n\}$, but again no simple conditions arrise. Thus when (9.30) is used as a test statistic, it is again best to assume $(f_1, f_2) \in \mathcal{F}_{ND}^{LSM}$.

Note, however, that $S_n$ is a $V$-statistic. The variance of $S_n$ can therefore be consistently estimated using standard theory of $V$-statistics. See, for example, Lee (1990). With such a variance estimator, provided the two medians are equal, the Shukatme test may be used for testing $H_0 : \pi^{(3)} = \frac{1}{4}$.

#### 9.5.3.4 The Mood Test

Mood's test (Mood (1954)) is based on the test statistic

$$T_{Mn} = \sqrt{n} \frac{M_n - \frac{1}{12}\frac{n^2-1}{n^2}}{\sigma_M} \tag{9.31}$$

with

$$\sigma_M^2 = \frac{n_2(n+1)(n^2-4)}{180 n_1 n^3}$$

and

$$M_n = \frac{1}{n_1 n^2} \sum_{i=1}^{n_1} \left( R(X_{1i}) - \frac{n+1}{2} \right)^2.$$

The factor $(n^2-1)/n^2$ in the numerator of (9.31) may (asymptotically) be dropped. This statistic is similar to the AB test statistic (9.27) and may be considered as the $L_2$-norm version of the AB test statistic. The $\delta$ parameter equals

$$\delta_M = \int_{\mathcal{S}} \left( \lambda F_1(x) + (1-\lambda) F_2(x) - \frac{1}{2} \right)^2 dF_1(x) - \frac{1}{12}. \tag{9.32}$$

In a balanced design, i.e., when $\lambda = \frac{1}{2}$, this reduces to

$$\begin{aligned}
\delta_M &= \frac{1}{4}\left\{\int_{\mathcal{S}} F_1(x)(1 - F_1(x))dF_2(x) - \int_{\mathcal{S}} F_2(x)(1 - F_2(x))dF_1(x)\right\} \\
&= \frac{1}{4}\left\{\Pr\left\{X_{11} \le X_2 \le X_{12}\right\} - \Pr\left\{X_{21} \le X_1 \le X_{22}\right\}\right\} \\
&= \frac{1}{4}\pi^{(4)}.
\end{aligned}$$

Although $\lambda = \frac{1}{2}$ is a design restriction rather than a distribution assumption, we consider a null hypothesis in terms of $\pi^{(4)}$ an *implied* null hypothesis for the Mood test. The *natural* null hypothesis in terms of $\delta_M$ has no simple interpretation. Thus, in a balanced design the Mood test may be used for testing the null hypothesis $H_0 : \pi^{(4)} = 0$, provided the variance $\sigma_M^2$ is asymptotically equivalent to $\mathrm{Var}\left\{\sqrt{n}M_n\right\}$ under this semiparametric null hypothesis. As before, no easily interpretable restrictions on $f_1$ and $f_2$ make this happen. Hence, the Mood test is again best considered as a test for the general two-sample null hypothesis which is consistent for $H_1 : \pi^{(4)} \neq 0$ as long as the variance of $\sqrt{n}M_n$ is bounded in probability, and provided the data come from a balanced design.

Because the Mood test reappears several times in subsequent chapters, an alternative interpretation is provided in the next paragraphs.

It was quite inconvenient to restrict the applicability of the test to balanced designs. The reason was that $\lambda$ appears in (9.32). On the other hand, $\lambda$ appears in (9.32) in the expression $\lambda F_1(x) + (1 - \lambda)F_2(x)$, which is recognised as the pooled distribution function $H(x)$. Now write

$$\begin{aligned}
&\int_{\mathcal{S}} \left(\lambda F_1(x) + (1 - \lambda)F_2(x) - \frac{1}{2}\right)^2 dF_1(x) \\
&\quad = \int_{\mathcal{S}} H^2(x)dF_1(x) + \frac{1}{4} - \int_{\mathcal{S}} H(x)dF_1(x).
\end{aligned}$$

When $Z$, $Z_1$, and $Z_2$ are independent random variables with distribution function $H$, then we can write

$$\begin{aligned}
&\int_{\mathcal{S}} H^2(x)dF_1(x) + \frac{1}{4} - \int_{\mathcal{S}} H(x)dF_1(x) \\
&= \Pr\left\{\max(Z_1, Z_2) \le X_1\right\} - \Pr\left\{Z \le X_1\right\} + \frac{1}{4} \\
&= \left(\Pr\left\{\max(Z_1, Z_2) \le X_1\right\} - \frac{1}{3}\right) - \left(\Pr\left\{Z \le X_1\right\} - \frac{1}{2}\right) + \frac{1}{12}.
\end{aligned}$$

Hence,

$$\delta_M = \left(\Pr\left\{\max(Z_1, Z_2) \le X_1\right\} - \frac{1}{3}\right) - \left(\Pr\left\{Z \le X_1\right\} - \frac{1}{2}\right),$$

which equals zero when $\Pr\{\max(Z_1, Z_2) \leq X_1\} = \frac{1}{3}$ and $\Pr\{Z \leq X_1\} = \frac{1}{2}$. This happens under the general two-sample null hypothesis. For later purposes it is also important to see that $\Pr\{Z \leq X_1\} = \frac{1}{2}$ is equivalent to the natural null hypothesis of the WMW test. The other part (i.e., $\Pr\{\max(Z_1, Z_2) \leq X_1\} = \frac{1}{3}$) expresses a *double likely equivalence* of $X_1$ and the marginal $Z$. From this representation, the Mood test may be seen as a test for the combined null hypothesis of the WMW natural null hypothesis and a double likely equivalence null hypothesis.

The discussion given in the previous paragraph suggests that the Mood test has a natural null hypothesis in terms of likely equivalences, without making any restrictive assumptions on $f_1$ and $f_2$. However, the validity of the Mood test statistic also depends on the variance used in the denominator of the test statistic. Under the general two-sample null hypothesis $\sigma_M^2$ is clearly correct, but under the natural null hypothesis a more general consistent variance estimator is required. Because $M_n$ is a simple linear rank statistic we refer to Section 9.3.4 for such estimators.

#### 9.5.3.5 The Lehmann Test

Lehmann (1951) suggested using the $U$ statistic

$$L_n = \frac{1}{\binom{n_1}{2}\binom{n_2}{2}} \sum_{i=1}^{n_1-1} \sum_{j=i+1}^{n_1} \sum_{k=1}^{n_2-1} \sum_{l=k+1}^{n_2} \phi\left(|X_{1i} - X_{1j}|, |X_{2k} - X_{2l}|\right),$$

with kernel $\phi(x, y) = 1$ if $x \leq y$ and $\phi(x, y) = 0$ otherwise. The test statistic $L_n$ is obviously an unbiased estimator of $\pi^{(2)}$. Despite the fact that $\pi^{(2)}$ satisfies the conditions C1 and C2, the simple interpretation of $\pi^{(2)}$ as a measure for scale differences, and the nice property that $L_n$ can be used in circumstances with arbitrary and unknown medians $m_1$ and $m_2$, the test is not distribution free (Sukhatme (1957)), even not asymptotically. In particular, the asymptotic variance depends on the unknown distribution $F_1 = F_2$ under the null hypothesis. Finally, also note that the Lehmann test statistic is not a rank statistic.

#### 9.5.3.6 The Fligner–Killeen Test

From the previous subsections we have learned that the AB and Shukatme tests are only consistent for scale differences when $F_1$ and $F_2$ have a common median, and when at least one of the two distributions is symmetric. Fligner and Killeen (1976) proposed tests that are also distribution free under the null hypothesis, but that are also consistent for scale differences without the assumption of common medians. In the development of their theory, however, they adopt the rather stringent location-scale model. Moreover, they assume that $F$ is symmetric. Despite this restrictive framework, their test statistics are of interest in their own right. The theory that we present here is slightly different from the original paper of Fligner and Killeen (1976).

First suppose that the medians $m_1$ and $m_2$ are known. Then define

$$FK_n = \frac{1}{n_1 n_2} \sum_{i=1}^{n_1} \sum_{j=1}^{n_2} \phi\left(|X_{1i} - m_1|, |X_{2j} - m_2|\right),$$

where $\phi$ is as for the Lehmann test. The statistic $FK_n$ is a $V$ statistic and is clearly an unbiased estimator of $\pi^{(1)}$. It is actually exactly the WMW test statistic (9.2) but with the observations replaced by their absolute deviations from the respective median. The test based on $FK_n$ is thus distribution free; the test statistic has the same null distribution as the WMW test statistic.

### *9.5.4 Conclusion*

Although the discussion presented in this section is far from complete, we hope that we have demonstrated that the rank tests for scale differences are very restrictive in their usefulness. Most of these tests have been proposed in the statistical literature as rank tests for testing scale differences, but all required very heavy distributional assumptions. Wasserstein and Boyer (1991) gave another important argument against the use of linear rank tests for testing for scale differences. They showed that the power of the linear rank tests do not approach one when the ratio of the two scale parameters goes to infinity. Note that this is a finite sample size property. Tests that show this defect are referred to as *nonresolving*. We think that many of these rank tests can still be usefully, but not necessarily for testing scale differences, but rather for testing informative hypothesis expressed in terms of probabilities. For testing such hypotheses no strong distributional assumptions are needed.

## 9.6 The Kruskal–Wallis Test and the ANOVA $F$-Test

In this section we present two tests for the $K$-sample problem. We start with the nonparametric Kruskal–Wallis (KW) rank test, which may be seen as an extension of the WMW test or as the rank statistic version of the $F$-test in an analysis of variance (ANOVA). The latter test is the parametric test for comparing equality of $K$ means. Just as with the WMW test the KW test can be treated in a semiparametric framework. However, because the arguments and methods are basically the same as for the two-sample setting, we do not elaborate on this. At the end of the section we only briefly comment on it.

### 9.6.1 The Hypotheses and the Test Statistic

Consider the general $K$-sample null hypothesis

$$H_0 : F_1(x) = F_2(x) = \cdots = F_K(x) \text{ of all } x.$$

We again use the notation $H(x) = (1/K)\sum_{s=1}^{K} F_s(x)$ for the pooled distribution function, and we use $Z$ to denote a random variable with distribution function $H$. As an alternative we consider the generalisation of the alternative of the WMW test; i.e.,

$$H_1 : \Pr\{Z \leq X_s\} \neq \frac{1}{2} \text{ for at least one } s = 1, \ldots, K. \tag{9.33}$$

A natural test statistic for this testing problem can be constructed by starting from estimators of the probabilities in $H_1$. The probability $\pi_s = \Pr\{Z \leq X_s\}$ is unbiasedly estimated by

$$\begin{aligned}
\hat{\pi}_s &= \frac{1}{nn_s}\sum_{i=1}^{n}\sum_{j=1}^{n_s} \mathrm{I}(Z_i \leq X_{sj}) \\
&= \frac{1}{n}\frac{1}{n_s}\sum_{j=1}^{n_s}\left(\sum_{i=1}^{n} \mathrm{I}(Z_i \leq X_{sj})\right) \\
&= \frac{1}{n}\frac{1}{n_s}\sum_{j=1}^{n_s} R(X_{sj}) \\
&= \frac{1}{n}\bar{R}_s,
\end{aligned}$$

where $R(X_{sj})$ is the rank of observation $X_{sj}$ in the pooled sample, and $\bar{R}_s$ is the average of the ranks of the observations in the $s$th sample. For each sample $s$ the statistic $W_s = \hat{\pi}_s - \frac{1}{2}$ is appropriate for measuring information against the null hypothesis in favor of the alternative. Combining all $W_s$ $(s = 1, \ldots, K)$ into one test statistic, and weighting the individual $W_s$ by the corresponding sample sizes $n_s$, results in

$$\sum_{s=1}^{K} n_s \left(\hat{\pi} - \frac{1}{2}\right)^2 = \frac{1}{n^2}\sum_{s=1}^{K} n_s \left(\bar{R}_s - \frac{n}{2}\right)^2.$$

The latter statistic comes very close to the original KW test statistic, which is usually defined as

$$KW = \frac{12}{n(n+1)}\sum_{s=1}^{K} n_s \left(\bar{R}_s - \frac{n+1}{2}\right)^2. \tag{9.34}$$

The difference between the two statistics is a factor 12 and an asymptotically vanishing factor $(n+1)/n$. The factor 12 is basically a scaling factor so that the KW test statistic has a convenient limiting null distribution; see the next section.

### 9.6.2 The Null Distribution

Because the general $K$-sample null hypothesis implies the randomisation hypothesis, the exact permutation null distribution may be enumerated or approximated using the methods of Section 7.1. Just as the WMW test the KW test is distribution free.

The asymptotic distribution under the general $K$-sample null hypothesis is stated in the next theorem. A sketch of the proof is given in Appendix A.10.

**Theorem 9.1.** *Suppose that all $\lambda_s = n_s/n$ are bounded away from 0 and 1 when $n \to \infty$. Then, under the general $K$-sample null hypothesis, as $n \to \infty$,*

$$KW \xrightarrow{d} \chi^2_{K-1}.$$

### 9.6.3 The Diagnostic Property

Because the KW test is a very direct extension of the WMW test, the discussion of Section 9.3 applies here too. We limit the discussion here to two remarks.

1. When it can be assumed that all $K$ distributions belong to the same location-scale model, i.e.,

   $$f_1(x - \Delta_1) = f_2(x - \Delta_2) = \ldots = f_K(x - \Delta_K) \text{ for all } x \in \mathcal{S}$$

   for some constants $\Delta_s$ $(s = 1, \ldots, K)$, the null hypothesis may also be expressed in terms of the means or the medians, without any changes to the form of the test statistic or its null distribution.
2. Suppose we can assume that all $K$ distributions are symmetric and have equal variances. Then again the null hypothesis may be formulated using means or medians. Evidently the exact permutation null distribution does not hold anymore, and for the asymptotic null distribution to hold, the test statistic must be rescaled first.

This brief discussion illustrates that the same type of assumptions as for the WMW has to be imposed for the KW test for making it a test for testing equality of means. The assumptions must now hold for all $K$ distributions simultaneously.

### 9.6.4 The $F$-Test in ANOVA

The analysis of variance is a very popular method for comparing means when normality can be assumed. We show in this section that the KW test is basically an $F$-test statistic, but applied to the rank-transformed observations.

Let $X_{si}$ denote the $i$th observation from the $s$th sample, and assume that $X_{si}$ i.i.d. $N(\mu_s, \sigma^2)$ $(s = 1, \ldots, K; i = 1, \ldots, n_s)$. It is thus assumed that the $K$ variances are equal. The null and alternative hypotheses are

$$H_0 : \mu_1 = \cdots = \mu_K \text{ and } H_1 : \text{ not } H_0.$$

Just as with the $t$-test, the null hypothesis together with the distributional assumptions imply the general $K$-sample null hypothesis. The $F$-test is defined in terms of the *between sum of squares* and the *total sum of squares*, denoted by SSB and SSTot:

$$\text{SSB} = \sum_{s=1}^{K} n_s \left(\bar{X}_s - \bar{X}\right)^2$$

$$\text{SSTot} = \sum_{s=1}^{K} \sum_{i=1}^{n_s} \left(X_{si} - \bar{X}\right)^2,$$

where $\bar{X}$ is the sample mean of all $n$ observations, and $\bar{X}_s$ is the sample mean of the $n_s$ observations in the $s$th sample. The $F$-test statistic is given by

$$F = \frac{\text{SSB}/(K-1)}{(\text{SSTot} - \text{SSB})/(n-K)}.$$

Under the null hypothesis $F$ has an $F_{K-1,n-K}$ distribution.

We now demonstrate that the KW test statistic (9.34) is related to the $F$-statistic. First we compute SSTot for the rank transformed data; i.e., we replace $X_{si}$ with its rank $R_{si}$ in the pooled sample. Note that

$$\bar{R} = \frac{1}{n} \sum_{s=1}^{K} \sum_{i=1}^{n_s} R_{si} = \frac{n+1}{2} \text{ and } \bar{R}_s = \frac{1}{n_s} \sum_{i=1}^{n_s} R_{si}.$$

Hence,

$$\text{SSTot} = \sum_{s=1}^{K} \sum_{i=1}^{n_s} \left(R_{si} - \bar{R}\right)^2$$

$$= \sum_{s=1}^{K}\sum_{i=1}^{n_s} R_{si}^2 - 2\bar{R}\sum_{s=1}^{K}\sum_{i=1}^{n_s} R_{si} + n\bar{R}^2$$
$$= \frac{1}{6}n(n+1)(2n+1) - 2\frac{n+1}{2}\frac{n(n+1)}{2} + n\frac{(n+1)^2}{4}$$
$$= \frac{n(n+1)^2}{12},$$

which is a constant that only depends on the total sample size. All relevant information in the rank-based $F$ statistic comes thus from

$$\text{SSB} = \sum_{s=1}^{K} n_s \left(\bar{R}_s - \bar{R}\right)^2 = \sum_{s=1}^{K} n_s \left(\bar{R}_s - \frac{n+1}{2}\right)^2,$$

which is indeed up to a factor the KW statistic of (9.34).

## 9.7 Some Final Remarks

### 9.7.1 Adaptive Tests

In Section 7.2.3 we have introduced the concept of *adaptive linear rank tests*. These are linear rank tests of which the scores are selected by a data-based selection rule so that the resulting rank test has good power. For example, earlier in this chapter we have shown that the WMW test is the LMPRT for testing the two-sample location shift hypotheses when the observations have a logistic distribution. When the observations have a normal distribution, the van der Waerden test is the LMPRT for location-shift alternatives. The two tests differ only in the scores defining the test statistics.

One of the first adaptive two-sample tests is due to Randles and Hogg (1973). They start from the observation that the power of rank tests is strongly influenced by the tail behavior of the distributions. They consider three types of tail behavior: light tails (e.g., uniform distribution), median tails (e.g., logistic), and heavy tails (e.g., double exponential). For each of these classes they propose a set of scores that have good power characteristics for distributions within that class. The data-based selection rule makes use of two statistics that only depend on the sample order statistics. Because order statistics and ranks are independently distributed, the score selection procedure has no effect on the distribution of the rank statistic. It is particularly this last property that makes this type of adaptive test attractive. Whether a good set of scores for a dataset at hand is selected by the statistician prior to looking at the data (i.e., the traditional way), or whether this set of scores is selected based on the order statistics-based selection rule, has no effect on the power. The adaptive tests thus increase the chance of testing with a good set of scores, and therefore the power of this adaptive testing procedure is expected to be better on average when applied to many different datasets. Many adaptive tests based on this general scheme have been proposed over the years, and even up to today new contributions to the statistical literature are added. Despite their simplicity, these tests appear not to have been implemented in the popular statistical software packages.

Finally we mention a shortcoming of this type of adaptive tests. It is only adaptive within the restrictive location-shift setting. The same idea can of course be translated to scale-difference models, but then again it is a very focused testing situation. In Chapter 10 we describe a more flexible adaptive test.

### *9.7.2 The Lepage Test*

Lepage (1971) proposed a two-sample rank test based on a simple combination of the WMW and the AB statistics,

$$L = U_1^2 + U_2^2,$$

where $U_1$ and $U_2$ denote the standardised WMW and AB statistics. He showed that $U_1$ and $U_2$ are independent and that the asymptotic null distribution of $L$ is thus $\chi_2^2$. In the context of the next chapter, $L$ is closely related to an order 2 smooth test statistic.

# Chapter 10
# Smooth Tests

This chapter is devoted to smooth tests for the two- and the $K$-sample problems. The literature on such tests may not be as vast as for the one-sample problem, though its applicability is very broad and often informative. Because many of the techniques and ideas used in this chapter rely heavily on what has been discussed in the previous chapters, this chapter is quite concise. The construction of the test is very similar to the one-sample smooth test of Chapter 4.

In Section 10.1 the construction of the smooth models and the test statistic is explained for the two-sample problem. Its null distribution is derived and a detailed discussion on the components is provided. The extension to comparing $K$ distributions is the topic of Section 10.3. The data-driven choice of the order of the smooth test is discussed in Section 10.4. We conclude the chapter with some practical recommendations in Section 10.7.

## 10.1 Smooth Tests for the 2-Sample Problem

### *10.1.1 Smooth Models and the Smooth Test*

#### 10.1.1.1 Smooth Models

Smooth tests for the two-sample problem, as we present them here, were first introduced by Janic-Wróblewska and Ledwina (2000), who immediately presented the smooth test in its data-driven order selection form. In this section, however, we assume that the order of the test is specified prior to looking at the data. The data-driven versions are discussed in Section 10.4.

As for the one-sample case, the test statistic arises as a score test statistic in an order $k$ smooth family of alternatives, which may also be referred to as the *smooth model*. We use again the notation $F_1$ and $F_2$ ($f_1$ and $f_2$) to denote the

O. Thas, *Comparing Distributions*, Springer Series in Statistics,
DOI 10.1007/978-0-387-92710-7_10, 

distribution functions (density functions) of the first and the second sample, respectively, and the notation $H$ (and $h$) for the pooled distribution function (density function). The latter is defined as

$$H(x) = \frac{n_1}{n} F_1(x) + \frac{n_2}{n} F_2(x) \text{ or } h(x) = \frac{n_1}{n} f_1(x) + \frac{n_2}{n} f_2(x). \tag{10.1}$$

The factor $n_1/n$ may also be replaced with $\lambda$ which is defined as the limit of $n_1/n$ as $n \to \infty$ and which is assumed to be bounded away from 0 and 1. Similarly, the factor $n_2/n$ may be replaced by $1 - \lambda$. Both definitions of $H$ and $h$ will asymptotically not make a difference. We therefore sometimes interchange the roles of $\lambda$ and $n_1/n$. Sometimes we write $\lambda_1$ for $\lambda$, and $\lambda_2$ for $1 - \lambda$.

The order $k$ family of alternatives that were considered by Janic-Wróblewska and Ledwina (2000) were first proposed by Neuhaus (1987). It is given by

$$f_{1k}(x) = C_1(\boldsymbol{\theta}) \exp\left(\frac{n_2}{n} \sum_{j=1}^{k} \theta_j h_j(H(x))\right) h(x) \tag{10.2}$$

$$f_{2k}(x) = C_2(\boldsymbol{\theta}) \exp\left(-\frac{n_1}{n} \sum_{j=1}^{k} \theta_j h_j(H(x))\right) h(x), \tag{10.3}$$

where $\boldsymbol{\theta}^t = (\theta_1, \ldots, \theta_k)$, $\{h_j\}$ is a set of orthonormal functions on the uniform distribution over $[0, 1]$, and $C_1$ and $C_2$ are two normalisation constants. (Note: our definition is slightly different from what has been used in the literature by considering different factors prior to the summation operator, but this will have no effect on further results as the factors may be resolved in the $\theta$ parameters.) The general two-sample null hypothesis reduces thus to $H_0 : \boldsymbol{\theta} = \mathbf{0}$. Because the models (10.2) and (10.3) use the exponential function, they are referred to as the *Neyman smooth models*. Just as in Chapter 4 the smooth tests based on these models appear to coincide with those constructed from the *Barton smooth models*, given by

$$f_{1k}(x) = \left(1 + \frac{n_2}{n} \sum_{j=1}^{k} \theta_j h_j(H(x))\right) h(x) \tag{10.4}$$

$$f_{2k}(x) = \left(1 - \frac{n_1}{n} \sum_{j=1}^{k} \theta_j h_j(H(x))\right) h(x). \tag{10.5}$$

Note that in both formulations of the smooth models the densities $f_{1k}$ and $f_{2k}$ contain the same set of $\theta$ parameters, but with different factors preceeding them. The factors, that depend on the sample sizes are a consequence of (10.1), which must also hold when $f_1$ and $f_2$ are replaced with $f_{1k}$ and $f_{2k}$.

The Barton model can be related to the comparison distributions of Section 7.4. Model (10.4) immediately gives

$$\frac{f_{1k}(H^{-1}(u)}{h(H^{-1}(u)} = 1 + \frac{n_2}{n}\sum_{j=1}^{k}\theta_j h_j(u), \tag{10.6}$$

which is exactly the comparison density function $r_1(u)$ of (7.20), and which basically shows that the Barton smooth model may be interpreted as an order $k$ orthogonal series expansion of $r_1(u)$. The comparison density function $r_2(u)$ becomes according to (10.5)

$$\frac{f_{2k}(H^{-1}(u)}{h(H^{-1}(u)} = 1 - \frac{n_1}{n}\sum_{j=1}^{k}\theta_j h_j(u). \tag{10.7}$$

Thus, when the $\theta$ parameters in the expansion of the comparison densities can be estimated, yet another estimation method of the comparison density arises. See Section 8.2.2 for more details on the estimation and use of the comparison density.

When we consider the Hilbert space $L_2(\mathcal{S}; H)$ the $\theta$ parameters have similar interpretations as in Section 4.1 for the one-sample smooth models. In particular,

$$\theta_j = \frac{n}{n_2}\left\langle h_j, \frac{f_1}{h}\right\rangle_h = \frac{n}{n_2}\left\langle h_j, r_1\right\rangle_h = -\frac{n}{n_1}\left\langle h_j, r_2\right\rangle_h,$$

in which $f_1$ and $f_2$ are represented by $f_{1k}$ and $f_{2k}$ with $k \to \infty$. The parameters may also be related to *Pearson's* $\phi^2$ *measure*, which was studied by Lancaster (1969), and particularly for the $K$-sample problem by Eubank and LaRiccia (1990). For $k = 2$, it becomes

$$\phi^2 = \sum_{s=1}^{2}\lambda_s\int_0^1 \frac{\left(f_s(H^{-1}(u)) - h(H^{-1}(u))\right)^2}{h(H^{-1}(u))}du = \sum_{s=1}^{2}\lambda_s\int_0^1 \left(r_s(u)-1\right)^2 du. \tag{10.8}$$

Similar calculations as in Section 4.1 give

$$\phi^2 = \frac{n_1 n_2}{n^2}\sum_{j=1}^{\infty}\theta_j^2. \tag{10.9}$$

Before continuing, note that the factor $n_1 n_2/n$ in (10.9) is not informative, and can be eliminated by redefining the densities $f_{1k}$ and $f_{2k}$ so that this factor gets resolved in the $\theta$s. Thus $\phi^2$ measures how far $f_1$ and $f_2$ are apart in terms of a squared norm in an appropriate Hilbert space. Because each $\theta_j$ is involved in the expansions of both $f_1$ and $f_2$, it must be interpreted in a slightly different way. It suggests that the distance interpretation goes over the pooled density $h$. However, by using the $\lambda_s$ and $n_s/n$ notation interchangeably, simple algebra immediately gives the equivalence between (10.8) and

$$\phi^2 = \frac{n_1 n_2}{n^2} \int_0^1 (r_1(u) - r_2(u))^2 \, du.$$

#### 10.1.1.2 Smooth Test Statistic and the Null Distribution

The first step in obtaining the order $k$ smooth test statistic is constructing the score test statistic for testing $H_0 : \boldsymbol{\theta} = \mathbf{0}$ in the Neyman or the Barton smooth models of the previous section. Both models give rise to the same score test statistic for the same reasons as made clear in the proof of Theorem 4.1. The following lemma is therefore restricted to the Neyman model.

**Lemma 10.1.** *Let $X_{s1}, \ldots, X_{sn_s}$ denote a sample of i.i.d. observations from $f_s$ $(s = 1, 2)$. Consider the order $k$ smooth family of alternatives (Neyman or Barton). The score test statistic for testing $H_0 : \theta_1 = \cdots = \theta_k$ is given by*

$$S_k = \sum_{j=1}^{k} \left\{ \frac{n_2}{n} \sum_{i=1}^{n_1} h_j(H(X_{1i}) - \frac{n_1}{n} \sum_{i=1}^{n_2} h_j(H(X_{2i})) \right\}^2 .$$

The score statistic $S_k$ can, however, not be used directly, because it depends on the unknown pooled distribution function $H$. It is replaced by its empirical version: the pooled empirical distribution function $\hat{H}_n$ of Equation (7.19): $\hat{H}_n(Z_i) = (R_i - 0.5)/n$, in which the conventional continuity correction is applied, and where $Z_1, \ldots, Z_n$ represent the pooled sample observations so that the first $n_1$ pooled sample observations correspond to $X_{11}, \ldots, X_{1n_1}$, and the last $n_2$ observations to $X_{21}, \ldots, X_{2n_2}$.

The order $k$ smooth test statistic and its asymptotic null distribution is presented in the following theorem.

**Theorem 10.1.** *Let $X_{s1}, \ldots, X_{sn_s}$ denote a sample of i.i.d. observations from $f_s$ $(s = 1, 2)$. Assume $\lambda = \lim_{n\to\infty}(n_1/n)$ is bounded away from 0 and 1. Consider the order $k$ smooth family of alternatives (Neyman or Barton). The order $k$ smooth test statistic for testing $H_0 : \theta_1 = \ldots = \theta_k$ is given by*

$$\begin{aligned} T_k &= \frac{n_1 n_2}{n} \sum_{j=1}^{k} \left\{ \frac{1}{n_2} \sum_{i=1}^{n_1} h_j \left( \frac{R_i - 0.5}{n} \right) - \frac{1}{n_2} \sum_{i=1}^{n_2} h_j \left( \frac{R_i - 0.5}{n} \right) \right\}^2 \\ &= \boldsymbol{U}^t \boldsymbol{U}, \end{aligned}$$

*where $\boldsymbol{U}$ is a $k$ vector with $j$th element equal to*

$$U_j = \sum_{i=1}^{n} c_{ni} h_j \left( \frac{R_i - 0.5}{n} \right),$$

*where*

$$c_{ni} = \sqrt{\frac{n_1 n_2}{n}} \begin{cases} \frac{1}{n_1} & \text{if } 1 \leq i \leq n_1 \\ -\frac{1}{n_2} & \text{if } n_1 + 1 \leq i \leq n. \end{cases}$$

*Under the null hypothesis $H_0 : F_1 = F_2$, as $n \to \infty$,*

$$\boldsymbol{U} \xrightarrow{d} MVN(\boldsymbol{0}, \boldsymbol{I}) \text{ and } T_k \xrightarrow{d} \chi^2_k.$$

First note that the factor $n/(n_1 n_2)$ which appears in $T_k$, and which was not yet part of the definition of $S_k$, is introduced here so that $T_k$ has a proper limiting null distribution. This factor is incorporated in the factor $c_{ni}$.

Under the general two-sample null hypothesis it is also possible to enumerate the exact permutation null distribution of the order $k$ smooth test statistic $T_k$. Moreover, $T_k$ is clearly a rank statistic, so that the exact null distribution is distribution free.

Before we move on to a more detailed discussion of the components we show that the components are again proportional to the estimators of the $\theta_j$ parameters in the order $k$ smooth models (10.6) and (10.7). Write

$$\begin{aligned} \mathrm{E}_k \left\{ \sqrt{\frac{n}{n_1 n_2}} U_j \right\} &= \mathrm{E}_k \left\{ \sum_{i=1}^{n_1} \frac{1}{n_1} h_j \left( \frac{R_i - 0.5}{n} \right) - \sum_{i=n_1+1}^{n} \frac{1}{n_2} h_j \left( \frac{R_i - 0.5}{n} \right) \right\} \\ &\approx \mathrm{E}_k \left\{ h_j \left( H(X_1) \right) \right\} - \mathrm{E}_k \left\{ h_j \left( H(X_2) \right) \right\} \\ &= \frac{n_2}{n} \theta_j + \frac{n_1}{n} \theta_j \\ &= \theta_j. \end{aligned}$$

Thus

$$\hat{\theta}_j = \sqrt{\frac{n}{n_1 n_2}} U_j \tag{10.10}$$

may be used as an estimator of $\theta_j$.

### 10.1.2 Components

Theorem 10.1 immediately shows that the test statistic $T_k$ is decomposed into $k$ components $U_j^2$ that are asymptotically mutually independent under the null hypothesis. In this section we further investigate the distributional properties of the components, as well as their interpretation and relation to other rank statistics.

The $j$th component equals

$$U_j = \sum_{i=1}^{n} c_{ni} h_j \left( \frac{R_i - 0.5}{n} \right), \tag{10.11}$$

which has exactly the form of a simple linear rank statistic (see Definition 7.3) with regression constants $c_i = c_{ni}$ (subscript $n$ is used to stress the dependence on the sample size) and scores $a_n(R_i) = h_j\left((R_i - 0.5)/n\right)$ determined by the system of orthonormal functions $\{h_j\}$. This important characterisation of the components implies that the properties of such linear rank statistics, as discussed in Section 7.2, directly apply to the components. For example, Theorem 7.2 shows that the $j$th component has asymptotically a standard normal distribution (this asymptotic property also follows from Theorem 10.1). The zero mean and unit variance follow for these particular statistics from the orthonormality of the $h_j$ functions.

At this point it is of interest to have a closer look at some of the lower-order components for a particular system of orthonormal functions. We consider here the Legendre polynomials.

#### 10.1.2.1 The First Component: WMW Statistic

With the first Legendre polynomial, $h_1(x) = \sqrt{12}(x - 0.5)$, the first component becomes (for notational simplicity we use $R_i$ instead of $R_i - 0.5$)

$$\begin{aligned} U_1 &= \sum_{i=1}^{n} c_{ni} h_1\left(\frac{R_i}{n}\right) \\ &= \sqrt{\frac{n_2}{n_1 n}} \sum_{i=1}^{n_1} \sqrt{12}\left(\frac{R_i}{n} - 0.5\right) - \sqrt{\frac{n_1}{n_2 n}} \sum_{i=n_1+1}^{n} \sqrt{12}\left(\frac{R_i}{n} - 0.5\right) \\ &= \sqrt{\frac{12}{n_1 n_2 n}} \left(\frac{n_2(n+1)}{2} - \sum_{i=n_1+1}^{n} R_i\right), \end{aligned}$$

where we have made use of the equality $\sum_{i=1}^{n} R_i = \sum_{i=1}^{n} i = n(n+1)/2$, and in which we recognise the standardised Wilcoxon rank sum test statistic of Section 9.2 up to an asymptotically neglectable factor $\sqrt{(n+1)/n}$. Using $\sum_{i=1}^{n} R_i = n(n+1)/2$ in the other direction gives

$$U_1 = \sqrt{\frac{12}{n_1 n_2 n}} \left(\sum_{i=n_1+1}^{n} R_i - \frac{n_1(n+1)}{2}\right).$$

This expression could also be obtained from (9.20), after appropriate rescaling with $\sqrt{(n_1 n_2)/n}$.

#### 10.1.2.2 The Second Component: Mood Statistic

The second Legendre polynomial is

$$h_2(x) = \sqrt{5}(6x^2 - 6x + 1) = 6\sqrt{5}\left[(x - 0.5)^2 - \frac{1}{12}\right].$$

Surely using $h_2$ in Equation (10.11) gives again a linear rank statistic. Here, however, we relate $U_2$ with a well known linear rank statistic. We therefore need to express $U_2$ in terms of ranks of only one of the two samples. This could again be obtained by using (9.20), but here we use the equality $\sum_{i=1}^n R_i^2 = \sum_{i=1}^n i^2 = (n+1)(2n+1)/(6n)$ to arrive at the identity

$$\sum_{i=1}^n \left(\frac{R_i}{n} - \frac{1}{2}\right)^2 = \frac{n^2+2}{12n}.$$

On the application of this equality we find

$$\begin{aligned}
U_2 &= \sum_{i=1}^n c_{ni} h_2\left(\frac{R_i}{n}\right) \\
&= \sqrt{\frac{n_2}{n_1 n}} \sum_{i=1}^{n_1} 6\sqrt{5}\left[\left(\frac{R_i}{n} - 0.5\right)^2 - \frac{1}{12}\right] \\
&\quad - \sqrt{\frac{n_1}{n_2 n}} \sum_{i=n_1+1}^{n} 6\sqrt{5}\left[\left(\frac{R_i}{n} - 0.5\right)^2 - \frac{1}{12}\right] \\
&= 6\sqrt{\frac{5}{n_1 n_2 n^3}} \sum_{i=1}^{n_1}\left[\left(R_i - \frac{n}{2}\right)^2 - \frac{n^2+2}{12}\right] \\
&= \frac{\sum_{i=1}^{n_1}\left[\left(R_i - \frac{n}{2}\right)^2 - \frac{n^2+2}{12}\right]}{\sqrt{\frac{1}{180} n_1 n_2 n^3}},
\end{aligned}$$

which is asymptotically equivalent to the standardised Mood statistic of (9.31).

#### 10.1.2.3 The Third Component: the SKEW Statistic

The third-order Legendre polynomial is

$$h_3(x) = \sqrt{7}(20x^3 - 30x^2 + 12x - 1) = \sqrt{7}\left[20(x-0.5)^3 - 3(x-0.5)\right].$$

The third component, $U_3$, is again related to a rank statistic that has been published in the statistical literature. Boos (1986) proposed a linear rank statistic, which he called SKEW, and which is exactly equal to $U_3$. He suggested that SKEW could be used to detect differences between $F_1$ and $F_2$ in their skewness. We give a more detailed discussion on the diagnostic properties of the $U_j$ components in Section 10.2.

#### 10.1.2.4 The Fourth Component: the KURT Statistic

The fourth-order Legendre polynomial is

$$h_3(x) = \sqrt{7}(20x^3 - 30x^2 + 12x - 1) = \sqrt{7}\left[20(x-0.5)^3 - 3(x-0.5)\right].$$

The fourth component, $U_4$, coincides with the KURT statistic of Boos (1986). He suggested that KURT could be used to detect differences between $F_1$ and $F_2$ in their kurtosis. We give a more detailed discussion on the diagnostic properties of the $U_j$ components in Section 10.2 following.

## 10.2 The Diagnostic Property

In the previous sections we have demonstrated that the (lower-order) components are related to well-known rank tests. Some of them have been described in more detail in Chapter 9. For example, the null and alternative hypothesis of the WMW test have been listed. A very important conclusion was that one should be very cautious when using the WMW test when conclusions about location shifts are wanted. A more correct view on the rank tests is to first find out for which population parameter the test statistic is an estimator, and use this population parameter in the formulation of the null and alternative hypotheses. The null hypothesis formulated in this way is less restrictive than the general two-sample null hypothesis. For the WMW test we argued that it actually tests hypotheses formulated in terms of $\pi = \Pr\{X_1 \leq X_2\}$. When the WMW test may be used for such hypotheses, we say that the test has the *diagnostic property*. In Section 9.3 we further argued that many of the rank tests are not diagnostic, unless the rank statistics are first scaled appropriately by using an estimator of the asymptotic variance that is consistent under the less restrictive null hypothesis. Two generic estimators were presented in Section 9.3.4.

Obviously the order $k$ smooth test can inherit the diagnostic property from its components. For the one-sample problem we showed in Section 4.5.6 briefly that the order $k$ smooth test can be rescaled too by replacing the covariance matrix $\boldsymbol{\Sigma}$ of the component vector $\boldsymbol{V}$ by an empirical estimator, say $\hat{\boldsymbol{C}}$ that is consistent under a weaker null hypothesis than the full parametric one-sample goodness-of-fit null hypothesis. A similar approach could be thought of here; thus instead of using the statistic $T_k = \boldsymbol{U}^t\boldsymbol{U}$ for the two-sample problem, the modified statistic $T_k = \boldsymbol{U}^{t-1}\hat{\boldsymbol{C}}\boldsymbol{U}$ could be used instead. Because no empirical results are available, at this moment we cannot advise positively or negatively on its use.

Instead of standardising by using an estimator of the covariance matrix of the $k$-dimensional vector $\boldsymbol{U}$, the components could be standardised individually and used in the data analysis after the general two-sample null hypothesis has been rejected. More precisely, the procedure could work as follows.

1. Test the general two-sample null hypothesis with an order $k$ smooth test (or one of its adaptive versions).
2. When the null hypothesis is not rejected, the procedure stops.
3. When the null hypothesis is rejected, it may of course be concluded that the two distributions are different. In a next phase, the individual components can be examined after, but now the components are standardised before being looked at.

### 10.2.1 Examples

In this section we apply the smooth tests to the gene expression data of the colorectal cancer study that was introduced in Section 6.2.1. At this point we only analyse the data of gene 1 with an order $k = 4$ smooth test, and we look at the individual component tests. Later, in Section 10.5 we redo the analysis by means of an adaptive version of the smooth test.

*Example 10.1 (The gene expression data: Gene 1).* We test the general two-sample null hypothesis with an order $k = 4$ smooth test, using the smooth.test function of the cd R package. It is the same function as for the one-sample problem; it recognises a two-sample problem by means of the formula argument.

```
> gene1.st<-smooth.test(expression~group,order=4,rescale=F,
+ B=NULL,probs=T,data=gene1)
> gene1.st
K-sample smooth goodness-of-fit test (K=2)

Smooth test statistic T_k =  21.6155  p-value =  0.0002
     1 st component =   2.4622  p-value =  0.0138
     2 nd component =  -3.8114  p-value =  0.0001
     3 rd component =  -0.4855  p-value =  0.6273
     4 th component =   0.8893  p-value =  0.3738

All p-values are obtained by the asymptotic approximation

Estimation of likely orderings
  Pr(X1<= X2) =  0.3269
  Pr(max(X11,X12)<= X2) = 0.2536
  Pr(max(X11,X12)<= X2) = 0.4587
```

The specification B=NULL implies that the asymptotic approximations are used, and the rescale=F argument specifies that the components are not rescaled with empirical variance estimators. From the output we read that the order $k = 4$ test statistic equals 21.6155 with a $p$-value of 0.0002. Thus at the 5% level of significance we reject the general two-sample null hypothesis. We now try to get a better understanding of how the two distributions differ by looking at the individual components. The output lists the first four components (not squared!) with their respective $p$-values. The latter seem to suggest that only the first two components are important.

Before interpreting the components, we explore the data. Normal QQ plots, a two-sample QQ plot, and boxplots are shown in Figure 10.1. Particularly the data from group 1 deviate from normality; this is confirmed by a one-sample Anderson–Darling test (results not shown). Moreover, the deviation is asymmetric. We therefore conclude that (1) a two-sample $t$-test for comparing means is not most appropriate, and (2) the WMW hypotheses in terms of $\Pr\{X_1 \leq X_2\}$ do not imply conclusions in terms of the means. The

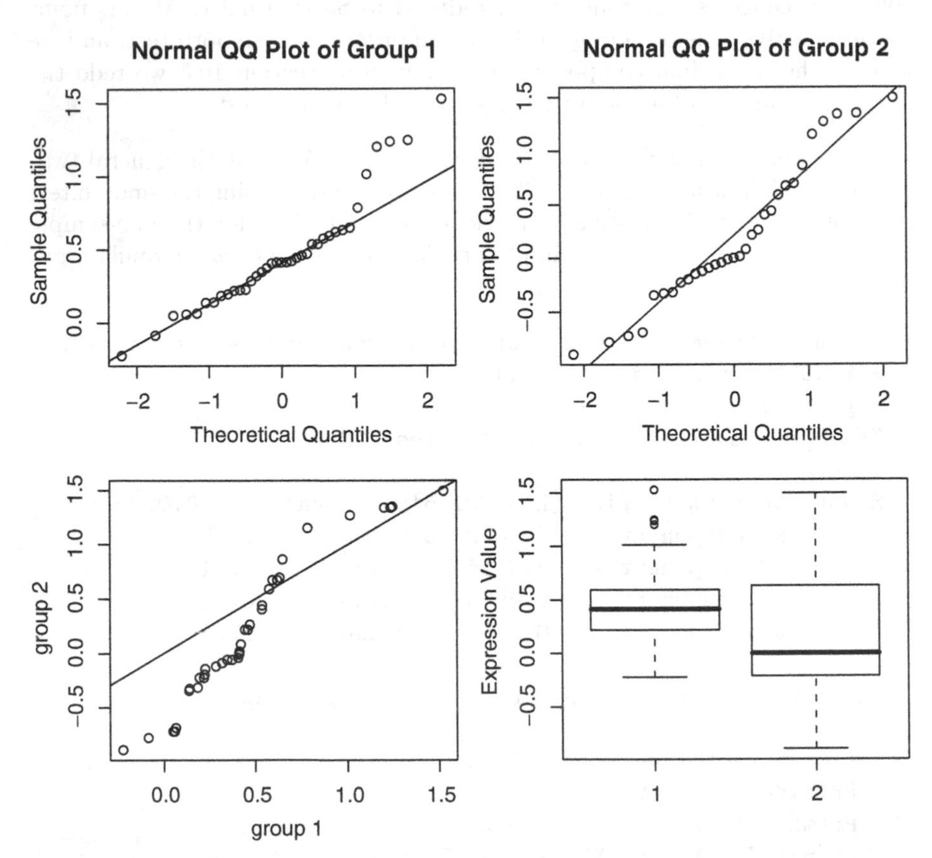


**Fig. 10.1** The normal QQ plots, two-sample QQ plot, and the boxplots of the expression values in the two groups for gene 1

two-sample QQ plot has points at the two sides of the 45 degree line. It is therefore not possible to use the concept of stochastic ordering in conjunction with the components.

The first component test, which is exactly the WMW test, thus cannot be used for testing equality of means. It may be used for the detection of likely ordering, provided that the null variance used in the WMW statistic is a good approximation of the true variance under the alternative. We therefore apply the wmw.diagnose function first.

```
> wmw.diagnose(expression~group,data=gene1)

Estimation of p112=Pr(max(X21,X22)<=X1)
  and p112=Pr(max(X11,X12)<=X2), and Var(MW)

  p112 =  0.2536
  p221 =  0.4587
  Estimated Var(MW) =  0.005368552
  Null Var(MW) =  0.005013078
  Ratio Estimated / Null =  1.07

  WMW test may be too liberal
```

Although the ratio of the two variances is larger than one, which could make the test liberal, the ratio is only slightly larger than 1. We therefore decide that the first-order component test may be used for testing likely ordering. The output shows the estimate of $\Pr\{X_1 \leq X_2\}$: 0.3269. It is thus likely that gene 1 is more expressed in a nonprogressed adenoma patient as compared to a carcinoma patient.

The interpretation of the second component, which is the Mood statistic, is much more difficult. As we clearly do not have a scale-shift model, it is not obvious to interpret a significant second-order component test in terms of scale differences. In Section 9.5.3.4 we first showed that it is only related to a measure for scale differences in balanced designs, but this is not the case here (37 and 31 observations). We further argued that the test statistic can be related to *double likely ordering* (see Section 7.6). In particular we have shown that the test may reject the null hypothesis in favour of

$$\delta_M = \left(\Pr\{\max(Z_1, Z_2) \leq X_1\} - \frac{1}{3}\right) - \left(\Pr\{Z \leq X_1\} - \frac{1}{2}\right) \neq 0, \quad (10.12)$$

but, as we have already concluded that there is a significant first-order ordering, the last term of $\delta_M$ is not zero. This further complicates the interpretation of the second component. It is a similar problem as that we encountered in the diagnostic interpretation of the components in the one-sample smooth test. Taking this caveat into account, we try to proceed, but we first check the appropriateness of the null variance used in this second-order component test. The smooth.diagnose function does the job (this is a more general function than wmw.diagnose).

```
> smooth.diagnose(gene1.st)

Ratio's of estimated / null variances:
  1 st component: 1.07  (liberal)
  2 nd component: 0.71  (conservative)
  3 rd component: 1.16  (liberal)
  4 th component: 0.94  (conservative)
```

Thus from a scaling point of view, there is not much to worry about. In conclusion, although the second order-component test is not formally diagnostic, the smooth test results seem to suggest that a likely ordering and a double likely ordering are present. It is even likely (i.e., a chance of more than 50%) that even the largest gene 1 expression of two carcinoma patients is still smaller than the gene 1 expression of a nonprogressed adenoma patient. It is, however, not straightforward to link this conclusion formally with a difference in scale, at least not with this rank-based smooth test.

Just for demonstration purposes, we also present the results of a two-sample Welch $t$-test.

```
> t.test(expression~group, data=gene1)

        Welch Two Sample t-test

data:  expression by group
t = 2.025, df = 46.276, p-value = 0.04865
alternative hypothesis: true difference in means is not
   equal to 0
95 percent confidence interval:
 0.001676391 0.542509120
sample estimates:
mean in group 1 mean in group 2
      0.4620662       0.1899734
```

This gives $p = 0.04865$, which is very close to the nominal significance level of 5%. Considering that Figure 10.1 showed that normality cannot be assumed, this $t$-test analysis is inconclusive.

## 10.3 Smooth Tests for the $K$-Sample Problem

### 10.3.1 Smooth Models and the Smooth Test

The extension from the two-sample to the $K$-sample setting is quite direct, though this $K$-sample smooth test has not been published yet. We use the notation introduced in Chapter 7. Thus $X_{s1}, \ldots, X_{sn_s}$ denotes a sample of

$n_s$ i.i.d. observations from $f_s$ ($s = 1, \dots, K$), and $Z_1, \dots, Z_n$ represent the $n$ observations from the pooled sample, with the convention that the first $n_1$ $Z$ observations are the observations from sample 1, the next $n_2$ observations are those from sample 2, etc., and the last $n_K$ observations are those from sample $K$. The pooled density becomes now $h(x) = \sum_{s=1}^{K}(n_s/n)f_s(x)$ (and so $H(x)$ is defined too). The asymptotically equivalent definition is $h(x) = \sum_{s=1}^{K} \lambda_s f_s(x)$ with $\lambda_s$ the limit of $n_s/n$ as the total sample size $n \to \infty$. As before it is assumed that all $\lambda_s$ stay away from 0 or 1.

Consider the following families of order $k$ alternatives in Neyman format ($s = 1, \dots, K$)

$$f_{sk}(x) = C_s(\boldsymbol{\theta}) \exp\left( \sum_{j=1}^{k} (\theta_{sj} - \bar{\theta}_j)\, h_j(H(x)) \right) h(x),$$

where $C_s(\boldsymbol{\theta})$ is a normalisation constant and

$$\bar{\theta}_j = \frac{1}{K} \sum_{s=1}^{K} \frac{n_s}{n} \theta_{sj}$$

are the average $j$th order effects. As a Barton smooth model we write ($s = 1, \dots, K$),

$$f_{sk}(x) = \left( 1 + \sum_{j=1}^{k} (\theta_{sj} - \bar{\theta}_j)\, h_j(H(x)) \right) h(x).$$

The construction of the families of alternatives guarantees that (10.1) holds for all $f_s$ replaced with their order $k$ Barton expansions ($s = 1, \dots, K$), so that the pooled density $h$ maintains its interpretation. Note that these densities have a more complicated system of $\theta$ parameters: it does not suffice anymore to have only one such parameter for each order $j$. There are now $K$ $\theta_{sj}$ parameters for each order $j$. It may be more convenient to reparameterise the alternatives in terms of parameters $\theta^*_{sj} = \theta_{sj} - \bar{\theta}_j$. This parameterisation shows $\sum_{s=1}^{K} \theta^*_{sj} \equiv 0$, implying that only $K - 1$ of the $K$ $\theta^*_{sj}$ parameters are linearly independent.

Before we give the order $k$ smooth test, we first focus on one $\theta$ parameter, say $\theta_{sj}$. The next lemma gives the score test statistic for testing $\theta_{sj} = 0$, assuming that the pooled distribution function $H$ is known.

**Lemma 10.2.** *Let $X_{s1}, \dots, X_{sn_s}$ denote a sample of i.i.d. observations from $f_s$ ($s = 1, \dots, K$), and let $Z_1, \dots, Z_n$ denote the pooled sample observations, and assume that $H$ is known. Assume $\lambda_s = \lim_{n\to\infty}(n_s/n)$ is bounded away from 0 and 1 ($s = 1, \dots, K$). Consider the order $k$ smooth family of alternatives (Neyman or Barton). The score test statistic for testing the null hypothesis $\theta_{sj} = 0$ is given by*

$$S_{sj}(H) = \sqrt{\frac{n-n_s}{n}} \sum_{i=1}^{n} c_{nsi} h_j(H(Z_i)),$$

*where*

$$c_{nsi} = \sqrt{\frac{n_s(n-n_s)}{n}} \begin{cases} \frac{1}{n_s} & \text{if observation } i \text{ is in sample } s \\ -\frac{1}{n-n_s} & \text{if observation } i \text{ is not in sample } s. \end{cases} \tag{10.13}$$

*Under the general two-sample null hypothesis, as* $n \to \infty$,

$$S_{sj}(H) \xrightarrow{d} N(0, 1-\lambda_s).$$

Surely in practice the pooled distribution function is again replaced by its EDF, giving, after continuity correction,

$$U_{sj} = S_{sj}(\hat{H}) = \sqrt{\frac{n-n_s}{n}} \sum_{i=1}^{n} c_{nsi} h_j \left( \frac{R_i - 0.5}{n} \right). \tag{10.14}$$

The interpretation of $U_{sj}$ is related to the comparison density. Straightforward calculations show that, as $n \to \infty$,

$$\begin{aligned} \frac{1}{\sqrt{n_s}} U_{sj} &\xrightarrow{p} \int_0^1 h_j(u)\,(r_s(u)-1)\,du \qquad (10.15) \\ &= \int_S h_j(H(x))\,(r_s(H(x))-1)\,dH(x) = \theta_{sj}. \end{aligned}$$

This demonstrates that $U_{sj}$ measures how far away the $s$th distribution is from the pooled distribution $H$ in the $h_j$ direction.

Testing for equality of the $K$ distributions in the $h_j$ direction may be formulated by $\theta_{1j} = \cdots = \theta_{sj} = 0$. This requires the joint limiting null distribution of the score statistics $U_{sj}$ $(s = 1, \ldots, K)$. The asymptotic joint multivariate normality may be demonstrated using the Cramér–Wald device and the covariances may, for example, be found by expressing the simple linear rank statistics $U_{sj}$ in terms of comparison distribution processes, similarly as in Section 9.3.4.1 for finding the asymptotic variance of such statistics. This very general procedure demands, however, many algebraic calculation steps. Instead we now just refer to Theorem 4 in Section 3.3.1 of Hájek et al. (1999) that gives the covariance between two simple linear rank statistics with different regression scores (here $\{c_{npi}\}$ and $\{c_{nqi}\}$ for $U_{pj}$ and $U_{qj}$, respectively $(p, q = 1, \ldots, K)$). For testing the general $K$-sample null hypothesis, we need to test $H_0 : \theta_{sj} = 0$ for all $s = 1, \ldots, K$ and all $j = 1, \ldots, k$. The score test statistic for this testing problem now requires the asymptotic joint multivariate normality of all $U_{sj}$ statistics. Again the Cramér–Wald device does the trick, and the additional asymptotic covariance between $U_{pi}$ and $U_{qj}$ $(p, q = 1, \ldots, K)$ with $i \neq j$ is zero because of the orthogonality of the polynomials. We summarise the most important properties in the following theorem.

**Theorem 10.2.** *Let $X_{s1}, \ldots, X_{sn_s}$ denote a sample of i.i.d. observations from $f_s$ ($s = 1, \ldots, K$). Assume $\lambda_s = \lim_{n\to\infty}(n_s/n)$ is bounded away from 0 and 1 ($s = 1, \ldots, K$). Consider the order $k$ smooth family of alternatives (Neyman or Barton). Let $\boldsymbol{U}_j^t = (U_{1j}, \ldots, U_{Kj})$, $j = 1, \ldots, k$, and $\boldsymbol{U}^t = (\boldsymbol{U}_1^t, \ldots, \boldsymbol{U}_k^t)$.*

*Under $H_0$, as $n \to \infty$, the following convergences hold.*
*(1) Let $\boldsymbol{\Sigma}_j$ a $K \times K$ matrix with at the sth diagonal position the element $1 - \lambda_s$, and at the $(p, q)$ off-diagonal position the element $-\sqrt{\lambda_p \lambda_q}$. Then,*

$$\boldsymbol{U}_j \xrightarrow{d} MVN(\boldsymbol{0}, \boldsymbol{\Sigma}_j).$$

*(2)*

$$T_{nj} = \boldsymbol{U}_j^t \boldsymbol{U}_j = \sum_{s=1}^{K} U_{sj}^2 \xrightarrow{d} \chi^2_{K-1}.$$

*(3) Let $\boldsymbol{\Sigma}$ be a $(kK) \times (kK)$ matrix with $\boldsymbol{\Sigma}_j$ ($j = 1, \ldots, k$) as diagonal blocks, and zeroes elsewhere. Then,*

$$\boldsymbol{U} \xrightarrow{d} MVN(\boldsymbol{0}, \boldsymbol{\Sigma}).$$

*(4)*

$$T_n = \boldsymbol{U}^t \boldsymbol{U} = \sum_{j=1}^{k} \sum_{s=1}^{K} U_{sj}^2 \xrightarrow{d} \chi^2_{k(K-1)}. \tag{10.16}$$

*A technical note*:
Although no formal proof of Theorem 10.2 is given here, we still want to clarify the perhaps less logical formulation of $U_{sj}$ in (10.14). The factor $\sqrt{(n - n_s)/n}$ may look strange, particularly because part (1) of Theorem 10.2 states that the asymptotic variance of $U_{sj}$ equals $1 - \lambda_s$. Hence, when the factor would have been dropped the asymptotic variance would be 1, as we usually want. However, the reason for not changing the definition of $U_{sj}$ is twofold. First, this factor appears naturally in the score statistic, and second, by recognising this variance and the covariances $-\sqrt{\lambda_p \lambda_q}$, Theorem 1 in Section 2.4.1 of Hájek et al. (1999) may be immediately applied to arrive at the asymptotic $\chi^2_{K-1}$ distribution. Finally, we also mention that the multivariate normal distributions have actually covariance matrices that are not of full rank.

There is again a connection between the $T_n$ statistic and Pearson's $\phi^2$ divergence. Extending (10.8) is obvious, and gives

$$\phi^2 = \sum_{s=1}^{K} \lambda_s \int_0^1 (r_s(u) - 1)^2 \, du = \sum_{s=1}^{K} \lambda_s \sum_{j=1}^{k} \left(\theta_{sj} - \bar{\theta}_j\right)^2.$$

On using (10.15), and writing (10.16) as

$$T_n = n \sum_{s=1}^{K} \frac{n_s}{n} \sum_{j=1}^{k} \left( \frac{1}{\sqrt{n_s}} U_{sj} \right)^2 ,$$

it follows that, as $n \to \infty$,

$$\frac{1}{n} T_n \xrightarrow{P} \phi^2.$$

## 10.3.2 Components

Just as with the two-sample order $k$ smooth test here also the test statistic $T_n$ has components that are related to well-known rank statistics. Equation (10.16) shows that $T_n$ decomposes into $k \times K$ $U_{sj}$ components, but we rather refer to a component as a summary statistic for the $j$th order deviations from the null hypothesis. Write $T_n = \sum_{j=1}^{k} T_{nj}$, and call $T_{nj}$ the $j$th component of $T_n$. Theorem 10.2 says that each component has asymptotically a $\chi^2_{K-1}$ distribution, and that the components are asymptotically mutually independent under the general $K$-sample null hypothesis. We briefly give some more details on the first few components.

When $j = 1$, the statistic $T_{n1}$ is asymptotically equivalent to the Kruskal–Wallis test statistic, and under suitable assumptions of the shapes of the $K$ distributions, it may be diagnostic for detecting location shifts. This is definitely true in location-shift models. The asymptotic equivalence between both statistics may be seen by rewriting $T_{n1}$ in a different form. We illustrate it more generally, i.e., for any $j$, because it is also useful for the other components. For notational comfort we use $h_j(s,i)$ for $h_j\left((R_{si} - 0.5)/n\right)$, $S_j = \sum_{s=1}^{K} \sum_{i=1}^{n_s} h_j(s,i)$, and let $d_s = \sqrt{n_s}/n$. Write $U_{sj}$ in (10.14) first as

$$\begin{aligned} U_{sj} &= \sqrt{n_s} \frac{n - n_s}{n} \left\{ \frac{1}{n_s} \sum_{i=1}^{n_s} h_j(s,i) - \frac{1}{n - n_s} \left( \sum_{t \neq s} \sum_{i=1}^{n_t} h_j(t,i) + \sum_{i=1}^{n_2} h_j(s,i) \right) \right. \\ &\quad \left. + \frac{1}{n - n_s} \sum_{i=1}^{n_s} h_j(s,i) \right\} \\ &= d_s \left\{ \frac{n}{n_s} \sum_{i=1}^{n_s} h_j(s,i) - S_j \right\}. \end{aligned} \tag{10.17}$$

Note that the $S_j$ term is not informative. Moreover, due to the characteristic $\int_0^1 h_j(u)du = 0$, $S_j$ vanishes asymptotically, and sometimes it is even exactly zero. For example, for $j = 1$,

$$S_1 = \sum_{s=1}^{K}\sum_{i=1}^{n_s}\left(\frac{R_{si}-\frac{1}{2}}{n}-\frac{1}{2}\right) = \frac{1}{n}\left(\frac{n(n+1)}{2}-\frac{n}{2}-\frac{n^2}{2}\right) = 0.$$

The $j$th component then becomes

$$\begin{aligned} T_j &= \sum_{s=1}^{K} U_{sj}^2 \\ &= \sum_{s=1}^{K} d_s^2 \frac{n^2}{n_s^2}\left(\sum_{i=1}^{n_s} h_j(s,i)\right)^2 \\ &\quad -2S_j \sum_{s=1}^{K} d_s^2 \frac{n}{n_s}\sum_{i=1}^{n_s} h_j(s,i) + S_j^2 \sum_{s=1}^{K} d_s^2 \\ &= \sum_{s=1}^{K}\left(\frac{1}{\sqrt{n_s}}\sum_{i=1}^{n_s} h_j(s,i)\right)^2 - \left(\frac{1}{\sqrt{n}}S_j\right)^2. \end{aligned} \tag{10.18}$$

This shows that a test based on our $j$th component is equivalent to a test based on the first term of this last equation. This term is exactly of the form of, for example, the Kruskal–Wallis statistic with $h_1(u) = \sqrt{12}(u-\frac{1}{2})$ the first-order Legendre polynomial:

$$\begin{aligned} &\sum_{s=1}^{K}\left(\frac{1}{\sqrt{n_s}}\sum_{i=1}^{n_s}\left[\sqrt{12}\left(\frac{R_{si}-\frac{1}{2}}{n}-\frac{1}{2}\right)\right]\right)^2 \\ &= \frac{12}{n^2}\sum_{s=1}^{K} n_s\left(\frac{1}{n_s}\sum_{i=1}^{n_s}\left(R_{si}-\frac{n+1}{2}\right)\right)^2 \\ &= \frac{12}{n^2}\sum_{s=1}^{K} n_s\left(\bar{R}_s-\frac{n+1}{2}\right)^2, \end{aligned}$$

which is up to a factor $(n+1)/n$ equal to the KW statistic of (9.34).

The second component, $T_{n2}$, is related to the $K$-sample Mood statistic, which is often considered as a nonparametric test for dispersion. The relation is again established using (10.18). However, the discussion given in Section 9.5.3.4 also applies here, and caution is thus in place here. The third- and fourth-order components, $T_{n3}$ and $T_{n4}$, are basically the SKEW and KURT test statistics of Boos (1986). Although the names SKEW and KURT suggest their relation to the skewness and the kurtosis, we do not have to repeat the arguments that these diagnostic conclusions in terms of the moments of the $K$ distributions are not always guaranteed. It is already quite complicated for relating the Mood test with dispersion, and with higher-order moments it becomes increasingly more depending on restrictive assumptions on the shapes of the distributions. We finish this brief overview of the components with remarking that the components described here are also related with those of Eubank and LaRiccia (1990).

## 10.4 Adaptive Smooth Tests

Adaptive data-driven smooth tests were discussed in detail in Section 4.3 for the one-sample goodness-of-fit tests. The general idea is here again to let the data decide which components to include in the test statistic. This will often result in a power improvement, or at least most frequently avoids the dilution effect. Just as in Section 4.3 several types of selection procedures are possible: order selection (with finite and infinite horizons) and subset selection. From a theoretical perspective we distinguish two different frameworks. First, when only finite horizons are allowed the methodology of Claeskens and Hjort (2004) is again applicable. No very restrictive conditions are imposed for this theory to hold. Second, with a horizon that grows with the sample size the data-driven smooth test is consistent. The theory of the data-driven test for the two-sample case has been presented by Janic-Wróblewska and Ledwina (2000). Other, but related adaptive tests exist, but as they are also related to the EDF tests of the next chapter, we postpone their description to Section 11.3.

Most of the arguments and methods are applicable here too, therefore we can limit the discussion here.

### *10.4.1 Order Selection and Subset Selection with a Finite Horizon*

In Section 4.3 the choice of the selection rule depended on how the nuisance parameter estimation was accounted for. This issue is not present here. From the selection rules presented before, we may consider here the analogues of the BIC-based $K2$ and $S2$ rules, or the AIC-based $M$ selection rule.

We use again the notation $S$ to denote the finite nonempty index set from which a subset is selected. For order selection, order $m < \infty$ is the largest order that may be selected by the selection rule. Furthermore, with $R \subseteq S$,

$$T_R = \sum_{j \in R} U_j^2 \text{ or } T_R = \sum_{j \in R} \sum_{s=1}^{K} U_{sj}^2$$

for the two- or $K$-sample smooth test statistics, respectively.

The selection rules become

$$\begin{aligned} K2_n &= \min\{k : 1 \leq k \leq m : T_k - k\log(n) \geq T_j - j\log(n), j = 1, \ldots, m\} \\ S2_n &= \{R \subseteq S : R \neq \phi \text{ and } T_R - |R|\log(n) \geq T_Q - |Q|\log(n), \forall Q \subseteq S\} \\ M_n &= \{R \subseteq S : R \neq \phi \text{ and } T_R - 2|R| \geq T_Q - 2|Q|, \forall Q \subseteq S\}. \end{aligned}$$

When $O_n$ is used to represent any of these three criteria, then $T_{O_n}$ is used to denote the corresponding adaptive test statistic. The appropriate lemmas and theorems of Section 4.3 remain essentially valid, after some minor obvious notational adaptations.

### 10.4.2 Order Selection with an Infinite Horizon

For the two-sample problem Janic-Wróblewska and Ledwina (2000) work with the $S2$ selection rule, but with the maximal order $m$ replaced by $d(n_1)$ which is an increasing function of $n_1$, and thus also of $n$ under the usual assumption that $n_1/n$ is kept away from 0 or 1. The next theorem combines Theorems 1 and 2 of Janic-Wróblewska and Ledwina (2000).

**Theorem 10.3.** *Suppose $d(n_1) = o\left((n_1/\log(n_1))^{1/9}\right)$, and $n_1/n$ is asymptotically in $(0,1)$. Then, under the general two-sample null hypothesis, as $n \to \infty$, $\Pr_0\{S2_n = 1\} \to 1$ and $T_{S2_n} \xrightarrow{d} \chi_1^2$. When the general two-sample null hypothesis is not true, $T_{S2_n}$ is no longer bounded in probability, and thus the data-driven test based on $T_{S2_n}$ is consistent.*

From a simulation study with $n_1 = n_2 = 50$ Janic-Wróblewska and Ledwina (2000) came to the following conclusions.

- The asymptotic $\chi_1^2$ approximation is not accurate. Note, however, that the data-driven smooth test is still a rank test and its exact permutation distribution can thus be enumerated, at least it can be well approximated using simulations.
- The null distribution depends on the maximal order $d$ (from $d(n_1)$) considered, but for $d \geq 3$ the $\alpha = 0.05$ critical values based on the null distribution do not change much anymore with further increasing $d$.
- The powers of the data-driven tests do not depend much on the choice of the maximal order $d$ for $d \geq 4$. Therefore any choice of $4 \leq k \leq 10$ seems appropriate for $n_1 = n_2 = 50$.
- For many alternatives studied, the data-driven two-sample smooth test has good power. Its power hardly breaks down. This has only been observed for alternatives with "high-order" deviations from the pooled distribution $H$. When the data analyst suspects this feature, another system of orthornormal functions may be used for constructing the test statistic.

## 10.5 Examples

*Example 10.2 (The gene expression data: Gene 1).* In Section 10.2.1 we analysed this dataset with an order four smooth test. Now we do the analysis over, but with a data-driven version of the smooth test. We have chosen here an order selection test with the BIC as an order selection criterion. The maximal order was set to 10. The output follows.

```
> smooth.test(expression~group,max.order=10,
+ adaptive=c("BIC","order"),rescale=F,B=10000,
+ probs=F,graph=T,data=gene1)

Adaptive K-sample smooth goodness-of-fit test (K=2)

Horizon: order selection (max. order = 10)
Order selection rule: BIC

Adaptive smooth test statistic T_k =  20.5884
   p-value < 0.0001
Selected order = 2

Components
      1 st component =    2.4622
      2 nd component =   -3.8114

All p-values are obtained by means of simulations
```

The test gives a $p$-value smaller than 0.0001, so that at the 5% level of significance we reject the general two-sample null hypothesis. The test has selected the first two components. See Example 10.1 for a detailed discussion on the interpretation of the components.

We go one step further now. Because the smooth tests are related to an orthogonal series expansion of the densities (the Barton representation), and because the order selection is basically a model selection criterion, orthogonal series density estimates can be plotted. To be more precise, it is more direct to plot the comparison densities of the two distributions. Consider the comparison densities $r_1$ and $r_2$ in (10.6) and (10.7) truncated at the order selected by the BIC selection criterion, and with the $\theta_j$ parameters replaced by their estimates. More precisely, when $o$ represents the order selected by BIC,

$$\hat{r}_1(u) = 1 + \frac{n_2}{n}\sum_{j=1}^{o}\hat{\theta}_j h_j(u) \text{ and } \hat{r}_2(u) = 1 - \frac{n_1}{n}\sum_{j=1}^{o}\hat{\theta}_j h_j(u).$$

As shown in Section 10.1.1.2 the parameters $\theta_j$ can be estimated by $\sqrt{n/(n_1n_2)}U_j$. Figure 10.2 shows the two estimated comparison densities;

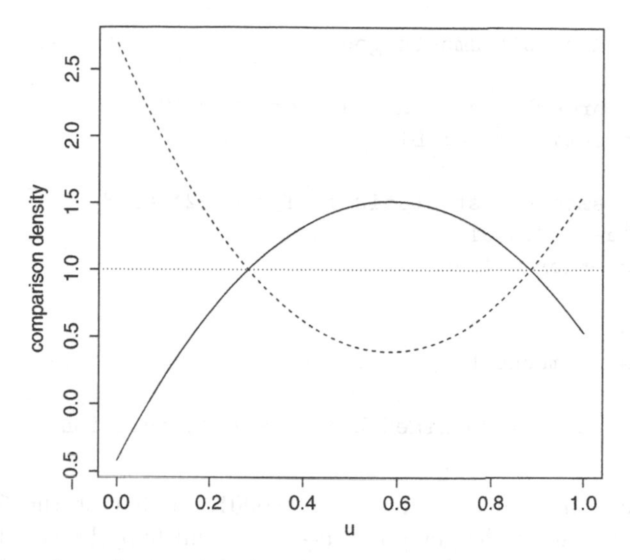


**Fig. 10.2** The estimated comparison densities of gene 1 based on the BIC selected orthogonal series expansion. The full and dashed lines correspond to the nonprogressed adenoma and carcinoma patients, respectively

this is obtained by specifying graph=T in the smooth.test function. The figure shows quite some symmetry at first sight. This is of course a consequence of the symmetry that is built in the two order $k$ smooth alternatives, particularly the use of the same set of $\theta_j$ parameters in the two densities. The plot demonstrates that the probability mass of the expression values of the carcinoma patients is relatively shifted to the smaller values.

In Chapter 4 we also used the relation between smooth tests and orthogonal series density estimates, and there it also allowed us to plot *improved density estimates*. In theory this is also possible here. It relies on (10.4) and (10.5), but, as these Barton models show, it requires an estimate of the pooled density $h$. Any nonparametric density estimate could be considered here, and an advantage could be that it may be calculated using all observations in the pooled sample. However, we do not proceed along these lines.

*Example 10.3 (The gene expression data: Gene 3).* In Example 9.6 we analysed this dataset with the WMW test. Now we do the analysis over, but with a data-driven smooth test. We have chosen here an order selection test with the BIC as an order selection criterion. The maximal order was set to 10. The output follows.

```
> smooth.test(expression~group,max.order=10,
+ adaptive=c("BIC","order"),rescale=F,B=10000,
+ probs=F,data=gene3)
```

```
Adaptive K-sample smooth goodness-of-fit test (K=2)

Horizon: order selection (max. order = 10)
Order selection rule: BIC

Adaptive smooth test statistic T_k =  21.4568
   p-value < 0.0001
Selected order = 1

Components
     1 st component =    -4.1186

All p-values are obtained by means of simulations
```

The test gives a $p$-value smaller than 0.0001, so that at the 5% level of significance we reject the general two-sample null hypothesis. The test has selected only the first component. See Example 9.6 for a detailed discussion on the interpretation of the WMW test.

*Example 10.4 (The traffic data).* We have analysed parts of the traffic data before in two-sample settings, but now we analyse the complete study by testing the general $K$-sample null hypothesis using an adaptive smooth test. The maximal order is set at $k = 5$ and the BIC model selection criterion is used for order selection.

```
>  traffic.st<-smooth.test(time~route,max.order=5,
+ adaptive=c("BIC","order"),rescale=F,B=10000,
+ probs=F,data=traffic)

Adaptive K-sample smooth goodness-of-fit test (K=5)

Horizon: order selection (max. order = 5)
Order selection rule: BIC

Adaptive smooth test statistic T_k =  192.6919
   p-value < 0.0001
Selected order = 4

Components
     1 st component = 53.5905
     2 nd component = 89.7120
     3 rd component = 25.4791
     4 th component = 23.9104

All p-values are obtained by means of simulations
```

From the very small $p$-value of the adaptive test we definitely reject the general $K$-sample null hypothesis strongly. It is, however, not immediately obvious from this analysis how the five samples differ among one another. We may shed some more light on it by looking at the very individual components, i.e., at the $k \times K$ individual $U_{sj}$ $(s = 1, \ldots, K, j = 1, \ldots, n_s)$ statistics. This is shown in the following output. The last column (behind the *) is the row sum of squares.

```
> components(traffic.st)

sample 1: -0.3782 -4.9015  0.2353  0.8876    *  6.2528
sample 2: -3.7203 -3.2362  3.3675 -0.4618    *  8.9667
sample 3:  5.4124 -0.8462 -3.6967 -2.0848    *  12.0055
sample 4: -2.8306  1.8308 -0.4076 -2.1036    *  3.9889
sample 5:  1.5167  7.1516  0.5022  3.7600    *  16.9591

comp:     :  53.5905 89.7120 25.4791 23.9104
```

Equations (10.17) and (10.15) show that the individual components $U_{sj}$ may be interpreted as the *jth-order effect* of the $s$th sample relative to the pooled (or marginal) distribution. Pairwise differences of the $U_{sj}$ between samples but within the same order $j$ are informative about $j$th-order differences between the samples. Note that we deliberately used the terminology *jth-order effects/differences* to avoid the difficult issues regarding their interpretation. The output also shows the row sums of squares,

$$R_s^2 = \sum_{j=1}^{o} U_{sj}^2,$$

with $o = 4$ the selected order. This allows us to write the order 4 test statistic as

$$T_4 = \sum_{s=1}^{K} R_s^2.$$

Each $R_s^2$ measures how far the $s$th sample distribution is away from the pooled distribution. A similar type of decomposition was also proposed by Boos (1986). The analysis thus suggests that particularly the distributions of the travel times with routes 3 and 5 deviate from the marginal distribution of travel times, and the distribution of travel times with route 4 comes closest to the marginal distribution. The marginal distribution is to be interpreted as the distribution that would arise when each taxi driver chose each route with a relative frequency of $\frac{1}{5}$.

Because route 1 (i.e., sample 1 in the output) is considered as the reference route, we compare it with the other routes. We only look at the first two order components.

```
> components(traffic.st,contrast=c("control",1),order=2)

 sample 1:  0        0
 sample 2: -3.3421  1.6653
 sample 3:  5.7906  4.0554
 sample 4: -2.4524  6.7323
 sample 5:  1.8949  1.2053
```

These results suggest that in terms of likely ordering, and compared to the reference route 1, it is more likely to have a faster taxi ride with routes 2 and 4. With route 3, on the other hand, it is more likely to spend more time in the taxi. These conclusions are consistent with what we have concluded before and with the boxplots of Figure 6.2. Note that we should actually first assess the diagnostic property of the components by calculating the empirical variances.

The second-order components should be interpreted with even more care. As $\delta_M$ in (10.12) demonstrates, these second-order components measure a combined effect of single and double likely ordering. It is also here not possible to use these components for formulating conclusions in terms of the scale difference measure $\pi^{(4)}$, because we have no reason to believe that the median travel times coincide (see Section 9.5.3.4). We believe, however, that it is more informative at this point to use the estimated double likely ordering probabilities. From the output we see that all estimates are positive. They are, however, slightly more difficult to interpret because they quantify a combined effect of first-order and second-order (or double) likely ordering, as can be seen from the form of the second-order Legendre polynomial (Section 2.6.2). The sign of the effects does thus not necessarily say something about the direction of the double likely ordering effect. At this point it would thus be more informative to estimate the double likely ordering probabilities so that the direction of the effect can be observed. However, we do not pursue this further here.

## 10.6 Smooth Tests That Are Not Based on Ranks

All smooth tests discussed in this chapter up to now are basically rank tests. The reason for this is they were defined starting from a smooth alternative to the pooled distribution $h$. This procedure required at some point the estimation of the pooled distribution function $H$ by the EDF $\hat{H}$, which introduced the ranks into the smooth test statistics.

In this section we describe a test for the $K$-sample problem that is not based on ranks. The method is due to Chervoneva and Iglewicz (2005) and it starts from orthogonal series expansions of the densities $f_1, \ldots, f_K$ (see Section 2.8.2). In particular, consider the order $k$ expansions ($s = 1, \ldots, K$)

$$f_{sk}(x) = \sum_{j=0}^{k} \theta_{sj} h_j(x), \tag{10.19}$$

where $\{h_j\}$ is a set of orthonormal functions on the uniform $[0,1]$ distribution, and where $\theta_{s0} = 1$ $(s = 1, \ldots, K)$ so that the order $k$ densities integrate to one. The $K$ models (10.19) are smooth alternatives to one another, and when it is assumed that the true $K$ densities are embedded in these order $k$ expansions, equality of the $K$ densities may be formulated by the null hypothesis

$$H_0 : \theta_{1j} = \theta_{2j} = \cdots = \theta_{Kj} \text{ for all } j = 1, \ldots, k.$$

Chervoneva and Iglewicz (2005) proposed to test this null hypothesis by means of a Wald test. This requires asymptotically normally distributed estimators of the $\theta_{sj}$ parameters, and a consistent estimator of their variance–covariance matrix.

Many times before we have seen that $\hat{\theta}_{sj} = (1/n)\sum_{i=1}^{n_s} h_j(X_{si})$ is an unbiased estimator of $\theta_{sj}$. Let $\boldsymbol{\theta}_s^t = (\theta_{s1}, \ldots, \theta_{sk})$ and $\hat{\boldsymbol{\theta}}_s^t = (\hat{\theta}_{s1}, \ldots, \hat{\theta}_{sk})$. Based on the $n_s$ sample observations of sample $s$, it is quite straightforward to show that, as $n_s \to \infty$,

$$\sqrt{n}(\hat{\boldsymbol{\theta}}_s - \boldsymbol{\theta}_s) \xrightarrow{d} MVN(\mathbf{0}, \boldsymbol{\Sigma}_s),$$

where the $(i,j)$th element of $\boldsymbol{\Sigma}_s$ is given by

$$\int_S h_i(x) h_j(x) f_s(x) dx - \theta_{si}\theta_{sj}.$$

This variance–covariance matrix may be estimated by a $U$-statistic, say $\hat{\boldsymbol{\Sigma}}_s$. In particular, the $(i,j)$th element of $\boldsymbol{\Sigma}_s$ may be unbiasedly estimated by

$$\frac{1}{n}\sum_{l=1}^{n_s} h_i(X_{sl}) h_j(X_{sl}) - \frac{1}{n_s(n_s-1)} \sum_{l=1}^{n_s} \sum_{m=1; m\neq l}^{n_s} h_i(X_{sl}) h_j(X_{sm}).$$

Because in the $K$-sample problem the $K$ samples consist of independently distributed observations, the $K$ vectors $\hat{\boldsymbol{\theta}}_s$ are also independently distributed. With this information, a Wald test statistic can be constructed.

## 10.7 Some Practical Guidelines for Smooth Tests

We list some of the most important features of smooth tests for the two- and the $K$-sample problem, as well as some practical guidelines.

- The smooth test statistics are basically rank statistics and easy to compute (only the tests described in Section 10.6 are not rank tests).

- The smooth tests have generally good power for detecting differences between distributions. This is illustrated in simulation studies. These studies suggest that an order of $k = 4$ for small sample sizes (e.g., $n = 50$ for $K = 2, 3$) and an order of up to $K = 6$ for larger datasets is sufficient for detecting many important deviations from the null hypothesis.
- For testing the general two- or $K$-sample null hypothesis, the exact null distribution may be enumerated (or approximated using Monte Carlo simulations). The convergence to the asymptotic $\chi^2$ null distribution is fairly slow when $k > 2$.
- The smooth test statistic decomposes into components that are again rank statistics. For the two-sample problem the first two components are basically the Wilcoxon rank sum statistic and the Mood statistics, respectively, and the for $K$-sample problem we find the Kruskal–Wallis and the generalised Mood statistics. Higher-order components ($k > 2$) may be considered as further generalisations of the Kruskal–Wallis and Mood statistics.
- The interpretation of the components should be done with great care. The components can only be related to differences in moments under some particular distributional assumptions. More generally they are related to likely orderings. From a theoretical point of view the components can be properly rescaled so that they possess a diagnostic property under less stringent conditions, but from simulation studies we have learnt that very large sample sizes are required before this rescaling does the job. We therefore actually do not recommend this procedure in general. See Chapter 9 for a detailed discussion.
- The smooth tests are related to the comparison distribution. This may be seen from the order $k$ smooth alternatives on which the construction of the smooth test statistic relies. This relation allows us to estimate the comparison densities using the components of the smooth test. A graphical display of the estimated comparison distribution may be very helpful in formulating the conclusions from the statistical analysis. Moreover, inasmuch as the graph and the test have such a close connection, the risk of finding contradictory conclusions is small.
- For choosing the order $k$ of the smooth test, data-driven selection rules can be used.
- The test of Chervoneva and Iglewicz (2005) may be considered as a smooth test which is not based on ranks. From a simulation study we have learned that it has good power when the densities are well approximated by a low-order linear series expansion. However, the basis functions must actually be chosen prior to looking at the data, so that there is a fair risk of ending up with a small power. Moreover, the test requires an empirical covariance estimate, which has a negative effect on the power.

# Chapter 11
# Methods Based on the Empirical Distribution Function

This chapter is devoted to tests for the two- and $K$-sample problems that are based on the empirical distribution functions (EDF) of the distributions to be compared. Such tests are generally known as *EDF tests.* The types of tests that are treated in this chapter are often of the same form of the EDF tests for the one-sample problem (Chapter 5). The Kolmogorov–Smirnov test is discussed in Section 11.1, and Section 11.2 concerns tests of the Anderson–Darling type. We conclude the chapter with some practical guidelines in Section 11.4. As in Chapter 5 we again prefer the use of empirical processes for studying the asymptotic properties of the tests.

## 11.1 The Two-Sample and $K$-Sample Kolmogorov–Smirnov Test

### *11.1.1 The Kolmogorov–Smirnov Test for the Two-Sample Problem*

#### 11.1.1.1 The Test Statistic

The Kolmogorov–Smirnov (KS) test for testing the general two-sample null hypothesis $H_0 : F_1 = F_2$ versus $H_1 : F_1 \neq F_2$ uses the test statistic

$$D_n = \sqrt{\frac{n_1 n_2}{n}} \sup_{x\in\mathcal{S}} \left| \hat{F}_{1n_1}(x) - \hat{F}_{2n_2}(x) \right| = \sup_{x\in\mathcal{S}} \left| \mathbb{C}_{n12}(x) \right|, \tag{11.1}$$

where $\mathbb{C}_{n12}(x) = \sqrt{n_1 n_2/n}(\hat{F}_{1n_1}(x) - \hat{F}_{2n_2}(x))$ is the contrast process (7.22).

O. Thas, *Comparing Distributions,* Springer Series in Statistics, 297
DOI 10.1007/978-0-387-92710-7_11, 

The rationale for the construction of $D_n$ is obvious. Just as for the one-sample KS test, the test statistic is related to the largest difference between the distribution functions $F_1$ and $F_2$. The statistic is rewritten as

$$D_n = \sqrt{\frac{n_1 n_2}{n}} \sup_{0<p<1} \left| \hat{F}_{1n_1}(\hat{F}_{2n_2}^{-1}(p)) - p \right| = \sqrt{\frac{n_2}{n}} \sup_{0<p<1} |\mathbb{C}_{n12}(p)| ,$$

where $\mathbb{C}_{n12}(p) = \sqrt{n_1}(\hat{F}_{1n_1}(\hat{F}_{2n_2}^{-1}(p)) - p)$ is now defined as the comparison distribution process (7.27). This representation shows that $D_n$ is the scaled maximal deviation of the PP plot from the 45 degree line. See Section 8.1 for more details on PP plots.

Similar as for the one-sample KS test, the test statistic $D_n$ may be formulated as $D_n = \sqrt{n_2/n} \max(D_n^+, D_n^-)$, where $D_n^+ = \sup_{0<p<1} \mathbb{C}_{n12}(p)$ and $D_n^- = \sup_{0<p<1}(-\mathbb{C}_{n12}(p))$. The statistics $D_n^+$ and $D_n^-$ are often suggested for testing the general two-sample null hypothesis versus the alternative of stochastic ordering, $F_1(x) > F_2(x)$ or $F_1(x) < F_2(x)$ for all $x \in \mathcal{S}$, but we do not recommend that usage here, unless some additional restrictions on the shapes of $F_1$ and $F_2$ are imposed (e.g., location-shift model).

#### 11.1.1.2 The Null Distribution

Because $D_n$ is basically a rank statistic, its exact permutation null distribution may be enumerated using the methods of Section 7.1.

The asymptotic null distribution follows directly from the weak convergence of the $\mathbb{C}_{n12}$ process (as a contrast process, or as comparison distribution process) and the continuous mapping theorem. It is, however, more convenient to use the comparison distribution process representation. Thus, on using (7.24) or Theorem 7.6, we find the following result. Under $H_0$, as $n \to \infty$,

$$D_n \xrightarrow{d} \sup_{p \in (0,1)} \left| \sqrt{1-\lambda}\mathbb{B}_1(p) - \sqrt{\lambda}\mathbb{B}_2(p) \right| ,$$

where $\mathbb{B}_1$ and $\mathbb{B}_2$ are two independent uniform Brownian bridges.

The asymptotic null distribution of $D_n$ was first studied by Smirnov (1939). He found an explicit form of the CDF of asymptotic null distribution of $D_n$. When $D$ denotes the random variable that possesses the asymptotic null distribution of $D_n$, then $D$ has distribution function

$$F_D(x) = \sum_{i=-\infty}^{\infty} (-1)^i \exp\left(-2i^2x^2\right) \quad \text{for } x > 0.$$

The test based on $D_n$ is consistent against any fixed alternative $F_1 \neq F_2$.

### 11.1.2 The Kolmogorov–Smirnov Test for the K-Sample Problem

The extension from two to $K \geq 2$ samples is quite straightforward, but there are several possibilities. A first solution exists in constructing all contrast or comparison distribution processes $\mathbb{C}_{nij}$, $i, j = 1, \ldots, K$ with $i \neq j$, and define the $K$-sample KS statistic as

$$DK_n = \max_{i \neq j} \sup_{p \in (0,1)} |\mathbb{C}_{nij}(p)| .$$

Another solution exists in the use of the process that contrasts each individual EDF $\hat{F}_{in_i}$ $(i = 1, \ldots, K)$ with the pooled EDF $\hat{H}_n$. This involves thus the comparison distribution processes $\mathbb{C}_{ni}$ of (7.26). A $K$-sample KS statistic may then be defined as

$$DKp_n = \max_{j=1,\ldots,K} \sup_{p \in (0,1)} \sqrt{n_j} \left| \hat{F}_{jn_j}(\hat{H}_n^{-1}(p)) - p \right| = \max_{j=1,\ldots,K} \sup_{p \in (0,1)} |\mathbb{C}_{nj}(p)| .$$

A related test uses the statistic

$$DK2p_n = \sup_{p \in (0,1)} \sum_{s=1}^{K} n_s \left( \hat{F}_{sn_s}(\hat{H}^{-1}(p)) - p \right)^2 .$$

As for the two-sample case, the exact distributions of these statistics under the general two-sample null hypothesis can be enumerated and the asymptotic approximations follow along the same lines. To our knowledge, however, these tests have actually never been used in real data analyses. We refer to Dwass (1960) and Fisz (1960) for some more details on these tests.

## 11.2 Tests of the Anderson–Darling Type

### 11.2.1 The Test Statistic

The Anderson–Darling (AD) test of Section 5.2.1 can be readily extended to the two- and the $K$-sample problem. Pettitt (1976) studied the two-sample AD statistic,

$$A_n = \frac{n_1 n_2}{n} \int_S \frac{\left( \hat{F}_{1n_1}(x) - \hat{F}_{2n_2}(x) \right)^2}{\hat{H}_n(x)(1 - \hat{H}_n(x))} d\hat{H}_n(x). \tag{11.2}$$

As the integrand does not exist for $x$ equal to or larger than the largest observation in the pooled sample, it is defined as zero for those $x$. An equivalent formulation is

$$A_n = \frac{n}{n_2}\int_S \frac{\left(\hat{F}_{1n_1}(x) - \hat{H}_n(x)\right)^2}{\hat{H}_n(x)(1-\hat{H}_n(x))} d\hat{H}_n(x). \tag{11.3}$$

The equivalent computational form of $A_n$ is then given by

$$A_n = \frac{1}{n_1 n_2}\sum_{i=1}^{n-1}\frac{(nM_i - n_1 i)^2}{i(n-i)},$$

where $M_i = n_1\hat{F}_{1n_1}(\hat{H}_n^{-1}(i/n))$ is the number of observations in the first sample that is not larger than the $i$th smallest in the pooled sample.

In passing we like to stress once more that we limit our presentation here to continuous distributions $F_1,\ldots,F_K$. This is necessary for the different forms of the Anderson–Darling statistics to be (at least asymptotically) equivalent. This is, for example, important when applying the transformation $x = \hat{H}_n^{-1}(p)$ so that the continuous version of the test statistic can be studied as well. Define the process

$$\mathbb{P}_n(p) = \frac{\mathbb{C}_{n1}(p)}{\sqrt{p(1-p)}}.$$

The continuous version is given by

$$A_n = \frac{n_1}{n_2}\int_0^1 \mathbb{P}_n^2(p)dp, \tag{11.4}$$

where again the domain of integration should be restricted, or where the process $\mathbb{P}_n$ should be modified for the statistic to be well defined. For example, Pettitt (1976) modified the process to $\mathbb{P}_n(p) = \sqrt{n}\{\hat{F}_{1n_1}(\hat{H}_n^{-1}(((n+1)/n)p) - p)\}$ when $p \leq n/(n+1)$ and $\mathbb{P}_n(p) = \sqrt{n}(1-t)/t$ when $p > n/(n+1)$.

For the one-sample AD test the transformation was less an issue, because there the variable of integration was $dG(x)$, in which $G$ was a specified distribution. Here, in (11.4), the variable of integration is $d\hat{H}_n(x)$ which is zero except when $x$ coincides with a sample observation.

For comparing $K$ distributions Scholz and Stephens (1987) proposed the statistic (both the original and the continuous version are given)

$$\begin{aligned} AK_n &= \sum_{s=1}^{K} n_j \int_S \frac{\left(\hat{F}_{sn_s}(x) - \hat{H}_n(x)\right)^2}{\hat{H}_n(x)(1-\hat{H}_n(x))} d\hat{H}_n(x) \\ &= \sum_{s=1}^{K}\frac{n_s}{n}\int_0^1 \frac{\mathbb{C}_{ns}^2(p)}{p(1-p)}dp, \end{aligned}$$

where the integrand is again defined as zero for $x$ for which the integrand does not exist (an equivalent way of circumventing this problem is to restrict the integration domain). The computational formula now becomes

$$AK_n = \frac{1}{n}\sum_{s=1}^{K}\frac{1}{n_s}\sum_{i=1}^{n-1}\frac{(nM_{si}-n_s i)^2}{i(n-i)}, \tag{11.5}$$

where $M_{si} = n_s\hat{F}_{sn_s}(\hat{H}_n^{-1}(i/n))$ is the number of observations in the $s$th sample that is not larger than the $i$th smallest observation in the pooled sample.

### 11.2.2 The Components

Just as in the one-sample case also here the AD test statistics decompose into components by applying a Kac–Siegert expansion. In Section 5.2.2 we have worked out those expansions in some detail for the one-sample AD test. For the AD tests of this chapter, basically the same machinery may be applied. There are actually two issues that arise only here, and were thus not discussed in previous chapters. We comment briefly on them, because they are important for arguing that the same principal component methodology of Section 5.2.2 may be applied.

1. The integrand of the AD statistic is not defined for all $p \in [0,1]$. This can be solved by restricting the domain of integration or by redefining the process $\mathbb{P}_n$ as Pettitt (1976) did. For the statistics that we study, these modifications do not have any effect, at least not asymptotically. We therefore choose not to make these modifications explicit throughout this chapter.
2. The process on which the Kac–Siegert expansion has to be applied must be sufficiently similar to those considered in the one-sample case. This is the case for the $\mathbb{P}_n$ process which is defined in terms of the comparison distribution function $\hat{F}_{1n_1}(\hat{H}_n^{-1}(p))$, but this does not hold for the contrast process $\hat{F}_{1n_1}(x) - \hat{H}_n(x)$, because of the discrete nature of $\hat{H}_n$. This is resolved in the former process.

Now that the similarity, and particularly the differences, with the one-sample AD test statistic have been clarified, we are ready to decompose the test statistic. We start with deriving the (asymptotic) covariance function of the $\mathbb{P}_n$ process. For notational comfort we first ignore the factor $1/\sqrt{p(1-p)}$. Let $c(p,q) = p \wedge q - pq$ denote the covariance function of a Brownian bridge and let $\mathbb{B}_1$ and $\mathbb{B}_2$ be two independent Brownian Bridges. The covariance function

$$c^*(p,q) = \text{Cov}_0\left\{\mathbb{C}_{n1}(p), \mathbb{C}_{n1}(q)\right\}$$

becomes for large sample sizes, on using Theorem 7.6,

$$\begin{aligned} c^*(p,q) &= \text{Cov}_0\left\{(1-\lambda)\left(\frac{1}{\sqrt{\lambda}}\mathbb{B}_1(p)+\frac{1}{\sqrt{1-\lambda}}\mathbb{B}_2(p)\right),\right. \\ &\left.\quad (1-\lambda)\left(\frac{1}{\sqrt{\lambda}}\mathbb{B}_1(q)+\frac{1}{\sqrt{1-\lambda}}\mathbb{B}_2(q)\right)\right\} \\ &= (1-\lambda)^2\left(\frac{1}{\lambda}c(p,q)+\frac{1}{1-\lambda}c(p,q)\right) \\ &= \lambda(1-\lambda)c(p,q). \end{aligned}$$

Thus the process $(1/\sqrt{\lambda(1-\lambda)})\mathbb{P}_n(p)$ or $(n/\sqrt{n_1n_2})\mathbb{P}_n(p)$ has the same covariance function as the process studied in Section 5.2.2.2 for the one-sample AD statistic. Thus the same eigenvalues and eigenfunctions as in (5.7) can be used here, and we do not repeat all details here. We only lift out the part of the calculations in which ranks will enter the components, using eigenvalues $\lambda_j = 1/(j(j+1))$ and eigenfunctions

$$l_j(p) = 2\sqrt{\frac{1}{j(j+1)}}\sqrt{p(1-p)}\frac{d}{dp}L_j(p).$$

Thus,

$$\begin{aligned} A_n &= \frac{n_1n_2}{n^2}\frac{n_1}{n_2}\int_0^1\left(\frac{n\mathbb{C}_{n1}(p)}{\sqrt{n_1n_2}\sqrt{p(1-p)}}\right)^2 dp \\ &= \frac{n_1n_2}{n^2}\frac{n_1}{n_2}\sum_{j=1}^{\infty}\frac{1}{j(j+1)}Z_{nj}^2, \end{aligned}$$

where

$$\begin{aligned} Z_{nj} &= \frac{1}{\sqrt{\lambda_j}}\int_0^1\left(\frac{n\mathbb{C}_{n1}(p)}{\sqrt{n_1n_2}\sqrt{p(1-p)}}\right)l_j(p)dp \\ &= 2\frac{n}{\sqrt{n_1n_2}}\int_0^1\mathbb{C}_{n1}(p)dL_j(p) \\ &= -2\frac{n}{\sqrt{n_1n_2}}\int_0^1 L_j(p)d\mathbb{C}_{n1}(p), \end{aligned}$$

where the last equation results from integration by parts. This becomes

$$Z_{nj} = -2\frac{n\sqrt{n}}{\sqrt{n_1n_2}}\int_0^1 L_j(p)d\hat{F}_{1n_1}(\hat{H}_n^{-1}(p)) + 2\frac{n\sqrt{n}}{\sqrt{n_1n_2}}\int_0^1 L_j(p)dp,$$

in which the last term is zero because of the orthogonality of the Legendre polynomials $L_j$. For the first term we know that $\hat{H}_n^{-1}(p)$ is a piecewise constant function with jumps in $i/n$ $(i = 1, \ldots, n)$, and $d\hat{F}_{1n_1}(x)$ is zero except when $x$ equals an observation from the first sample. Define $x_i$ as an indicator so that $x_i$ is 1 when observation $i$ comes from the first sample, and $x_i = 0$ otherwise. Then we may write

$$\begin{aligned} Z_{nj} &= -2\frac{n\sqrt{n}}{\sqrt{n_1 n_2}} \sum_{i=1}^{n} L_j\left(\frac{i}{n}\right)\frac{x_i}{n_1} \\ &= -2\frac{n\sqrt{n}}{\sqrt{n_1 n_2}}\frac{1}{n_1} \sum_{i=1}^{n_1} L_j\left(\frac{R_i}{n}\right), \end{aligned}$$

where $R_i$ is the rank of observation $X_{1i}$ in the pooled sample. With this expression for $Z_{nj}$ we find for the AD test statistic

$$A_n = 4\frac{n}{n_1 n_2} \sum_{j=1}^{\infty} \frac{1}{j(j+1)} \left( \sum_{i=1}^{n_1} L_j\left(\frac{R_i}{n}\right) \right)^2. \tag{11.6}$$

From this representation we see that $A_n$ decomposes into a series of weighted components, $\sum_{i=1}^{n_1} L_j\left(R_i/n\right)$ that are simple linear rank statistics. The functions $L_j$ are $j$th-order Legendre polynomials, so that these components equal the components of the smooth test statistics of Section 10.1.2, where we used the notation $h_j$ instead of $L_j$ for the $j$th-order polynomial. The weight of the $j$th component equals $1/(j(j+1))$ $(j = 1, \ldots)$, just as for the one-sample AD statistic. Consequently, the first component is proportional to the WMW statistic, the second to the Mood statistic, etc. We can thus simply refer to Section 10.1.2 for a discussion of the interpretation of the components.

In the calculations resulting in (11.6) we have chosen not to adopt any particular continuity correction, but surely it allows for any correction of choice.

### *11.2.3 The Null Distribution*

Apart from the exact permutation null distribution for testing the general two- or $K$-sample null hypothesis, also the asymptotic null distributions may be used. Although the convergence is not very good, there are some minor modifications to $A_n$ and $AK_n$ suggested to highly improve the approximation. When $A_n$ is replaced by

$$A_n^* = (A_n - 1)\left(1 + \frac{1.55}{n}\right) + 1$$

the asymptotic null distribution may be used to get quite accurate results in the upper tail of the distribution (Pettitt (1976)). The rationale for this transformation is that it makes the first two moments of $A_n^*$ and the asymptotic null distribution fit better. A similar modification has been suggested for the $K$-sample $AK_n$ statistic. The standardising transformation is, however, more complicated. We refer to Scholz and Stephens (1987) for details.

The asymptotic null distributions of $A_n$ and $AK_n$ result again from the weak convergence of the $\mathbb{C}_n$ processes (Theorem 7.6) and from the continuous mapping theorem, but there are some complications that are caused by the typical Anderson–Darling weight function $1/(p(1-p))$ that causes the integrand to not exist for $p$ too close to 1. The trick exists in proving that the statistics defined as $\int_\delta^{1-\delta}$, with small $\delta > 0$, has a limiting distribution and that the remainder is asymptotically neglectable.

The asymptotic null distributions of the two statistics are provided in the following theorem.

**Theorem 11.1.** *Under $H_0$, as $n \rightarrow \infty$,*

$$A_n \xrightarrow{d} \int_0^1 \frac{\mathbb{B}_1^2(p)}{p(1-p)} dp,$$

*and*

$$AK_n \xrightarrow{d} \int_0^1 \frac{\sum_{j=2}^K \mathbb{B}_j^2(p)}{p(1-p)}, \tag{11.7}$$

*where $\mathbb{B}_1, \ldots, \mathbb{B}_K$ are $K$ mutually independent uniform Brownian bridges.*

It is worthwhile to focus for a moment on the limit distribution of $AK_n$ in (11.7). Note that it involves only $K-1$ mutually independent Brownian bridges. The arguments that result in the loss of one term are similar to those that we used in Section 10.3.2 for the $K$-sample Kruskal–Wallis test statistic.

### *11.2.4 Examples*

We illustrate the use of the Anderson–Darling test on the gene expression data. In Example 11.2 following we also provide a general discussion on the practical use of EDF tests, and on the relevance of the conclusions for the gene expression data in the colorectal cancer study.

*Example 11.1 (The gene expression data: Gene 1).* The two-sample Anderson–Darling test is also available through the EDF.test function in the cd R package.

```
> gene1.AD<-EDF.test(expression~group,type="AD",B=10000,
+ data=gene1)

          K-sample Anderson-Darling Test
```

```
data:  gene1
T = 5.61, number simulations = 10000, p-value = 0.0013
```

Based on the permutation null distribution, which is here approximated by 10,000 simulation runs, the Anderson–Darling test gives a $p$-value of 0.0013. Hence, at the 5% level of significance the general two-sample null hypothesis is rejected. Because the AD test allows a decomposition in terms of the same components as the two-sample smooth test, we advise looking at those components, or to use the components for calculating the estimates of the $\theta_j$ parameters of the orthogonal series expansion of the comparison densities. The components may again be obtained with the components function (see below). We refer to Example 10.2 for the interpretation of the components and the estimated comparison densities.

```
> components(gene1.AD,order=2)

Components
1 st component = 2.4622
2 nd component = -3.8114
```

*Example 11.2 (The gene expression data: All genes).* In the previous chapters we have done several analyses on the gene expression data. We first give the results of the two-sample Anderson–Darling tests for all four genes, and next we summarise our conclusions. Without presenting the R code and output, we report the $p$-values of the AD tests for genes 1 to 4: 0.0013, 0.0002, $< 0.0001$, and 0.0085. Thus for all genes the general two-sample null hypothesis is rejected at the 5% level of significance. At the rejection of the null hypothesis our general recommendation is

1. To look at the individual components and try to interpret them. For the first-order component, which is the WMW statistic, the interpretation is usually not too hard, but interpretation of higher-order components may become complicated, particularly when they have to be related to differences in scale, skewness, etc. In Chapter 10 we have illustrated several times the difficulties that may arise. The components may be interpreted in terms of likely orderings, but for the higher-order components the interpretation is not always very clear.
2. To use the components for the calculation of the estimates of the $\theta_j$ parameters in the orthogonal expansion of the comparison densities, and to plot the estimated comparison densities. Conclusions may then be formulated based on these graphs. Because the graphs and the AD test (or smooth tests) are based on the same components, they both contain the same information.

For illustrative purposes we have also performed Welch $t$-tests for the four genes. The $p$-values are: 0.04865, 0.05482, 0.07379, and 0.051. They are all very close to the nominal significance level of 5%. Moreover, because most of the gene expression distributions appear to be nonnormal, the $p$-values from the Welch $t$-tests may not be trusted. This leaves us inconclusive in terms of potential differences in mean expression values. On the other hand, however, for all four genes the first-order component (WMW statistic) turns out to be important. After careful assessment of the validity of the WMW test (i.e., empirical variance ratio close to one), conclusions may thus be formulated in terms of likely orderings. Likely orderings are also biologically very informative in the context of the present colorectal cancer study. For example, the probability $\Pr\{X_1 \leq X_2\} > 0.5$ indicates that among many pairs of patients, one of which is a nonprogressed adenoma patient and the other one is a carcinoma patient, there are relatively more pairs for which the carcinoma patient has a higher gene expression value than the nonprogressed adenoma patient. Thus even when the means of the two gene expression distributions are not very different, this stochastic difference may have a biological explanation, and such a gene can thus play an important role in a biological pathway that is related to colorectal cancer.

## 11.3 Adaptive Tests of Neuhaus

We describe here an apparently different framework for the construction of adaptive tests, but eventually we show that both the smooth tests and the EDF tests fit into it. It is based on a series of papers by Behnen and Neuhaus (1983), Behnen et al. (1983), Behnen and Husková (1984), and Neuhaus (1987). We refer to it as the adaptive tests of Neuhaus, because it is particularly the 1987 paper that comes to the general form that is introduced here. Although Neuhaus constructed the tests very formally, we have chosen to give here merely an intuitive construction of his method.

### *11.3.1 The General Idea*

In Section 7.2.2 we have given a taste of the proof Theorem 7.5, which gives the optimal scores for the LMPRT. This result is based on the fundamental Neyman–Pearson lemma, but it was applied to a paramcterised local alternative, so that a first-order Taylor approximation was appropriate. Here we actually start with an even simpler situation. Suppose we want to test the general two-sample null hypothesis versus the alternative

$$H_1 : f_1 = f_1^* \neq f_2^* = f_2,$$

where $f_1^*$ and $f_2^*$ are two different fixed densities. This is basically a *simple* alternative, and not a *composite* hypothesis as we always have considered before. The fundamental Neyman–Pearson lemma gives the most powerful test, which is the log-likelihood ratio (LRR) statistic

$$LLR \propto \sum_{i=1}^{n_1} \log \frac{f_1^*(Z_i)}{f_2^*(Z_i)} - \sum_{i=n_1+1}^{n} \log \frac{f_1^*(Z_i)}{f_2^*(Z_i)},$$

where $Z_1, \ldots, Z_n$ denotes again the order sample observations so that the first $n_1$ observations come from the first sample. For later convenience we rewrite $LLR$ using the regression constants $c_i = c_{ni}$ as defined in Theorem 10.1. These constants appear typically in two-sample rank statistics. With this notation we may write

$$LLR \propto \sum_{i=1}^{n} c_i \log \frac{f_1^*(Z_i)}{f_2^*(Z_i)}.$$

When $h$ denotes the pooled density

$$h = \frac{n_1}{n} f_1^* + \frac{n_2}{n} f_2^*,$$

and using the definition of the comparison density of (7.20), we may write

$$f_1^*(H^{-1}(u)) = r_1^*(u)h(H^{-1}(u)) \text{ and } f_2^*(H^{-1}(u)) = r_2^*(u)h(H^{-1}(u)),$$

where $r_1^*$ and $r_2^*$ denote the fixed comparison densities. At this point we still assume that $h$ and $H$ are known. The $LLR$ now becomes

$$LLR \propto \sum_{i=1}^{n} c_i \log \frac{r_1^*(H(Z_i))}{r_2^*(H(Z_i))}.$$

Up to now we have considered $f_1^*$ and $f_2^*$ (or $r_1^*$ and $r_2^*$) as fixed. Now we assume that they are very close to each other so that the $LLR$ statistic may be approximated to first order by

$$LLR \approx \sum_{i=1}^{n} c_i \left\{ r_1^*(H(Z_i)) - r_2^*(H(Z_i)) \right\},$$

and we further use the notation $b(u) = r_1^*(u) - r_2^*(u)$, which is basically a score function. Furthermore, the null hypothesis may actually also be expressed as $b(u) = 0$. Thus $LLR \approx \sum_{i=1}^{n} c_i b(H(Z_i))$. The first important conclusion is that an optimal test can be constructed if the function $b$ were known. Unfortunately this function depends on fixed alternatives whereas in reality we are actually interested in testing the general two-sample null hypothesis against all such fixed alternatives. The problem could be solved if we were in the position to estimate the function $b$. In the next two sections we discuss two types of solutions.

### 11.3.2 Smooth Tests

Consider the order $k$ Barton type expansions of the densities $f_1^*$ and $f_2^*$ in (10.4) and (10.5). The $LLR$ statistic then becomes

$$LLR \approx \sum_{i=1}^{n} c_i \sum_{j=1}^{k} \theta_j h_j(H(Z_i)).$$

This statistic can still not be directly used because the $\theta_j$ and $H$ are unknown. We now replace them by their estimators (see (10.10) for the estimator $\hat{\theta}_j$ of $\theta_j$). This results in the test statistic

$$N_k = \sum_{i=1}^{n} c_i \sum_{j=1}^{k} \left[ \sqrt{\frac{n}{n_1 n_2}} \sum_{l=1}^{n} c_l h_j(\hat{H}(Z_i)) \right] h_j(\hat{H}(Z_i)).$$

On using the convention that $\hat{H}(Z_i)$ is replaced by $\frac{R_i - 0.5}{n}$ we find

$$\begin{aligned} N_k &= \sqrt{\frac{n}{n_1 n_2}} \sum_{j=1}^{k} \sum_{i=1}^{n} \sum_{l=1}^{n} c_i c_l h_j \left( \frac{R_i - 0.5}{n} \right) h_j \left( \frac{R_l - 0.5}{n} \right) \\ &= \sqrt{\frac{n}{n_1 n_2}} \sum_{j=1}^{k} \left\{ \sum_{i=1}^{n} c_i h_j \left( \frac{R_i - 0.5}{n} \right) \right\}^2 \\ &= \sqrt{\frac{n}{n_1 n_2}} \sum_{j=1}^{k} U_j^2, \end{aligned}$$

with $U_j$ the $j$th component of the two-sample smooth test statistic as in Theorem 10.1. Hence, the test based on $N_k$ is equivalent to the order $k$ two-sample smooth test.

### 11.3.3 EDF tests

The general approach of Neuhaus (1987) consists in replacing $b(u)$ in the $LLR$ statistic by a nonparametric estimator. A very rough estimate may be obtained by first recognising that (dropping the * notation)

$$b(u) = r_r(u) - r_2(u) = \frac{d}{du} \left\{ F_1(H^{-1}(u)) - F_2(H^{-1}(u)) \right\}$$

(see Section 2.4). When we define $B(u) = F_1(H^{-1}(u)) - F_2(H^{-1}(u))$, we have $b(u) = dB(u)/du$. The function $B$ has a nonparametric estimator by simply replacing the CDFs with their empirical counterparts. This gives

$$\hat{B}(u) = \hat{F}_1(\hat{H}^{-1}(u)) - \hat{F}_2(\hat{H}^{-1}(u)).$$

On using this empirical estimator, the function $b$ may be roughly estimated by

$$\bar{b}(u) = n\left\{\hat{B}\left(\frac{i}{n}\right) - \hat{B}\left(\frac{i-1}{n}\right)\right\} \text{ for } \frac{i-1}{n} \leq u < \frac{i}{n}$$

$(i = 1, \ldots, n)$.

Neuhaus (1987) continues by smoothing $\bar{b}(u)$ using kernel functions. In particular, when $K(u,v)$ $((u,v) \in [0,1]^2)$ is a proper square integrable convolution kernel, the smoothed score function becomes

$$\tilde{b}(u) = \int_0^1 \bar{b}(v)K(u,v)dv.$$

A test statistic for testing $H_0 : b(u) = 0$ may be constructed as the squared norm in the Hilbert space, or, as Neuhaus (1987) proposed,

$$T_n = < \tilde{b}, \bar{b} > .$$

Without going into further detail here, it is interesting to note that $T_n$ is a rank statistic and that it has a principal component decomposition of the form

$$T_n = \sum_{j=1}^{\infty} \lambda_j < \psi_j, \bar{b} >^2,$$

where the $\lambda_j$ and the $\psi_j$ are eigenvalues and eigenfunctions related to the kernel function $K$. For a particular choice of $K$, Neuhaus (1987) showed that $T_n$ becomes the two-sample CvM statistic. This illustrates that the construction of Neuhaus is quite general.

## 11.4 Some Practical Guidelines for EDF Tests

We list some of the most important features of the EDF tests for the two- and the $K$-sample problem, as well as some practical guidelines.

- The EDF tests of the Kolmogorov–Smirnov type are related to PP plots, but they generally have not very good powers. They may, however, be useful for detecting stochastic ordering, though better and more specialised tests are available.
- The EDF tests of the Anderson–Darling type have overall good power properties.
- The EDF tests are rank tests.
- Under the general two- and $K$-sample null hypothesis, the exact null distributions may be enumerated (or at least approximated by means of Monte Carlo simulations).

- The Anderson–Darling statistic decomposes into components that coincide with those of the smooth test statistics. Thus they possess the same properties as discussed in the two previous chapters. We therefore recommend to first apply an Anderson–Darling test and when the general two-sample or $K$-sample null hypothesis is rejected, we recommend using the individual components in exactly the same manner as with the smooth tests of the previous chapter (e.g., interpretability and estimation of the comparison density).
- An advantage of the Anderson–Darling test over the smooth test is that the former does not require a data-driven order selection rule. It weights the components of order $j$ with weights of the order $1/j^2$ so that particularly the lower-order components determine the power of the test. Because in many real situations the most interesting and relevant differences between the distributions can be described by lower-order basis functions, experience has snown that the Anderson–Darling test has overall good powers.

# Chapter 12
# Two Final Methods and Some Final Thoughts

A seemingly completely different approach to arrive at two- or $K$-sample EDF test statistics is described in Section 12.1. This method is known as the *contingency table approach*, as proposed by Rayner and Best (2001). The *sample space partition tests* of Section 12.2 can be looked at as a combination of the contingency table approach and the tests of the EDF type. Although both type of tests are basically EDF tests, we have chosen to present them in a separate chapter, because the manner in which they are constructed deviates from what is seen in the previous chapters. This chapter, and the book, is concluded with Section 12.3 with some final thoughts.

## 12.1 A Contigency Table Approach

Rayner and Best (2001) proposed a quite general method of constructing nonparametric rank tests. Their method consists in constructing an appropriate contingency table which is filled up with counts related to the sample observations, and once this table is set up, they calculate the statistic of Pearson's chi-squared test for indepedence. The components that result from a decomposition of the Pearson statistic turn out to be well-known rank statistics and generalisations of them. In this section we give an outline of their method for the $K$-sample problem.

As before, we use the notation $Z_i$ for denoting an observation from the pooled sample, and the $Z_{(i)}$ are the corresponding order statistics. Consider a $K \times n$ table with $(s, i)$th element $N_{si}$ defined as

$$N_{si} = \begin{cases} 1 & \text{if observation } Z_{(i)} \text{ is in sample } s \\ 0 & \text{otherwise.} \end{cases}$$

O. Thas, *Comparing Distributions*, Springer Series in Statistics,
DOI 10.1007/978-0-387-92710-7_12, 

This table has row sums equal to $n_s$ ($s = 1, \ldots, K$) and all column totals are equal to $1/n$. The Pearson chi-squared test statistic may then be written as

$$X^2 = \sum_{s=1}^{K} \sum_{i=1}^{n} \frac{\left(N_{si} - n\frac{1}{n}\frac{n_s}{n}\right)^2}{n\frac{1}{n}\frac{n_s}{n}}$$
$$= \sum_{s=1}^{K} \boldsymbol{Y}^t \boldsymbol{Y}$$

where $\boldsymbol{Y}$ is a vector with $i$th element equal to $i = 1, \ldots, n$

$$\frac{N_{si} - \frac{n_s}{n}}{\sqrt{\frac{n_s}{n}}} = \frac{\sqrt{n}}{\sqrt{n_s}}\left(N_{si} - \frac{n_s}{n}\right).$$

Let $\boldsymbol{D} = \text{diag}(1/n)$ be an $n \times n$ diagonal matrix and let $\boldsymbol{H}$ be a $k \times n$ ($k < n$) orthonormal matrix, with $(j, i)$th element $h_{ji}$ ($j = 1, \ldots, k$, $i = 1, \ldots, n$) so that $\boldsymbol{H}^t\boldsymbol{D}\boldsymbol{H} = \boldsymbol{I}$ and so that each column vector is orthogonal to $(1, \ldots, 1)$. The columns of $\boldsymbol{H}$ are thus orthonormal vectors on the discrete uniform distribution on $n$ support points. With this notation the Pearson statistic may be written as

$$X^2 = \sum_{s=1}^{K} \boldsymbol{Y}_s^t \boldsymbol{Y}_s = \sum_{s=1}^{K} \boldsymbol{Y}_s^t \boldsymbol{H}^t \boldsymbol{D}\boldsymbol{H}\boldsymbol{Y}_s = \left(\boldsymbol{D}^{1/2}\boldsymbol{H}\boldsymbol{Y}_s\right)^t \left(\boldsymbol{D}^{1/2}\boldsymbol{H}\boldsymbol{Y}_s\right),$$

in which the vectors $\boldsymbol{D}^{1/2}\boldsymbol{H}\boldsymbol{Y}_s$ have the $j$th element equal to ($j = 1, \ldots, k$),

$$\sum_{i=1}^{n} h_{ji} \frac{1}{\sqrt{n_s}} \left(N_{si} - \frac{n_s}{n}\right) = \frac{1}{\sqrt{n_s}} \sum_{i=1}^{n} h_{ji} N_{si},$$

where we made use of the property $\sum_{i=1}^{n} h_{ji} = \sum_{i=1}^{n} h_{ji} \times 1 = 0$. Let $W_{sj}$ denote this element. Hence,

$$X^2 = \sum_{s=1}^{K} \sum_{j=1}^{k} W_{sj}^2. \tag{12.1}$$

In Chapter 10 we have seen a similar decomposition of the order $k$ smooth $K$-sample test statistic (Equation (10.16)). There the components had the expression (see (10.14))

$$U_{sj} = \sqrt{\frac{n - n_s}{n}} \sum_{i=1}^{n} c_{nsi} h_j \left(\frac{R_i - 0.5}{n}\right),$$

with $R_i$ the rank of observation $i$ in the pooled sample, and the constants $c_{nsi}$ as defined in (10.13). We next show that the components $W_{sj}$ and $U_{sj}$ are equivalent. Write

$$
\begin{aligned}
U_{sj} &= \frac{n-n_s}{n}\sqrt{n_s} \\
&\times \sum_{i=1}^{n}\left\{\frac{1}{n_s}h_j\left(\frac{R_i-0.5}{n}\right)N_{si} - \frac{1}{n-n_s}h_j\left(\frac{R_i-0.5}{n}\right)(1-N_{si})\right\} \\
&= \frac{n-n_s}{n}\sqrt{n_s} \\
&\times \sum_{i=1}^{n}\left\{h_j\left(\frac{R_i-0.5}{n}\right)N_{si}\left(\frac{1}{n_s}+\frac{1}{n-n_s}\right) - h_j\left(\frac{R_i-0.5}{n}\right)\frac{1}{n-n_s}\right\} \\
&\approx \frac{1}{\sqrt{n_s}}\sum_{i=1}^{n} h_{ji}N_{si},
\end{aligned}
$$

where in the final step we used two properties of the orthonormal polynomials. First, because the $h_j$ are orthonormal on the continuous uniform $[0,1]$ distribution and because they are evaluated in the scaled ranks $(R_i - 0.5)/n$, they can be very well approximated by the orthonormal polynomial vectors $(h_{j1}, \ldots, h_{jn})$, which form a set of orthonormal vectors on the discrete uniform distribution. The second property we used is again $\sum_{i=1}^{n} h_{ji} = \sum_{i=1}^{n} h_{ji} \times 1 = 0$. These calculations demonstrate that $U_{sj} = W_{sj}$ for all $s = 1, \ldots, K$ and $j = 1, \ldots, k$, and we therefore may conclude that $X^2$ statistic from the contingency table approach of Rayner and Best (2001) coincides with an order $k = n$ smooth $K$-sample statistic, and both statistics have the same decomposition. Finally note that for each sample size $n$, the statistic $X^2 \equiv n(K-1)$, so that it has no proper distribution, but its components of course have.

## 12.2 The Sample Space Partition Tests

In Section 5.4 we have generalised the Anderson–Darling test to the *sample space partition test* SSP, or, more specifically the SSPc test, in which the letter $c$ stands for the number of intervals on which the *localised* Pearson statistics are calculated. In this section we describe a similar type of extension of the $K$-sample Anderson–Darling statistic of Section 11.2 to a sample space partition test. This class of tests was first introduced by Thas and Ottoy (2004).

First we write the $AK_n$ statistic (11.5) as

$$
\begin{aligned}
AK_n &= \frac{1}{n}\sum_{s=1}^{K}\frac{1}{n_s}\sum_{i=1}^{n-1}\frac{(nM_{si}-n_s i)^2}{i(n-i)} \\
&= \sum_{s=1}^{K}\frac{1}{n}\sum_{i=1}^{n-1}\left\{\frac{\left(M_{si}-n_s\frac{i}{n}\right)^2}{n_s\frac{i}{n}} + \frac{\left((n_s-M_{si})-n_s(1-\frac{i}{n})\right)^2}{n_s\frac{n-i}{n}}\right\},
\end{aligned}
$$

in which

$$X_s^2(i) = \frac{\left(M_{si} - n_s \frac{i}{n}\right)^2}{n_s \frac{i}{n}} + \frac{\left((n_s - M_{si}) - n_s(1 - \frac{i}{n})\right)^2}{n_s \frac{n-i}{n}}$$
$$= n_s \left\{ \frac{\left(\hat{F}_s(\hat{H}^{-1}(i/n)) - \frac{i}{n}\right)^2}{\frac{i}{n}} + \frac{\left(1 - \hat{F}_s(\hat{H}^{-1}(i/n)) - (1 - \frac{i}{n})\right)^2}{1 - \frac{i}{n}} \right\}$$

may be recognised as the Pearson chi-squared statistic for testing the null hypothesis

$$H_0 : F_s\left(H^{-1}\left(\frac{i}{n}\right)\right) = H\left(H^{-1}\left(\frac{i}{n}\right)\right) = \frac{i}{n}.$$

This null hypothesis is basically a hypothesis about the probability parameter of a binomial distribution. The probability $F_s(H^{-1}(i/n))$ is implied by categorising the data into two groups or cells, and the cell boundary is determined by the observation $Z_{(i)}$. We say that the Pearson statistic $X_s^2(i)$ is *localised* at $Z_{(i)}$. This is completely analogous to how the one-sample AD test statistic has been interpreted in Section 5.4. The $K$-sample AD statistic is thus basically an average of such *localised* Pearson statistics, averaged over the first $n-1$ order statistics $Z_{(i)}$. The largest order statistic may not be used, as explained in Section 11.2.

The SSPKc test statistic is obtained by extending the localised Pearson statistic $X_s^2(i)$ to Pearson statistics that are localised at more than one observation, say at $c-1$ ($c > 1$) distinct observations $Z_{(i_1)}, \ldots, Z_{(i_{c-1})}$, and subsequently averaging these localised Pearson statistics. The observations $Z_{(i_1)}, \ldots, Z_{(i_{c-1})}$ serve as cell boundaries, implying $c$ cells and $c$ probabilities of a multinomial distribution. The constant $c$ is referred to as the SSP size. Each localised Pearson statistic is thus the Pearson chi-squared statistic for testing the multinomial null hypothesis

$$H_0 : F_s(H^{-1}(i_1/n)) = \frac{i_1}{n} \text{ and } F_s(H^{-1}(i_2/n)) - F_s(H^{-1}(i_1/n)) = \frac{i_2 - i_1}{n}$$
$$\text{and } \cdots \text{ and } 1 - F_s(H^{-1}(i_{c-1}/n)) = 1 - \frac{i_{c-1}}{n}.$$

More specifically, let $D_c = \{i_1, \ldots, i_{c-1}\}$, with the convention that $1 \leq i_1 < i_2 < \cdots < i_{c-1} < n$. The localised Pearson statistic is then given by

$$X_s^2(D_c) = n \sum_{j=1}^{c} \frac{\left(\hat{F}_s(\hat{H}^{-1}(i_j/n)) - \hat{F}_s(\hat{H}^{-1}(i_{j-1}/n)) - \frac{i_j - i_{j-1}}{n}\right)^2}{\frac{i_j - i_{j-1}}{n}}, \quad (12.2)$$

where $i_0 \equiv 0$ and $i_c \equiv n$. The SSPKc test statistic then becomes

$$T_{K,c} = \frac{1}{m_c} \sum_{s=1}^{K} \sum_{D_c} X_s^2(D_c),$$

where $\sum_{D_c}$ is over all $c-1$ tuples of ordered distinct integers $1, 2, \ldots, n-1$, and where $m_c$ is the number of such sets.

The asymptotic null distribution of the SSPKc test statistic may be found using empirical processes, similarly as in Theorem 11.1. However, because the SSPKc test statistic is a rank test, its exact null distribution under the general $K$-sample null hypothesis may also be enumerated.

Just as for the SSPc test in Section 5.4 for the one-sample problem, the SSPKc test can be made adaptive by applying a data-based SSP size selection rule which is of the form

$$C_n = \text{ArgMax}_{c \in \Gamma} \{T_{K,c} - 2(c-1)(K-1) \log a_n\},$$

where $\Gamma$ and $a_n$ are as in Section 5.4. Simulation studies in Thas (2001) demonstrated that the data-driven SSPKc test has overall very good powers under various alternatives. The powers observed for this test are often larger than for the $K$-sample AD test, which is equivalent to the SSPk2 (i.e., $c = 2$) test. As in the one-sample problem, the adaptiveness is not required for making the test omnibus consistent, but it generally improves the power of the test.

Finally, we have a look at a limiting case of the SSPKc test: suppose $c = n$. In this case $i_1 = 1$, $i_2 = 2, \ldots, i_{c-1} = n-1$ and the estimated probabilities in (12.2) reduce to (multiplied by $n$)

$$n\hat{F}_s(\hat{H}^{-1}(i/n)) - n\hat{F}_s(\hat{H}^{-1}((i-1)/n)) = N_{si},$$

where $N_{si}$ is as in the contingency table approach as in Section 12.1. The SSPKn (i.e., $c = n$) statistic is thus equivalent to the $X^2$ statistic (12.1), and so are the components. This interesting observation brings us to the conclusion that the class of SSPKc tests may be considered as a bridge between the AD test, which is traditionally classified as an EDF test, and the $K$-sample smooth test.

## 12.3 Some Final Thoughts and Conclusions

- I hope that I have succeeded in demonstrating throughout the book that the one-sample problem and the two-sample problem are basically two settings belonging to the same archetype, an archetype that I essentially consider as "comparing distributions".
- In both parts of the book we came across the same types of tests. I have mainly focussed on smooth and EDF tests, but many other types of tests are very closely related to those two classes. Although the smooth and EDF tests have a completely different origin, it turns out that they both are related to the same components. Also tests of different types (e.g., ECF tests) are often related to these components.

- The components are defined in terms of functions that form a basis in an appropriate Hilbert space. The Hilbert space view of the *comparing-distributions* problem is very valuable for studying the properties of the tests. Many methods for comparing distributions reduce to finding an efficient way of representing the density functions in a low-dimensional subspace of the infinite dimensional Hilbert space, i.e., a subspace spanned by as few as possible basis functions.
- Many of the tests described in this book have components that are related to moment deviations between the distributions as specified under the null hypothesis and the true distributions, particularly for the one-sample problem. However, the components should be interpreted with care. It is not only the expectation of a component that determines its interpretation, but because its (asymptotically normal) distribution is what is used in the construction of the hypothesis test, it is just as important to know how the variance of the component behaves under the hypotheses. This has led to the concept of the *diagnostic property* and the use of variance estimators of the components, that are also consistent under the alternative hypothesis, for standardising so that the diagnostic property is regained. Despite the theoretical correctness of this approach, simulation studies have demonstrated that the theoretical properties only kick in for very large sample sizes (often >10,000). Many components are well-known test statistics that were well known long before the decompositions of the smooth or EDF statistics were studied, and because these tests are very popular among many data analysts, I believe that this deserves more attention by statisticians so that perhaps better solutions will be proposed to circumvent the caveats that exist nowadays.
- As a side remark related to the previous point, I want to add that many simulation studies have suggested that the components frequently are diagnostic in realistic settings with moderately large sample sizes ($n \approx 50$). This happens particularly in the one-sample problem for testing goodness-of-fit for distributions that belong to the exponential family.
- When components are used in an informative statistical analysis, it may be wise to first verify the distributional assumptions that allow them to be used as diagnostic components. For example, when the WMW statistic is used for detecting shifts in location, the location-shift model should be verified first. Or, when the WMW statistic is used for testing likely orderings, it should be assessed first whether the variance of WMW under the null hypothesis and a consistent estimator of the variance do not differ too much. This procedure resembles the way the two-sample $t$-test is used in daily practice: apart from the assessment of normality, one will often use boxplots for assessing the equality of variance assumption. When no large differences are observed the two-sample $t$-test is used with the pooled variance estimator, and otherwise the Welch modified $t$-test is used.
- Closely related to the previous point is the interplay among the distributional assumption, the formulation of the hypotheses, and the null

distribution. This is particularly important for the two- and $K$-sample problems. I stressed several times that one should think very carefully about how the hypotheses should be formulated: they should reflect the substantial research question. One should always be aware of the distributional assumptions that sometimes have to be made. Assumptions must always at least be verifiable. I have introduced the terms *natural null hypothesis* and *implied null hypothesis* to simplify the process.

- When the components are rescaled using a more generally valid consistent variance estimator, the test is in fact no longer a score test. Its construction resembles more that of a Wald test, with the only difference that the numerator of the test statistic is not a maximum likelihood estimator.
- Throughout the book power comparisons have often only been mentioned in terms of simulation experiments. On the other hand, much research has focussed on the theoretical (asymptotical) power properties of goodness-of-fit tests. See, for example, Janssen (1995, 2000) and Janic-Wróblewska (2004).
- Particularly the construction of the smooth tests shows that there is a very close relationship between goodness-of-fit hypothesis testing and non-parametric density estimation (orthogonal series expansions). Of course this link exists for many statistical applications, but it has often been ignored in the context of goodness-of-fit testing. This observation is one of the reasons why I prefer the term "comparing distributions" rather than goodness-of-fit testing. In both parts of the book it has been illustrated that informative conclusions may be derived based on a plot of the *improved density*, which is the (truncated) orthogonal series expansion with the parameters replaced by their estimates, and these estimates are basically the components of the smooth and EDF tests. Because these graphs use the same statistics and thus also the same sample information as the accompanying goodness-of-fit test, the conclusions from both are expected to be consistent. Thus when the diagnostic property of the components is in doubt I recommend using the improved density estimate when the null hypothesis is rejected.
- The density functions mentioned in the previous point also include the comparison density. Throughout the book the importance of the comparison density has been stressed. It appears that many techniques for comparing distributions have a close connection to it. A comparison density is a very convenient and informative way of summarising the difference between two distributions.
- The importance of the comparison density also follows from Neuhaus (1987). He showed basically that the comparison density summarises the differences between two distributions in the most efficient way. It is related to the optimal score function for detecting differences between the two distributions. From this point of view both the (data-driven) smooth tests and the EDF tests may be considered as *adaptive* score tests; adaptive in the sense that they try to approximate the optimal score function

based on the sample data at hand. This is again related to the Hilbert space representation, in which this score function is essentially the most informative dimension.

- Not only smooth and EDF tests have components and are related to a Hilbert space representation. Other goodness-of-fit tests also appear to be related to it.
- The *contingency table approach* presented earlier in this chapter is a seemingly unrelated way of constructing hypothesis tests, but eventually it gives exactly the same components as the $K$-sample smooth test and the $K$-sample Anderson–Darling test. The method demonstrates that rank tests may be obtained by first applying the most extreme categorisation of the data and subsequently applying Pearson's chi-squared test, which is arguably the oldest hypothesis test in statistical history. Thus it looks like proceeding along very old statistical practice methodology, but avoiding the arbitrary choices of the cell boundaries in the categorisation step by putting each observation into exactly one cell. It has also been shown that the Anderson–Darling test may be related to a Pearson test applied to a categorised sample, but now the sample space is only partitioned into two cells. There are $n$ such possible categorisations. To avoid the arbitrariness of choosing one cell boundary, the Anderson–Darling statistic averages the Pearson chi-squared statistics over all $n$ choices for the cell boundary. Again the same components arise, but now with a particular weighting scheme. The SSP tests fill the gap between the two previous methods. Instead of considering all partitionings with 2 cells, it considers all partitionings with $c$ cells, and the SSP test statistic is defined as the average of Pearson chi-squared statistics computed for all these categorisations. In the extreme cases of $c = 2$ and $c = n$ the SSP statistic reduced to the Anderson–Darling and the order $n$ smooth test statistic, respectively. The choice of $c$ is not important from an omnibus consistency point of view, but a data-driven choice of $c$ improves the small sample powers.
- Most of the (nonparametric) tests that were presented for the two- and the $K$-sample problems are rank tests. Apart from the parametric $t$-test I only presented the test of Chervoneva and Iglewicz (2005) as a test not based on ranks. Again it is an example of a test that is constructed from Hilbert space arguments.
- Most of the methods described in the book are available in the cd R package. The package also contains diagnostic methods that help in assessing the assumptions underlying some of the tests. This is particularly helpful for arriving at informative conclusions.
- One of the major emphases throughout the book is that a hypothesis test for comparing distributions should be *informative*. This means that when the null hypothesis is rejected, the statistical method should suggest what the reason was for rejection. In the context of the comparison of distributions this means that the method should indicate in what sense the distributions are different. For example, when a $t$-test is used and the

null hypothesis is rejected, the statistician concludes that the means are different. However, only focussing on the means does not tell the whole story. Differences between populations may occur in different characteristics of the distributions, for example, differences in variance, skewness, or kurtosis. This is an example in which moments are used for expressing differences. When rank tests are used, informative conclusions may also be obtained by looking at likely orderings, depending on which additional distributional assumptions can be made. Particularly the first-order likely ordering is very informative, and in many settings it is even more interpretable than the difference between two means. Many of the tests described in this book allow such informative analyses, and many of the methods can be made *adaptive* so that the data analyst does not have to specify a priori what aspect of the distribution he or she wants to investigate. To some extent the adaptive methods will point the data analyst to the characteristics of the distribution that are most important for understanding the differences between the distributions.

- Although most of the methods for comparing distributions that are included in this book are described as hypothesis tests, they are often related to parameterised densities or comparison density functions (orthogonal series expansions). Parameterised statistical models can usually be extended easily to cope with more complicated study designs, therefore I think it should be possible to extent some of the existing methods for comparing distributions to more complicated study designs as well. For example, the smooth tests for the $K$-sample problem can perhaps be extended so that blocked experiments can be analysed. These methods shall then basically be an extension of the Friedman rank test. Rayner and Best (2001) succeeded in such an extension by using their contingency table approach, which was, however, not developed starting from a parameterised model. Similar solutions, and thus also more extensions, should surely be possible using appropriately parameterised orthogonal series expansions and using some of the methodology treated in this book. This would result in statistical analysis methods that are more informative than methods which only focus on means.

# Appendix A
# Proofs

## A.1 Proof of Theorem 1.1

We first provide a lemma (see, e.g., Lemma 17.1 in van der Vaart (1998)).

**Lemma A.1.** *Assume $\boldsymbol{Z} \sim MVN(\boldsymbol{0}, \boldsymbol{\Sigma})$, where the $p \times p$ matrix $\boldsymbol{\Sigma}$ has eigenvalues $\lambda_1, \ldots, \lambda_p$. Let $X_1, \ldots, X_p$ denote i.i.d. standard normal variates. The quadratic form $\boldsymbol{Z}^t \boldsymbol{\Sigma}^{-1} \boldsymbol{Z}$ is then equivalent in distribution with the random variable*

$$\sum_{i=1}^{p} \lambda_i X_i^2.$$

Throughout the proof, we assume that $H_0$ holds true.

First, write

$$X_n^2 = n \sum_{j=1}^{k} \frac{(\hat{p}_j - \pi_{0j})^2}{\pi_{0j}},$$

where $\hat{p}_j = N_j/n$ is an unbiased and consistent estimator of $\pi_{0j}$, and let $\hat{\boldsymbol{p}}_n^t = (\hat{p}_1, \ldots, \hat{p}_k)$. Let $\boldsymbol{D}_{\pi_0} = \text{diag}(\boldsymbol{\pi}_0)$. With this new notation, we may write $X_n^2 = n(\hat{\boldsymbol{p}}_n - \boldsymbol{\pi}_0)^t D_{\pi_0}^{-1}(\hat{\boldsymbol{p}}_n - \boldsymbol{\pi}_0)$, which is a quadratic form in $\boldsymbol{Z}_n = \sqrt{n}(\hat{\boldsymbol{p}}_n - \boldsymbol{\pi}_0)$. By the multivariate central limit theorem (see, e.g., Theorem 5.4.4 in Lehmann (1999)), as $n \to \infty$,

$$\boldsymbol{Z}_n \xrightarrow{d} \text{MVN}(\boldsymbol{0}, \boldsymbol{\Sigma}),$$

where $\boldsymbol{\Sigma} = \boldsymbol{D}_{\pi_0} - \boldsymbol{\pi}_0 \boldsymbol{\pi}_0^t$. Because $X_n^2$ is a quadratic form in $\boldsymbol{Z}_n$, Lemma A.1 gives, as $n \to \infty$,

$$X_n^2 \xrightarrow{d} \sum_{j=1}^{k} \lambda_j Z_j^2,$$

where $Z_1, \ldots, Z_k$ are i.i.d. $N(0,1)$, and $\lambda_1 \leq \cdots \leq \lambda_k$ are the eigenvalues of

$$\boldsymbol{L} = \boldsymbol{D}_{\pi_0}^{-1/2} \boldsymbol{\Sigma} \boldsymbol{D}_{\pi_0}^{-1/2} = \boldsymbol{I}_k - \sqrt{\boldsymbol{\pi}_0} \sqrt{\boldsymbol{\pi}_0^t}$$

with $\sqrt{\boldsymbol{\pi}_0^t} = (\sqrt{\pi_{01}}, \ldots, \sqrt{\pi_{0k}})$. It can be shown that $\lambda_1 = 0$ and $\lambda_2 = \cdots = \lambda_k = 1$. This completes the proof. □

## A.2 Proof of Theorem 1.2

Because $\hat{\boldsymbol{\beta}}$ is BAN, we obtain

$$\hat{\boldsymbol{\beta}} = \boldsymbol{\beta} + (\hat{\boldsymbol{p}}_n - \boldsymbol{\pi}_0(\boldsymbol{\beta}))\boldsymbol{D}_{\pi_0}^{-1/2}\boldsymbol{A}(\boldsymbol{A}^t\boldsymbol{A})^{-1} + o_p(n^{-1/2}),$$

where the matrix $\boldsymbol{A}$ has $(i,j)$th element $(i = 1, \ldots, k; j = 1, \ldots, p)$,

$$\frac{1}{\sqrt{\pi_{0i}}}\frac{\partial \pi_{0i}}{\partial \beta_j}(\boldsymbol{\beta}).$$

By Birch's regularity conditions, we find

$$\boldsymbol{\pi}_0(\hat{\boldsymbol{\beta}}) - \boldsymbol{\pi}_0(\boldsymbol{\beta}) = (\hat{\boldsymbol{\beta}} - \boldsymbol{\beta})\frac{\partial \boldsymbol{\pi}_0}{\partial \boldsymbol{\beta}}(\boldsymbol{\beta}) + o_p(n^{-1/2}).$$

Hence,

$$\boldsymbol{\pi}_0(\hat{\boldsymbol{\beta}}) - \boldsymbol{\pi}_0(\boldsymbol{\beta}) = (\hat{\boldsymbol{p}}_n - \boldsymbol{\pi}_0(\boldsymbol{\beta}))\boldsymbol{L} + o_p(n^{-1/2}),$$

where $\boldsymbol{L} = \boldsymbol{D}_{\pi_0}^{-1/2}\boldsymbol{A}(\boldsymbol{A}^t\boldsymbol{A})^{-1}\boldsymbol{A}^t\boldsymbol{D}_{\pi_0}^{1/2}$. Write

$$\boldsymbol{M}_n = \begin{bmatrix} \hat{\boldsymbol{p}}_n - \boldsymbol{\pi}_0 \\ \boldsymbol{\pi}_0(\hat{\boldsymbol{\beta}}) - \boldsymbol{\pi}_0 \end{bmatrix} = (\hat{\boldsymbol{p}}_n - \boldsymbol{\pi}_0(\boldsymbol{\beta}))\begin{bmatrix} \boldsymbol{I}_k \\ \boldsymbol{L} \end{bmatrix} + o_p(n^{-1/2}).$$

Because $\sqrt{n}(\hat{\boldsymbol{p}}_n - \boldsymbol{\pi}_0)$ is asymptotically multivariate normal, we may conclude that $\boldsymbol{M}_n$ is also asymptotically multivariate normal with zero mean and variance–covariance matrix equal to

$$\begin{bmatrix} \boldsymbol{D}_{\pi_0} - \boldsymbol{\pi}_0\boldsymbol{\pi}_0^t & (\boldsymbol{D}_{\pi_0} - \boldsymbol{\pi}_0\boldsymbol{\pi}_0^t)\boldsymbol{L} \\ \boldsymbol{L}^t(\boldsymbol{D}_{\pi_0} - \boldsymbol{\pi}_0\boldsymbol{\pi}_0^t) & \boldsymbol{L}^t(\boldsymbol{D}_{\pi_0} - \boldsymbol{\pi}_0\boldsymbol{\pi}_0^t)\boldsymbol{L} \end{bmatrix}.$$

Hence, $\sqrt{n}(\hat{\boldsymbol{p}}_n - \boldsymbol{\pi}_0(\hat{\boldsymbol{\beta}})) = \sqrt{n}(\hat{\boldsymbol{p}}_n - \boldsymbol{\pi}_0) - \sqrt{n}(\boldsymbol{\pi}_0(\hat{\boldsymbol{\beta}}) - \boldsymbol{\pi}_0)$ is also asymptotically zero mean multivariate normal with variance–covariance matrix (after some simple algebra)

$$\boldsymbol{\Sigma} = \boldsymbol{D}_{\pi_0} - \boldsymbol{\pi}_0\boldsymbol{\pi}_0^t - \boldsymbol{D}_{\pi_0}^{1/2}\boldsymbol{A}(\boldsymbol{A}^t\boldsymbol{A})^{-1}\boldsymbol{A}^t\boldsymbol{D}_{\pi_0}^{1/2}. \tag{A.1}$$

The asymptotic null distribution of $\hat{X}_n^2$ is now again obtained by applying Lemma A.1. This time we need the eigenvalues of

$$\boldsymbol{D}_{\pi_0}^{-1/2}\boldsymbol{\Sigma}\boldsymbol{D}_{\pi_0}^{-1/2} = \boldsymbol{I} - \sqrt{\boldsymbol{\pi}_0}\sqrt{\boldsymbol{\pi}_0^t} - \boldsymbol{A}(\boldsymbol{A}^t\boldsymbol{A})^{-1}\boldsymbol{A}^t.$$

It can be shown that this matrix has $k-p-1$ eigenvalues equal to 1, and the remaining $p+1$ eigenvalues equal to 0. □

## A.3 Proof of Theorem 4.1

(1) To obtain the score statistic, we first need to specify the log-likelihood function. From Equation (4.1), we find

$$\begin{aligned} l(\boldsymbol{\theta}) &= \log\left(\prod_{i=1}^{n} g_k(X_i;\boldsymbol{\theta})\right) \\ &= n\log C(\boldsymbol{\theta}) + \sum_{i=1}^{n}\log g(X_i) + \sum_{j=1}^{k}\theta_j\sum_{i=1}^{n}h_j(X_i). \end{aligned}$$

The score function for parameter $\theta_j$ is given by

$$\begin{aligned} u_j(\boldsymbol{\theta}) &= \frac{\partial l(\boldsymbol{\theta})}{\partial\theta_j} \\ &= n\frac{\partial\log C(\boldsymbol{\theta})}{\partial\theta_j} + \sum_{i=1}^{n}h_j(X_i). \end{aligned}$$

For the construction of the score test statistics, we need to evaluate the score function under the null hypothesis. This gives $u_j(\boldsymbol{\theta})|_{\boldsymbol{\theta}=\mathbf{0}} = \sum_{i=1}^{n}h_j(X_i)$, where we have used

$$\frac{\partial\log C(\boldsymbol{\theta})}{\partial\boldsymbol{\theta}}|_{\boldsymbol{\theta}=\mathbf{0}} = 0.$$

As for all score functions, $\mathrm{E}_0\{u_j\} = 0$. This could also be directly seen from the orthogonality property of the $h_j$; i.e., $\mathrm{E}_0\{u_j\} = n\int_S h_j(x)g(x)\,dx = n < h_1, 1 >_g = 0$.

The variance–covariance matrix of the vector $\boldsymbol{U}^t = (1/\sqrt{n})\,(u_1,\ldots,u_k)_{\boldsymbol{\theta}=\mathbf{0}}$ involves the covariances

$$\begin{aligned} \mathrm{Cov}_0\{h_i(X),h_j(X)\} &= \int_S h_i(x)h_j(x)g(x)dx \\ &= < h_i, h_j >_g \\ &= \delta_{ij}, \end{aligned}$$

where $\delta_{ij}$ is Kronecker delta. Hence, $\mathrm{Var}_0\{\boldsymbol{U}\} = \boldsymbol{I}$, the $k\times k$ identity matrix. The multivariate central limit theorem gives that

$$\boldsymbol{U} \xrightarrow{d} MVN(\mathbf{0},\boldsymbol{I}),$$

and we therefore have also the convergence of the quadratic form (Lemma 17.1 in van der Vaart (1998))

$$T_k = \boldsymbol{U}^t \boldsymbol{U} \xrightarrow{d} \chi^2_k.$$

(2) To prove the second part of the theorem, we only have to show that the score function based on the Barton model is the same as $u_j$ when restricted under the null hypothesis $\boldsymbol{\theta} = \mathbf{0}$.

The log-likelihood function becomes

$$l(\boldsymbol{\theta}) = \sum_{i=1}^{n} \log g_k(X_i; \boldsymbol{\theta}) = \sum_{i=1}^{n} \log g(X_i) + \sum_{i=1}^{n} \log \left(1 + \sum_{j=1}^{k} \theta_j h_j(X_i)\right),$$

and the score function for $\theta_j$

$$u_j(\boldsymbol{\theta}) = \frac{\partial l}{\partial \theta_j} = \sum_{i=1}^{n} \frac{h_j(X_i)}{1 + \sum_{j=1}^{k} \theta_j h_j(X_i)}.$$

Hence,

$$u_j|_{\boldsymbol{\theta}=\mathbf{0}} = \sum_{i=1}^{n} h_j(X_i),$$

which is exactly the same as what we found in part (1) of the proof. □

## A.4 Proof of Lemma 4.1

Straightforward calculations give

$$\begin{aligned}
\mathrm{E}_k\left\{(X-\mu)^m\right\} &= \int_{\mathcal{S}} (x-\mu)^m \left(1 + \sum_{j=1}^{k} \theta_j h_j(x)\right) g(x)dx \\
&= \mu_m + \sum_{j=1}^{k} \theta_j < (x-\mu)^m, h_j(x) >_g \\
&= \mu_m + \sum_{j=1}^{k} \theta_j < (x-\mu)^m - \mu_m, h_j(x) >_g, \qquad \text{(A.2)}
\end{aligned}$$

where the last step makes use of $< \mu_m 1, h_j(x) >_g = \mu_m < 1, h_j(x) >_g = 0$. Because $\mathrm{E}_0\left\{(X-\mu)^m - \mu_m\right\} =< (x-\mu)^m - \mu_m, 1 >_g = 0$, we may write the degree $m$ polynomial $(x-\mu)^m - \mu_m$ in terms of the $m$ base functions $h_1, \ldots, h_m$,

$$(x-\mu)^m - \mu_m = \sum_{j=1}^{m} c_j h_j(x), \qquad \text{(A.3)}$$

where $c_1, \ldots, c_m$ are constants. After substituting (A.3) into (A.2), we get

$$\begin{aligned} \mathrm{E}_k\left\{(X-\mu)^m\right\} &= \mu_m + \sum_{j=1}^{k} \theta_j < \sum_{i=1}^{m} c_i h_i, h_j >_g \\ &= \mu_m + \theta_j c_j. \end{aligned}$$

Hence, if $\theta_j = 0$ then $\mathrm{E}_k\left\{(X-\mu)^m\right\} = \mu_m$. The $\Leftarrow$ part of the proof is also true because $(x-\mu)^m - \mu_m$ is a polynomial of exactly degree $m$, and thus $c_m \neq 0$, and, therefore, $\mathrm{E}_k\left\{(X-\mu)^m\right\} = \mu_m$ if and only if $\theta_j = 0$. □

## A.5 Proof of Lemma 4.2

Because $h_0(x) = 1$, we get $\mathrm{E}_0\left\{h_j(X)\right\} =< h_j, 1 >_g = 0$ for all $j$. To stress that the lemma imposes a restriction on the polynomials, we use the notation $h_j^\star$ whenever they are of the form of Equation (4.13). Under the null hypothesis, $\mu_j = \mathrm{E}_0\left\{(X-\mu)^j\right\}$. Hence, also $\mathrm{E}_0\left\{h_j^\star(X)\right\} =< h_j^\star, 1 >_g = 0$.

It is always possible to write

$$h_j(x) = h_j^\star(x) + z(x),$$

where $z$ is a polynomial of degree $\leq j$. The lemma is proven if we can show that $z(x) \equiv 0$.

We know that $< h_i, h_j >_g = 0$ for all $i \neq j$, thus we get

$$\begin{aligned} 0 &= < h_i, h_j >_g \\ &= < h_i, h_j^\star + z >_g \\ &= < h_i, h_j^\star >_g + < h_i, z >_g. \end{aligned}$$

Hence, $< h_i, z >_g = - < h_i, h_j^\star >_g$. This holds for all $i \neq j$, and since the $h_i$ form a base in a Hibert space, therefore we may conclude that $z = -h_j^\star$ or $z = 0$. However, the former implies $h_j(x) = 0$ which is a contradiction. Therefore, $z = 0$. □

## A.6 Proof of Lemma 4.3

(1) Because all first $j$ moments agree with $g$, Lemma 4.2 implies that $\mathrm{E}\left\{h_j(X)\right\} = 0$. Hence,

$$\begin{aligned} \mathrm{Var}\left\{U_j\right\} &= \mathrm{Var}\left\{h_j(X)\right\} \\ &= \mathrm{E}\left\{\left(h_j(X) - \mathrm{E}\left\{h_j(X)\right\}\right)^2\right\} \qquad \text{(A.4)} \\ &= \mathrm{E}\left\{h_j^2(X)\right\}, \end{aligned}$$

where $h_j^2$ is a polynomial of degree $2j$ which may be written as $h_j^2(x) = \sum_{l=0}^{2j} c_l h_l(x)$. Note that this is a sum of polynomials of degrees corresponding to moments which all agree with $g$, and, again according to Lemma 4.2, these polynomials have expectation equal to zero. Hence, $\mathrm{E}\left\{h_j^2(X)\right\} = c_0$. Because the same result would have been found under the null hypothesis, we find $c_0 = \mathrm{Var}_0\left\{U_j\right\} = 1$.

(2) We start from (A.4), in which now $\mathrm{E}\left\{h_j(X)\right\}$ is not necessarily zero. Write

$$\begin{aligned}\mathrm{Var}\left\{U_j\right\} &= \mathrm{E}\left\{h_j^2(X)\right\} - \left(\mathrm{E}\left\{h_j(X)\right\}\right)^2 \\ &= \mathrm{E}\left\{1 + \sum_{l=1}^{2j} c_l h_l(X)\right\} - \left(\mathrm{E}\left\{h_j(X)\right\}\right)^2 \\ &= 1 + \sum_{l=m}^{2j} c_l \,\mathrm{E}\left\{h_l(X)\right\} - \left(\mathrm{E}\left\{h_j(X)\right\}\right)^2 . \end{aligned} \tag{A.5}$$

Lemma 4.2 tells again when the last or the two last terms in (A.5) are zero. This gives the statement in (4.14). □

## A.7 Proof of Theorem 4.10

First we introduce some matrix notation. Let $\boldsymbol{H}^t$ the $m \times k$ matrix with the $(i,j)$th element equal to $h_{ij}$; i.e., the $j$th column corresponds to the $j$th orthonormal vector. We may now write $\boldsymbol{U} = (1/\sqrt{n})\boldsymbol{H}\boldsymbol{N}$, and the orthonormality condition becomes $\boldsymbol{H}\boldsymbol{D}_{\pi_0}\boldsymbol{H}^t = \boldsymbol{I}$, where $\boldsymbol{D}_{\pi_0} = \mathrm{diag}(\boldsymbol{\pi}_0)$. The restriction $\sum_{i=1}^m h_{ij}N_i = 0$ for all $j = 1, \ldots, k$ now becomes $\boldsymbol{H}\boldsymbol{\pi}_0 = \boldsymbol{0}$. This latter restriction allows us to write the equality

$$\boldsymbol{U} = \frac{1}{\sqrt{n}}\boldsymbol{H}\boldsymbol{N} = \frac{1}{\sqrt{n}}\boldsymbol{H}\left(\boldsymbol{N} - n\boldsymbol{\pi}_0\right) = \sqrt{n}\boldsymbol{H}\left(\hat{p} - \boldsymbol{\pi}_0\right). \tag{A.6}$$

With this notation the order $k$ smooth test statistic becomes

$$T_k = \boldsymbol{U}^t\boldsymbol{U} = n\left(\hat{p} - \boldsymbol{\pi}_0\right)^t \boldsymbol{H}^t\boldsymbol{H}\left(\hat{p} - \boldsymbol{\pi}_0\right).$$

Because $k = m - 1$ and because $\boldsymbol{H}\boldsymbol{D}_{\pi_0}\boldsymbol{H}^t = \boldsymbol{I}$, we find $\boldsymbol{H}^t\boldsymbol{H} = \boldsymbol{D}_{\pi_0}^{-1}$. Substituting this equality in Equation (A.6) completes the proof. □

## A.8 Proof of Theorem 4.2

In this section we acually give the proof of a more general theorem which states the asymptotic distribution of $\hat{\boldsymbol{V}}$ under a sequence of local alternatives. First some notation is introduced.

As a sequence of local alternatives to $g$, we consider model (4.1) with

$$\boldsymbol{\theta} = \boldsymbol{\theta}_n = n^{-1/2}\boldsymbol{\delta}, \tag{A.7}$$

where $\boldsymbol{\delta}$ is a vector of $k$ positive nonzero constants and $\delta^2 = \boldsymbol{\delta}^t\boldsymbol{\delta} < \infty$. The null hypothesis corresponds to $\boldsymbol{\delta} = \mathbf{0}$. The density or model of the local alternatives is now denoted as

$$g_{nk}(x) = g_{nk}(x;\boldsymbol{\theta}_n,\boldsymbol{\beta}) = C(\boldsymbol{\theta}_n,\boldsymbol{\beta})\exp\left(\sum_{j=1}^{k}\theta_{nj}h_j(x;\boldsymbol{\beta})\right)g(x;\boldsymbol{\beta}). \tag{A.8}$$

The next two lemmas are needed.

**Lemma A.2 (Local Asymptotic Normality (LAN)).** *Consider the sequence of alternatives given in (A.7) and model (A.8). Then, the log-likelihood ratio admits the following asymptotic expansion*

$$\log\left(\frac{g_{nk}(x;\boldsymbol{\theta}_n;\boldsymbol{\beta})}{g(x;\boldsymbol{\beta})}\right) = \frac{1}{\sqrt{n}}\boldsymbol{\delta}^t\boldsymbol{h}(x;\boldsymbol{\beta}) - \frac{1}{2}\frac{1}{n}\boldsymbol{\delta}^t\boldsymbol{\delta} + o(\delta^2/n), \tag{A.9}$$

*and, as $n \to \infty$,*

$$\log\prod_{i=1}^{n}\left(\frac{g_{nk}(X;\boldsymbol{\theta}_n;\boldsymbol{\beta})}{g(X;\boldsymbol{\beta})}\right) \xrightarrow{d} N\left(-\frac{1}{2}\boldsymbol{\delta}^t\boldsymbol{\delta},\boldsymbol{\delta}^t\boldsymbol{\delta}\right) \tag{A.10}$$

*Proof.* To prove Equation (A.9) we start with substituting $g_{nk}$ and $g$ into the log-likelihood ratio

$$\begin{aligned}\log\frac{g_{nk}(x;\boldsymbol{\theta}_n;\boldsymbol{\beta})}{g(x;\boldsymbol{\beta})} &= \log C(\boldsymbol{\theta}_n) - \log(C(\mathbf{0})) + \frac{1}{\sqrt{n}}\boldsymbol{\delta}^t\boldsymbol{h}(x)\\ &= \log C(\boldsymbol{\theta}_n) + \frac{1}{\sqrt{n}}\boldsymbol{\delta}^t\boldsymbol{h}(x).\end{aligned}$$

This can be further simplified by applying a Taylor series expansion on $\log C(\boldsymbol{\theta}_n)$,

$$\begin{aligned}\log C(\boldsymbol{\theta}_n) &= \log C(\mathbf{0}) + \frac{1}{\sqrt{n}}\boldsymbol{\delta}^t\left.\frac{\partial\log C(\boldsymbol{\theta})}{\partial\boldsymbol{\theta}}\right|_{\boldsymbol{\theta}=\mathbf{0}} + \frac{1}{2}\frac{1}{n}\boldsymbol{\delta}^t\left.\frac{\partial^2\log C(\boldsymbol{\theta})}{\partial\boldsymbol{\theta}\partial\boldsymbol{\theta}^t}\right|_{\boldsymbol{\theta}=\mathbf{0}}\boldsymbol{\delta} + o(\delta^2/n)\\ &= \frac{1}{2}\frac{1}{n}\boldsymbol{\delta}^t\,\mathrm{E}_0\left\{-\boldsymbol{h}(X)\boldsymbol{h}^t(X)\right\}\boldsymbol{\delta} + o(\delta^2/n)\\ &= -\frac{1}{2}\frac{1}{n}\boldsymbol{\delta}^t\boldsymbol{I}\boldsymbol{\delta} + o(\delta^2/n)\\ &= -\frac{1}{2}\frac{1}{n}\boldsymbol{\delta}^t\boldsymbol{\delta} + o(\delta^2/n).\end{aligned}$$

The convergence in Equation (A.10) follows from

$$\log \prod_{i=1}^{n} \frac{g_{nk}(X;\boldsymbol{\theta}_n;\boldsymbol{\beta})}{g(X;\boldsymbol{\beta})} = \frac{1}{\sqrt{n}} \sum_{i=1}^{n} \boldsymbol{\delta}^t \boldsymbol{h}(X_i;\boldsymbol{\beta}) - \frac{1}{2}\boldsymbol{\delta}^t\boldsymbol{\delta} + o_P(1), \tag{A.11}$$

where $(1/\sqrt{n})\sum_{i=1}^{n} \boldsymbol{h}(X_i;\boldsymbol{\beta})$ converges according to the multivariate central limit theorem to a multivariate normal distribution with mean

$$\mathrm{E}_0\{\boldsymbol{h}(X)\} = \mathbf{0},$$

and variance–covariance matrix

$$\mathrm{Var}_0\{\boldsymbol{h}(X)\} = \mathrm{E}_0\{\boldsymbol{h}(X)\boldsymbol{h}^t(X)\} = \boldsymbol{I}.$$

Using this result and applying Slutsky's lemma completes the proof.

**Lemma A.3.** *Let $\boldsymbol{w}(x;\boldsymbol{\beta})$ be a vector-valued function that satisfies the regularity conditions, and for which* $\mathrm{E}_0\{\boldsymbol{w}(X;\boldsymbol{\beta})\} = \mathbf{0}$. *Then*

$$\mathrm{E}_0\left\{\frac{\partial \boldsymbol{w}}{\partial \boldsymbol{\beta}}(X;\boldsymbol{\beta})\right\} = -\mathrm{Cov}_0\{\boldsymbol{w}(X;\boldsymbol{\beta}), \boldsymbol{u}_{\beta}(X;\boldsymbol{\beta})\} = <\boldsymbol{w}, \boldsymbol{u}_{\beta}>.$$

*Proof.* It is assumed that

$$\mathrm{E}_0\{\boldsymbol{w}(X;\boldsymbol{\beta})\} = \mathbf{0} = \int_{-\infty}^{+\infty} \boldsymbol{w}(x;\boldsymbol{\beta})g(x;\boldsymbol{\beta})dx.$$

Differentiating both sides of this equation yields

$$\begin{aligned}
\int_{-\infty}^{+\infty} \frac{\partial \boldsymbol{w}}{\partial \boldsymbol{\beta}}(x;\boldsymbol{\beta})f(x;\boldsymbol{\beta})dx + \int_{-\infty}^{+\infty} \boldsymbol{w}(x;\boldsymbol{\beta})\frac{\partial g}{\partial \boldsymbol{\beta}}(x;\boldsymbol{\beta})dx &= 0\\
\mathrm{E}_0\left\{\frac{\partial \boldsymbol{w}}{\partial \boldsymbol{\beta}}(X;\boldsymbol{\beta})\right\} + \int_{-\infty}^{+\infty} \boldsymbol{w}(x;\boldsymbol{\beta})\frac{\partial \log g}{\partial \boldsymbol{\beta}}(x;\boldsymbol{\beta})g(x;\boldsymbol{\beta})dx &= 0\\
\mathrm{E}_0\left\{\frac{\partial \boldsymbol{w}}{\partial \boldsymbol{\beta}}(X;\boldsymbol{\beta})\right\} + \mathrm{E}_0\left\{\boldsymbol{w}(X;\boldsymbol{\beta})\frac{\partial \log g}{\partial \boldsymbol{\beta}}(X;\boldsymbol{\beta})\right\} &= 0.
\end{aligned}$$

Because

$$\mathrm{E}_0\{\boldsymbol{w}(X;\boldsymbol{\beta})\} = \mathrm{E}_0\left\{\frac{\partial \log g}{\partial \boldsymbol{\beta}}(X;\boldsymbol{\beta})\right\} = \mathbf{0},$$

we obtain

$$\mathrm{E}_0\left\{\frac{\partial \boldsymbol{w}}{\partial \boldsymbol{\beta}}(X;\boldsymbol{\beta})\right\} = -\mathrm{Cov}_0\left\{\boldsymbol{w}(X;\boldsymbol{\beta}), \frac{\partial \log g}{\partial \boldsymbol{\beta}}(X;\boldsymbol{\beta})\right\},$$

which completes the proof. □

**Theorem A.1.** *Under the sequence of local alternatives given in (A.7), the vector $\boldsymbol{V}(\hat{\boldsymbol{\beta}})$ converges, as $n \to \infty$, in distribution to a multivariate normal distribution with variance–covariance matrix*

$$\boldsymbol{\Sigma}_{\hat{v}} = \boldsymbol{\Sigma}_v + \boldsymbol{\Sigma}_{v\beta}\boldsymbol{\Sigma}_{b\beta}^{-1}\boldsymbol{\Sigma}_{bb}\boldsymbol{\Sigma}_{\beta b}^{-1}\boldsymbol{\Sigma}_{\beta v} - \boldsymbol{\Sigma}_{vb}\boldsymbol{\Sigma}_{\beta b}^{-1}\boldsymbol{\Sigma}_{\beta v} - \boldsymbol{\Sigma}_{v\beta}\boldsymbol{\Sigma}_{b\beta}^{-1}\boldsymbol{\Sigma}_{bv}, \quad \text{(A.12)}$$

*and mean*

$$\boldsymbol{\mu}_{\hat{v}} = \left(\boldsymbol{\Sigma}_{vh} - \boldsymbol{\Sigma}_{v\beta}\boldsymbol{\Sigma}_{b\beta}^{-1}\boldsymbol{\Sigma}_{bh}\right)\boldsymbol{\delta}.$$

*Proof.* The proof consists of two parts. First the asymptotic null distribution of $\boldsymbol{V}(\hat{\boldsymbol{\beta}})$ is found. Then the joint null distribution of $\boldsymbol{V}(\hat{\boldsymbol{\beta}})$ and the log-likelihood ratio statistic is proven, from which by means of Le Cam's third lemma the theorem immediately follows.

1. A first-order Taylor expansion of $\boldsymbol{v}(\hat{\boldsymbol{\beta}})$ gives

$$\boldsymbol{v}(x;\hat{\boldsymbol{\beta}}) = \boldsymbol{v}(x;\boldsymbol{\beta}) + \frac{\partial \boldsymbol{v}}{\partial \boldsymbol{\beta}}(x;\boldsymbol{\beta})(\hat{\boldsymbol{\beta}} - \boldsymbol{\beta}) + o_P(n^{-1/2}).$$

Substituting this into $\boldsymbol{V}(\hat{\boldsymbol{\beta}})$ and recognising that $\boldsymbol{\beta}$ is an asymptotic linear estimator it becomes

$$\boldsymbol{V}(\hat{\boldsymbol{\beta}}) = \frac{1}{\sqrt{n}}\sum_{i=1}^{n}\boldsymbol{v}(X_i;\boldsymbol{\beta}) + \left(\frac{1}{n}\sum_{i=1}^{n}\frac{\partial \boldsymbol{v}}{\partial \boldsymbol{\beta}}(X_i;\boldsymbol{\beta})\right)\left(\frac{1}{\sqrt{n}}\sum_{i=1}^{n}\boldsymbol{\Psi}(X_i;\boldsymbol{\beta})\right) + o_P(1).$$

This is further simplified by applying the law of large numbers on

$$\frac{1}{n}\sum_{i=1}^{n}\frac{\partial \boldsymbol{v}}{\partial \boldsymbol{\beta}}(X_i;\boldsymbol{\beta}),$$

resulting in

$$\boldsymbol{V}(\hat{\boldsymbol{\beta}}) = \frac{1}{\sqrt{n}}\sum_{i=1}^{n}\boldsymbol{v}(X_i;\boldsymbol{\beta}) + \mathrm{E}_0\left\{\frac{\partial \boldsymbol{v}}{\partial \boldsymbol{\beta}}(X;\boldsymbol{\beta})\right\}\left(\frac{1}{\sqrt{n}}\sum_{i=1}^{n}\boldsymbol{\Psi}(X_i;\boldsymbol{\beta})\right) + o_P(1). \quad \text{(A.13)}$$

Under the null hypothesis, the multivariate central limit theorem gives

$$\frac{1}{\sqrt{n}}\sum_{i=1}^{n}\boldsymbol{v}(X_i;\boldsymbol{\beta}) \xrightarrow{d} N(\boldsymbol{0},\boldsymbol{\Sigma}_v),$$

and

$$\frac{1}{\sqrt{n}}\sum_{i=1}^{n}\boldsymbol{\Psi}(X_i;\boldsymbol{\beta}) \xrightarrow{d} N(\boldsymbol{0},\boldsymbol{\Sigma}_{\Psi}),$$

where

$$\boldsymbol{\Sigma}_{\Psi} = \boldsymbol{\Sigma}_{b\beta}^{-1} \boldsymbol{\Sigma}_{b} \boldsymbol{\Sigma}_{\beta b}^{-1}.$$

The joint distribution of these two random vectors is obtained by applying the Cramér–Wald device. In particular it is a multivariate normal distribution with mean $\mathbf{0}$ and variance–covariance matrix

$$\begin{bmatrix} \boldsymbol{\Sigma}_v & \boldsymbol{\Sigma}_{v\Psi} \\ \boldsymbol{\Sigma}_{\Psi v} & \boldsymbol{\Sigma}_{\Psi} \end{bmatrix},$$

where $\boldsymbol{\Sigma}_{v\Psi} = \mathrm{Cov}_0\left\{\boldsymbol{v}(X;\boldsymbol{\beta}), \boldsymbol{\Psi}(X;\boldsymbol{\beta})\right\}$. Using Lemma A.3 we find

$$\mathrm{E}_0\left\{\frac{\partial \boldsymbol{v}}{\partial \boldsymbol{\beta}}(X;\boldsymbol{\beta})\right\} = -\,\mathrm{Cov}_0\left\{\boldsymbol{v}(X), \boldsymbol{u}_{\beta}\right\} = -\boldsymbol{\Sigma}_{v\beta},$$

and using Slutsky's lemma, we find that the limiting null disitribution of $\boldsymbol{V}(\hat{\boldsymbol{\beta}})$ is a multivariate normal distribution with mean $\mathbf{0}$ and variance–covariance matrix $\boldsymbol{\Sigma}_{\hat{v}}$ as stated in Equation (A.12).

2. The proof of the joint null distribution of $\boldsymbol{V}(\hat{\boldsymbol{\beta}})$ and the log-likelihood ratio statistic is along the same lines as van der Vaart (1998), p. 219. We only need to calculate the covariance between the two random vectors,

$$\mathrm{Cov}_0\left\{\boldsymbol{V}(\hat{\boldsymbol{\beta}}), \log\prod_{i=1}^{n}\frac{g_{nk}(X_i;\boldsymbol{\theta}_n;\boldsymbol{\beta})}{g(X_i;\boldsymbol{\beta})}\right\}.$$

The solution is obtained by substituting $\boldsymbol{V}(\hat{\boldsymbol{\beta}})$ and the log-likelihood ratio statistic by their respective asymptotic expansions (Equations (A.13) and (A.11)):

$$\begin{aligned}
&\mathrm{Cov}_0\left\{\boldsymbol{V}(\hat{\boldsymbol{\beta}}), \log\prod_{i=1}^{n}\frac{g_{nk}(X_i;\boldsymbol{\theta}_n;\boldsymbol{\beta})}{g(X_i;\boldsymbol{\beta})}\right\} \\
&= \mathrm{Cov}_0\left\{\frac{1}{\sqrt{n}}\sum_{i=1}^{n}\boldsymbol{v}(X_i;\boldsymbol{\beta}) + \mathrm{E}_0\left\{\frac{\partial \boldsymbol{v}}{\partial \boldsymbol{\beta}}(X;\boldsymbol{\beta})\right\}\left(\frac{1}{\sqrt{n}}\sum_{i=1}^{n}\boldsymbol{\Psi}(X_i;\boldsymbol{\beta})\right)\right. \\
&\quad \left. + o_P(1), \frac{1}{\sqrt{n}}\sum_{i=1}^{n}\boldsymbol{\delta}^t\boldsymbol{h}(X_i;\boldsymbol{\beta}) - \frac{1}{2}\delta^2 + o_P(1)\right\} \\
&= \mathrm{Cov}_0\left\{\boldsymbol{v}(X;\boldsymbol{\beta}) + \mathrm{E}_0\left\{\frac{\partial \boldsymbol{v}}{\partial \boldsymbol{\beta}}(X;\boldsymbol{\beta})\right\}\boldsymbol{\Psi}(X;\boldsymbol{\beta}), \boldsymbol{\delta}^t\boldsymbol{h}(X;\boldsymbol{\beta})\right\} + o(1) \\
&= \mathrm{Cov}_0\left\{\boldsymbol{v}(X;\boldsymbol{\beta}), \boldsymbol{h}(X;\boldsymbol{\beta})\right\}\boldsymbol{\delta} \\
&\quad + \mathrm{E}_0\left\{\frac{\partial \boldsymbol{v}}{\partial \boldsymbol{\beta}}(X;\boldsymbol{\beta})\right\}\mathrm{Cov}_0\left\{\boldsymbol{\Psi}(X;\boldsymbol{\beta}), \boldsymbol{h}(X;\boldsymbol{\beta})\right\}\boldsymbol{\delta} + o(1) \\
&= \boldsymbol{\Sigma}_{vh}\boldsymbol{\delta} + \mathrm{E}_0\left\{\frac{\partial \boldsymbol{v}}{\partial \boldsymbol{\beta}}(X;\boldsymbol{\beta})\right\}\boldsymbol{\Sigma}_{\Psi h}\boldsymbol{\delta} + o(1) \\
&= \boldsymbol{\Sigma}_{vh}\boldsymbol{\delta} + \mathrm{E}_0\left\{\frac{\partial \boldsymbol{v}}{\partial \boldsymbol{\beta}}(X;\boldsymbol{\beta})\right\}\mathrm{E}_0\left\{-\dot{\boldsymbol{b}}(X)\right\}^{-1}\boldsymbol{\Sigma}_{bh}\boldsymbol{\delta} + o(1).
\end{aligned}$$

Applying Lemma A.3 to the last equation gives

$$\mathrm{Cov}_0\left\{\boldsymbol{V}(\hat{\boldsymbol{\beta}}), \log\prod_{i=1}^{n}\frac{g_{nk}(X_i;\boldsymbol{\theta}_n;\boldsymbol{\beta})}{g(X_i;\boldsymbol{\beta})}\right\} = \boldsymbol{\Sigma}_{vh}\boldsymbol{\delta} - \boldsymbol{\Sigma}_{v\beta}\boldsymbol{\Sigma}_{b\beta}^{-1}\boldsymbol{\Sigma}_{bh}\boldsymbol{\delta} + o(1).$$

Now that the joint distribution of $\boldsymbol{V}(\hat{\boldsymbol{\beta}})$ and the log-likelihood ratio statistic are known, we can directly apply Le Cam's third lemma which immediately completes the proof.

□

## A.9 Heuristic Proof of Theorem 5.2

(1) Because both $\{h_j \circ G\}$ and $\{k_j^a\}$ are systems of orthonormal functions in $L_2(\mathcal{S}, G)$, there exists a set of constants $\{a_{ij}\}$ so that for all $x \in \mathcal{S}$,

$$\sum_i a_{ij} v_i(x) = k_j^a(x). \qquad \text{(A.14)}$$

Let $\boldsymbol{A}$ denote the matrix with $(i,j)$th element equal to $a_{ij}$, and assume that $\boldsymbol{A}$ has an inverse $\boldsymbol{A}^{-1}$. Equation (A.14) may now be written as $\boldsymbol{A}^t\boldsymbol{v}(x) = \boldsymbol{k}_a(x)$. We now project both sides of the equation onto $\boldsymbol{v}$, resulting in $\boldsymbol{A}^t\boldsymbol{\Sigma}_{\hat{v}} =< \boldsymbol{k}_a, \boldsymbol{v} >_g$, from which we find

$$\boldsymbol{A} = \boldsymbol{\Sigma}_{\hat{v}}^{-1} < \boldsymbol{v}, \boldsymbol{k}_a >_g . \qquad \text{(A.15)}$$

We now simplify this expression for $\boldsymbol{A}$ by looking for an alternative representation of $\boldsymbol{\Sigma}_{\hat{v}}$.

Denote the $(i,j)$th element of $\boldsymbol{A}^{-1}$ as $a^{ij}$. From $\boldsymbol{A}^t\boldsymbol{v}(x) = \boldsymbol{k}_a(x)$ we find $\boldsymbol{v}(x) = \boldsymbol{A}^{-t}\boldsymbol{k}_a(x)$, or $v_i(x) = \sum_j a^{ji}k_j^a(x)$.

The $(i,j)$th element of $\boldsymbol{\Sigma}_{\hat{v}} =< \boldsymbol{v}, \boldsymbol{v} >_g$ is given by

$$< v_i, v_j >_g = \sum_m\sum_n a^{mi}a^{nj}\int_{\mathcal{S}} k_m^a(x)k_n^a(x)dG(x) = \sum_m a^{mi}a^{mj},$$

which is the $(i,j)$th element of $\boldsymbol{A}^{-t}\boldsymbol{A}^{-1}$. Hence,

$$\begin{aligned}\boldsymbol{\Sigma}_{\hat{v}} &= \boldsymbol{A}^{-t}\boldsymbol{A}^{-1}\\ &= \left(< \boldsymbol{v}, \boldsymbol{l} >_g^{-1} \boldsymbol{\Sigma}_{\hat{v}}\right)^t \left(< \boldsymbol{v}, \boldsymbol{k}_a >_g^{-1} \boldsymbol{\Sigma}_{\hat{v}}\right)\\ &= \boldsymbol{\Sigma}_{\hat{v}} < \boldsymbol{v}, \boldsymbol{k}_a >_g^{-t} < \boldsymbol{v}, \boldsymbol{k}_a >_g \boldsymbol{\Sigma}_{\hat{v}}.\end{aligned}$$

Solving this equation for $\boldsymbol{\Sigma}_{\hat{v}}$ gives $\boldsymbol{\Sigma}_{\hat{v}} =< \boldsymbol{v}, \boldsymbol{k}_a >_g < \boldsymbol{k}_a, \boldsymbol{v} >_g$. We now substitute this expression into (A.15),

$$\boldsymbol{A} = \boldsymbol{\Sigma}_{\hat{v}}^{-1} < \boldsymbol{v}, \boldsymbol{k}_a >_g = < \boldsymbol{v}, \boldsymbol{k}_a >_g^{-1} = \boldsymbol{\Sigma}_{\hat{v}}^{-1/2}.$$

(2) By the definition of the $\{l_j\}$ and the $\{\gamma_j\}$, we have

$$\int_S c(x,y)l_j(x)dG(x) = \gamma_j l_j(y).$$

We now project both sides of the equation onto $l_j$,

$$\int_S l_j(y) \int_S c(x,y)l_j(x)dG(x)dG(y) = \gamma_j.$$

Equation (5.11) is found by substituting $l_j(x) = \boldsymbol{a}_j^t \boldsymbol{v}(x)$.

## A.10 Proof of Theorem 9.1

We provide only a sketch of the proof.

Write

$$\boldsymbol{R}^t = (R_{11}, R_{12}, \ldots, R_{1n_2}, R_{21}, \ldots, R_{Kn_K}),$$

which is the vector of ranks, ordered according the usual convention. From Lemma 7.3 we know that

$$\mathrm{Var}\{\boldsymbol{R}\} = \boldsymbol{\Sigma}_R = \frac{n+1}{12}\left(n\boldsymbol{I} - \boldsymbol{J}\boldsymbol{J}^t\right)$$

with $\boldsymbol{I}$ and $\boldsymbol{J}$ the $n \times n$ identity matrix and the $n$-unit vector, respectively. Define the $n$-dimensional vectors $\boldsymbol{c}_s$ as vectors with all entries equal to zero, except the entries at the positions corresponding to the elements of the $s$th sample in $\boldsymbol{R}$; these entries are equal to

$$\frac{\sqrt{12}}{\sqrt{n_s n(n+1)}}$$

$(s = 1, \ldots, K)$. Let $\boldsymbol{C}$ denote an $n \times K$ matrix with $s$th column equal to $\boldsymbol{c}_s^t$. Note that the columns of this matrix are orthogonal. In a similar fashion we also construct a matrix $\boldsymbol{D}$ which only differs from $\boldsymbol{C}$ by the absence of the factor $\sqrt{12}/\sqrt{n(n+1)}$. The columns of this matrix are orthonormal.

Let $W_s = \boldsymbol{c}_s^t(\boldsymbol{R} - ((n+1)/2)\boldsymbol{J})$. With this notation the KW statistic becomes

$$KW = \sum_{s=1}^{K} W_s^2 = \left(\boldsymbol{R} - \frac{n+1}{2}\boldsymbol{J}\right)^t \boldsymbol{C}\boldsymbol{C}^t \left(\boldsymbol{R} - \frac{n+1}{2}\boldsymbol{J}\right).$$

Asymptotic multivariate normality of $\boldsymbol{W} = \boldsymbol{C}^t \left(\boldsymbol{R} - \frac{n+1}{2}\boldsymbol{J}\right)$ can be shown easily. It has mean zero and its covariance matrix equals

$$\begin{aligned}\text{Var}\{\boldsymbol{W}\} &= \boldsymbol{C}^t \boldsymbol{\Sigma}_R \boldsymbol{C} \\ &= \boldsymbol{D}^t \left(\boldsymbol{I} - (\boldsymbol{J}/\sqrt{n})(\boldsymbol{J}/\sqrt{n})^t\right) \boldsymbol{D} \\ &= \boldsymbol{D}^t \boldsymbol{E} \boldsymbol{\Gamma} \boldsymbol{E}^t \boldsymbol{D},\end{aligned}$$

where $\boldsymbol{E}$ has rows equal to the eigenvectors of $\boldsymbol{I} - (\boldsymbol{J}/\sqrt{n})(\boldsymbol{J}/\sqrt{n})^t$, and $\boldsymbol{\Gamma}$ is the diagonal matrix with the eigenvalues. Note that this particular matrix has exactly one zero eigenvalue and $n-1$ eigenvalues equal to one (see also Appendix A.1). For convenience we set this zero at the first diagonal position of $\boldsymbol{\Gamma}$. We now write

$$\text{Var}\{\boldsymbol{W}\} = \boldsymbol{D}^t(\boldsymbol{E}\boldsymbol{\Gamma}^{1/2})(\boldsymbol{E}\boldsymbol{\Gamma}^{1/2})^t\boldsymbol{D}.$$

The zero eigenvalue implies that all entries in the first column of $\boldsymbol{E}\boldsymbol{\Gamma}^{1/2}$ are zero. Moreover, by the orthonormality of $\boldsymbol{D}$ and the $n-1$ eigenvalues 1 in $\boldsymbol{\Gamma}$, we may conclude that $\text{Var}\{\boldsymbol{W}\}$ has also one eigenvalue equal to zero, and $K-1$ eigenvalues equal to one. On using Lemma A.1 we may conclude that $KW$ has asymptotically a $\chi^2_{K-1}$ distribution under the general $K$-sample null hypothesis. □

# Appendix B
# The Bootstrap and Other Simulation Techniques

## B.1 Simulation of EDF Statistics Under the Simple Null Hypothesis

In traditional univariate statistics, many test statistics have a limiting standard normal null distribution. For instance, let $T_n$ denote such a test statistic; then the asymptotic results may be denoted by $T_n \xrightarrow{d} N(0,1)$, or $T_n \xrightarrow{d} Z$, where $Z \sim N(0,1)$. A one-sided $\alpha$-level test may be performed by comparing the observed test statistic with the $1-\alpha$ quantile of the standard normal distribution, which can be found in tables in many textbooks. When working with empirical processes, however, we will often encounter test statistics which have a limiting distribution that has no explicit distribution function. The limiting distribution is often expressed as a function of a Gaussion process. In this case, the criticial values will often have to be esimated by means of simulations of the empirical process. The next R-code generates a realization of a Brownian bridge at frequency=1000 equally spaced points between 0 and 1. The larger the frequency, the better the realization approximates a true continuous process.

```
> B<-rbridge(frequency=1000)
```

The asymptotic null distribution of the KS test can now be simulated by the following lines.

```
> ks<-rep(NA,10000)
> for(i in 1:10000) {
+   ks[i]<-max(abs(rbridge(frequency=1000)))
+ }
```

We can use ks for instance to compute the $p$-value of the PRG example.

```
> length(ks[ks>sqrt(100000)*0.0029])/10000
[1] 0.3394
```

A better approximation can be obtained by increasing the frequency and the number of Monte Carlo simulation runs.

## B.2 The Parametric Bootstrap for Composite Null Hypotheses

The parametric bootstrap may be used for testing a full parametric null hypothesis, whether simple or composite. Here we describe the method for testing for a composite null hypothesis, but it can be applied to simple null hypotheses too by simply fixing the $\boldsymbol{\beta}$ nuisance parameter throughout the algorithm.

Consider the null hypothesis

$$H_0 : F \in \{G(.;\boldsymbol{\beta}) : \boldsymbol{\beta} \in B\}$$

(see Section 4.2.2 for more details on this type of composite null hypothesis). Let $\boldsymbol{X}^t = (X_1, \ldots, X_n)$ denote the sample of $n$ i.i.d. observations, and $\hat{\boldsymbol{\beta}}$ is a $\sqrt{n}$-consistent estimator of $\boldsymbol{\beta}$ under $H_0$. Suppose the test statistic is denoted by $T = T(\boldsymbol{X}, \hat{\boldsymbol{\beta}})$.

The parametric bootstrap procedure consists in sampling $B$ times $n$ i.i.d. observations from the distribution $G(.;\hat{\boldsymbol{\beta}})$. The $j$th sample is denoted by $\boldsymbol{X}_j^*$, and the estimator of $\boldsymbol{\beta}$ by $\hat{\boldsymbol{\beta}}_j^*$. For each bootstrap sample the test statistic is recalculated, which is denoted by $T_j^* = T(\boldsymbol{X}_j^*, \hat{\boldsymbol{\beta}}_j^*)$. The empirical distribution of the $B$ bootstrapped test statistics, $T_1^*, \ldots, T_B^*$, serves as an approximation of the asymptotic null distribution of $T$.

## B.3 A Modified Nonparametric Bootstrap for Testing Semiparametric Null Hypotheses

The method described here was proposed by Bickel and Ren (2001). See also Bickel et al. (2006).

Let $\mathcal{F}$ denote a class of density functions for which the distribution of the test statistic behaves well, and let $\boldsymbol{X}^t = (X_1, \ldots, X_n)$ denote the vector of the $n$ i.i.d. sample observations. Let $\boldsymbol{U} = \boldsymbol{U}(\boldsymbol{X})$ denote a $k$-dimensional statistic. Consider test statistics of the form

$$T = \boldsymbol{U}^t(\boldsymbol{X})\hat{\boldsymbol{\Sigma}}^{-1}(\boldsymbol{X})\boldsymbol{U}(\boldsymbol{X}),$$

where $\hat{\boldsymbol{\Sigma}}(\boldsymbol{X})$ is an estimator of $\text{Var}\{\boldsymbol{U}\}$ that is $\sqrt{n}$-consistent for all $f \in \mathcal{F}$. Consider a semiparametric null hypothesis formulated as

$$H_0 : f \in \mathcal{F}_0,$$

where

$$\mathcal{F}_0 = \{f \in \mathcal{F} : \mathrm{E}_f \{\boldsymbol{U}\} = 0\}.$$

Consider now a nonparametric bootstrap procedure in which $\boldsymbol{X}_j^*$ denotes the $j$th bootstrap sample. For each bootstrap sample the test statistic is calculated as

$$T_j^* = \left(\boldsymbol{U}(\boldsymbol{X}_j^*) - \boldsymbol{U}(\boldsymbol{X})\right)^t \hat{\boldsymbol{\Sigma}}^{-1}(\boldsymbol{X}_j^*) \left(\boldsymbol{U}(\boldsymbol{X}_j^*) - \boldsymbol{U}(\boldsymbol{X})\right).$$

When $B$ bootstrap simulations are performed, the empirical distribution of $T_1^*, T_2^*, \ldots, T_B^*$ is used as an approximation of the null distribution of $T$.

# References


L. Acion, J. Peterson, S. Temple, and S. Arndt. Probabilistic index: An intuitive non-parametric approach to measuring the size of treatment effects. *Statistics in Medicine*, 25:591–602, 2006.

N. Aguirre and M. Nikulin. Goodness-of-fit tests for the family of logistic distributions. *Q uesti'o*, 18:317–335, 1994.

H. Akaike. Information theory and an extension of the maximum likelihood principle. In B. Petrov and F. Csàki, editors, *Second International Symposium on Inference Theory*, pages 267–281, Budapest, 1973. Akadémiai Kiadó.

H. Akaike. A new look at statistical model identification. *I.E.E.E. Transactions on Automatic Control*, 19:716–723, 1974.

M. Akritas and E. Brunner. A unified approach to rank tests for mixed models. *Journal of Statistical Planning and Inference*, 61:249–277, 1997.

W. Alexander. *Boundary Kernel Estimation of the Two-Sample Comparison Density Function*. PhD thesis, Texas A& M University, College Station, Texas, USA, 1989.

D. Allison, G. Page, T. Beasley, and J. E. Edwards. *DNA Microarrays and Related Genomics Techniques : Design, Analysis, and Interpretation of Experiments*. Chapman and Hall, Boca Raton, Florida, USA, 2006.

T. Anderson and D. Darling. Asymptotic theory of certain "goodness of fit" criteria based on stochastic processes. *Annals of Mathematical Statistics*, 23:193–212, 1952.

T. Anderson and D. Darling. A test of goodness-of-fit. *Journal of the American Statistical Association*, 49:765–769, 1954.

A. Ansari and R. Bradley. Rank-sum tests for dispersion. *Annals of Mathematical Statistics*, 31:1174–1189, 1960.

A. Atkinson. Tests of pseudo-random numbers. *Applied Statistics*, 29:164–171, 1980.

G. Babu and A. Padmanabhan. Resampling methods for the non-parametric Behrens-Fisher problem. *Sankhya, Series A*, 64:678–692, 2002.

G. Babu and C. Rao. Goodness-of-fit tests when parameters are estimated. *Sankhya, series A*, 2004.

D. Bamber. The area above the ordinal dominance graph and the area below the receiver operator characteristic graph. *Journal of Mathematical Psychology*, 12:287–415, 1975.

L. Baringhaus and N. Henze. A class of tests for exponentiality based on the empirical Laplace transform. *annals of the institute of statistical mathematics*, 43:551–564, 1991.

L. Baringhaus, N. G Urtler, and N. Henze. Weighted integral test statistics and components of smooth tests of fit. *Australian and New Zealand journal of statistics*, 42:179–192, 2000.

D. Barton. On Neyman's smooth test of goodness of fit and its power with respect to a particular system of alternatives. *Skandinavisk Aktuarietidskrift*, 36:24–63, 1953.

D. Bauer. Constructing confidence sets using rank statistics. *Journal of the American Statistical Association*, 67:687–690, 1972.

T. Bednarski and T. Ledwina. A note on biasedness of tests of fit. *Mathematische Operationsforschung und Statistik, Series Statistics*, 9:191–193, 1978.

K. Behnen and M. Husková. A simple algorithm for the adaptation of scores and power behavior of the corresponding rank tests. *Communications in Statistics - Theory and Methods*, 13:305–325, 1984.

K. Behnen and G. Neuhaus. Galton's test as a linear rank test with estimated scores and its local asymptotic efficiency. *Annals of Statistics*, 11:588–599, 1983.

K. Behnen, G. Neuhaus, and F. Ruymgaart. Two sample rank estimators of optimal nonparametric score-functions and corresponding adaptive rank statistics. *Annals of Statistics*, 11:1175–1189, 1983.

A. Bernard and E. Bos-Levenbach. The plotting of observations on probability paper. *Statistica Neerlandica*, 7:163–173, 1953.

P. Bickel and D. Freedman. Some asymptotic theory for the bootstrap. *annals of statistics*, 9:1196–1217, 1981.

P. Bickel and J. Ren. The *Bootstrap in Hypothesis Testing*. In M. de Gunst, C. Klaassen, and A. van der Vaart, editors, *Festschrift for Willem R. van Zwet*, pages 91–112. IMS, Beachwood, USA, 2001.

P. Bickel, Y. Ritov, and T. Stoker. Tailor-made tests of goodness of fit to semiparametric hypotheses. *Annals of Statistics*, 34:721–741, 2006.

M. Birch. A new proof of the Pearson-Fisher theorem. *AMS*, 35:817–824, 1964.

Z. Birnbaum and O. Klose. Bounds for the variance of the Mann-Whitney statistic. *Annals of Mathematical Statistics*, 23:933–945, 1957.

Y. Bishop, S. Fienberg, and P. Holland. *Discrete Multivariate Analysis: Theory and Practice*. MIT Press, Cambridge, MA, USA, 1975.

G. Blom. *Statistical Estimates and Transformed Beta Variables*. Wiley, New York, 1958.

M. Bogdan. Data driven version of Pearson's chi-square test for uniformity. *Journal of Statistical Computation and Simulation*, 52:217–237, 1995.

D. Boos. Comparing $k$ populations with linear rank statistics. *Journal of the American Statistical Association*, 81:1018–1025, 1986.

D. Boos. On generalized score tests. *The American Statistician*, 46:327–333, 1992.

J. Box. *R. A. Fisher, the Life of a Scientist*. Wiley, New York, USA, 1978.

S. Buckland. Fitting density functions with polynomials. *Applied Statistics*, 41:63–76, 1992.

C. Carolan and J. Tebbs. Nonparametric tests for and against likelihood ratio ordening in the two-sample problem. *Biometrika*, 92:159–171, 2005.

B. Carvalho, C. Postma, S. Mongera, E. Hopmans, S. Diskin, M. van de Wiel, W. Van Criekinge, O. Thas, A. Matth Ai, M. Cuesta, J. Terhaar, M. Craanen, E. Schr Ock, B. Ylstra, and G. Meijer. Integration of dna and expression microarray data unravels seven putative oncogenes on 20q amplicon involved in colorectal adenoma to carcinoma progression. *Cellular Oncology*, 2008.

N. Cencov. Evaluation of an unknown distribution density from observations. *Soviet. Math.*, 3:1559–1562, 1962.

H. Chernoff and E. Lehmann. The use of maximum-likelihood estimates in $\chi^2$ tests for goodness of fit. *Annals of Mathematical Statistics*, 25:579–586, 1954.

H. Chernoff and I. Savage. Asymptotic normality and efficiency of certain non-parametric test statistics. *Annals of Mathematical Statistics*, 29:972–994, 1958.

I. Chervoneva and B. Iglewicz. Orthogonal basis approach for comparing nonnormal continuous distributions. *Biometrika*, 92:679–690, 2005.

Y. Cheung and J. Klotz. The Mann Whitney Wilcoxon distribution using linked lists. *Statistica Sinica*, 7:805–813, 1997.

G. Claeskens and N. Hjort. Goodness of fit via non-parametric likelihood ratios. *Scandinavian Journal of Statistics*, 31:487–513, 2004.

A. Cohen and H. Sackrowitz. Unbiasedness of the chi-squared, likelihood ratio, and other goodness of fit tests for the equal cell case. *Annals of Statistics*, 3:959–964, 1975.

H. Cramér. On the composition of elementary errors. *Skandinavisk Aktuarietidskrift*, 11: 13–74, 141–180, 1928.

N. Cressie and T. Read. Multinomial goodness-of-fit tests. *Journal of the Royal Statistical Society, Series B*, 46:440–464, 1984.

M. Csörgö. *Quantile Processes with Statistical Applications*. SIAM, Philadelphia, USA, 1983.

M. Csörgö and L. Horváth. *Weighted Approximations in Probability and Statistics*. Wiley, New York, USA, 1993.

M. Csörgö and P. Révész. Strong approximations of the quantile process. *The Annals of Statistics*, 6:822–894, 1978.

M. Csörgö, S. Csörgö, L. Horváth, and D. Mason. Weighted empirical and quantile process. *Annals of Probability*, 14:31–85, 1986.

M. Csörgö, L. Horváth, and Q. Shao. Convergence of integrals of uniform empirical and quantile processes. *Stochastic Processes and Their Applications*, 45:283–294, 1993.

S. Csörgö. Weighted correlation tests for scale families. *test*, 11:219–248, 2002.

S. Csörgö and J. Faraway. The exact and asymptotic distribuitons of Cramér-von Mises statistics. *JRSSB*, 58:221–234, 1996.

C. Cunnane. Unbiased plotting positions - a review. *Journal of Hydrology*, 37:205–222, 1978.

J. Cwik and J. Mielniczuk. Data-dependent bandwidth choice for a grade kernel estimate. *Statistics and Probability Letters*, 16:397–405, 1993.

R. D'Agostino and M. Stephens. *Goodness-of-Fit Techniques*. Marcel Dekker, New York, USA, 1986.

G. Dallal and L. Wilkinson. An analytic approximation to the distribution of lilliefors' test for normality. *The American Statistician*, 40:294–296, 1986.

T. de Wet. Goodness-of-fit tests for location and scale families based on a weighted $l_2$-Wasserstein distance measure. *Test*, 11:89–107, 2002.

E. del Bario, J. Cuesta-Albertos, and C. Matrán. Contributions of empirical and quantile processes to the asymptotic theory of goodness-of-fit tests. *Test*, 9:1–96, 2000.

E. del Barrio, J. Cuesta Albertos, and C. Matrán Y J. Rodríguez Rodríguez. Tests of goodness of fit based on the L2-Wasserstein distance. *Annals of Statistics*, 27:1230–1239, 1999.

E. Del Barrio, E. Giné, and F. Utzet. Asymptotics for $l_2$ functionals of the empirical quantile process, with applications to tests of fit based on the weighted Wasserstein distances. *Bernoulli*, 11:131–189, 2005.

K. Doksum. Empirical probability plots and statistical inference for nonlinear models in the two sample case. *Annals of Statistics*, 2:267–277, 1974.

H. Doss and R. Gill. An elementary approach to weak convergence for quantile processes, with applications to censored survival data. *JASA*, 87:869–877, 1992.

F. Drost. Asymptotics for generalized chi-square goodness-of-fit tests. In *CWI Tract*. Centrum voor Wiskunde en Informatica, Amsterdam, The Netherlands, 1988.

J. Durbin. Weak convergence of the sample distribution function when parameters are estimated. *Annals of Statistics*, 1:279–290, 1973.

J. Durbin and M. Knott. Components of Cramér - von Mises statistics. *Journal of the Royal Statistical Society, Series B*, 34:290–307, 1972.

J. Durbin, M. Knott, and C. Taylor. Components of Cramér - von Mises statistics: II. *Journal of the Royal Statistical Society, Series B*, 37:216–237, 1975.

M. Dwass. Some $k$-sample rank-order tests. In *Contributions to Probabilitiy and Statistics, Essays in Honor of H. Hotelling*, pages 198–202. Stanford University Press, Stanford, USA, 1960.

B. Efron and R. Tibshirani. Using specially designed exponential families for density estimation. *Annals of Statistics*, 24:2431–2461, 1996.

J. Einmahl and D. Mason. Strong limit theorems for weighted quantile processes. *Annals of Probability*, 16:1623–1643, 1988.

J. Einmahl and I. McKeague. Empirical likelihood based hypothesis testing. *Bernoulli*, 9:267–290, 2003.

K. Entacher and H. Leeb. Inversive pseudorandom number generators: Empirical results. In *Proceedings of the Conference Parallel Numerics 95*, pages 15–27, Sorrento, Italy, 1995.

T. Epps and L. Pulley. A test for normality based on the empirical characteristic function. *Biometrika*, 70:723–726, 1983.

R. Eubank and V. LaRiccia. Components of pearson's phi-squared distance measure for the $k$-sample problem. *Journal of the American Statistical Association*, 85:441–445, 1990.

R. Eubank, V. LaRiccia, and R. Rosenstein. Test statistics derived as components of Pearson's phi-squared distance measure. *Journal of the American Statistical Association*, 82:816–825, 1987.

J. Fan and I. Gijbels. *Local Polynomial Modelling and its Applications*. Chapman and Hall, London, UK, 1996.

R. Farrell. On the best obtainable asymptotic rates of convergence in estimation of a density function at a point. *Annals of Mathematical Statistics*, 43:170–180, 1972.

P. Feigin and C. Heathcore. The empirical characteristic function and the Cramér-von Mises statistic. *Sankhya A*, 38:309–325, 1977.

J. Filliben. The probability plot coefficient test for normality. *Technometrics*, 17:111–117, 1975.

M. Fisz. Some non-parametric tests for the $k$-sample problem. *Colloquium Math.*, 7:289–296, 1960.

M. Fligner and T. Killeen. Distribution-free two-sample tests for scale. *Journal of the American Statistical Association*, 71:210–213, 1976.

M. Fligner and G. Policello. Robust rank procedures for the Behrens-Fisher problem. *JASA*, 76:162–168, 1981.

G. Gajek. On improving density estimators which are not bona fide functions. *Annals of Statistics*, 14:1612–1618, 1986.

R. Gentleman, R. Irizarry, V. Carey, S. Dutoit, and W. E. Huber. *Bioinformatics and Computational Biology Solutions Using R and Bioconductor*. Springer, New York, USA, 2005.

D. Gillen and S. Emerson. Nontransitivity in a class of weighted logrank statistics under nonproportional hazards. *Statistics and Probability Letters*, 77:123–130, 2007.

I. Glad, N. Hjort, and N. Ushakov. Correction of density estimators that are not densities. *Scandinavian Journal of Statistics*, 30:415–427, 2003.

P. Good. *Resampling Methods: a Practical Guide to Data Analysis*. Birkhauser, Boston, USA, 3rd edition, 2005.

P. Greenwood and M. Nikulin. *A Guide to Chi-Squared Testing*. Wiley, New York, USA, 1996.

I. Gringorten. A plotting rule for extreme probability paper. *Journal of Geophysical Research*, 68:813–814, 1963.

N. Gürtler and N. Henze. Goodness-of-fit tests for the Cauchy distribution based on the empirical characteristic function. *Annals of the Institute of Statistical Mathematics*, 52:267–286, 2000.

J. Hájek, Z. Šidák, and P. Sen. *Theory of Rank Tests*. Academic Press, San Diego, USA, 2nd edition, 1999.

P. Hall. Orthogonal series distribution function estimation, with applications. *Journal of the Royal Statistical Society, Series B*, 45:81–88, 1983.

P. Hall. On the rate of convergence of orthogonal series density estimators. *Journal of the Royal Statistical Society, Series B*, 48:115–122, 1986.

P. Hall. On Kullback-Leibler loss and density estimation. *Annals of Statistics*, 15: 1491–1519, 1987.

P. Hall and R. Murison. Correcting the negativity of high-order kernel density estimators. *Journal of Multivariate Analysis*, 47:103–122, 1993.

W. Hall and D. Mathiason. On large-sample estimation and testing in parametric models. *International Statistical Review*, 58:77–97, 1990.

M. Halperin, P. Gilbert, and J. Lachin. Distribution-free confidence intervals for $pr(x_1 < x_2)$. *Biometrics*, 43:71–80, 1987.

F. Hampel, E. Ronchetti, P. Rousseeuw, and W. Stahel. *Robust Statistics: The Approach Based on the Influence Function.* Springer-Verlag, New York, USA, 1986.

D. Hand, F. Daly, A. Lunn, K. McConway, and E. Ostrowsky. *A Handbook of Small Data Sets.* Chapman and Hall, London, UK, 1994.

M. Handcock and M. Morris. *Relative Distribution Methods in Social Siences.* Springer-Verlag, New York, USA, 1999.

J. Hart. *Nonparametric Smoothing and Lack-of-Fit Tests.* Springer, Berlin, Germany, 1997.

A. Hazen. *Flood Flows.* Wiley, New York, USA, 1930.

C. Heathcore. A test of goodness of fit for symmetric random variables. *Australian Journal of Statistics*, 14:172–181, 1972.

N. Henze. A new flexible class of tests for exponentiality. *Communications in Statistics - Theory and Methods*, 22:115–133, 1993.

N. Henze. Do components of smooth tests of fit have diagnostic properties? *Metrika*, 45:121–130, 1997.

N. Henze and B. Klar. Properly rescaled components of smooth tests of fit are diagnostic. *Australian Journal of Statistics*, 38:61–74, 1996.

N. Henze and S. Meintanis. Goodness-of-fit tests based on a new characterization of the exponential distribution. *Communications in Statistics - Theory and Methods*, 31:1479–1497, 2002.

R. Hilgers. On the Wilcoxon-Mann-Whitney-test as nonparametric analogue and extension of *t*-test. *Biometrical Journal*, 24:1–15, 2007.

N. Hjort and I. Glad. Nonparametric density estimation with a parametric start. *Annals of Statistics*, 23:882–904, 1995.

J. Hodges and E. Lehmann. Some problems in minimax point estimation. *Annals of Mathematical Statistics*, 21:182–197, 1956.

J. Hodges and E. Lehmann. *Hodges-Lehmann Estimators.* In S. Kotz, L. Johnson, and C. Read, editors, *Encyclopedia of Statistical Sciences, Volume 3.* Wiley, New York, USA, 1983.

P. Holland. A variation on the minimum chi-square test. *Journal of Mathematical Psychology*, 4:377–413, 1967.

M. Hollander and D. Wolfe. *Nonparametric Statistical Methods.* Wiley, New York, USA, 1999.

E. Holmgren. The P-P plot as a method for comparing treatment effects. *JASA*, 90:360–365, 1995.

T. Hothorn, K. Hornik, M. van de Wiel, and A. Zeileis. A lego system for conditional inference. *The American Statistician*, 60:257–263, 2006.

F. Hsieh. The empirical process approach for semiparametric two-sample models with heterogeneous treatment effect. *JRSSB*, 57:735–748, 1995.

F. Hsieh and B. Turnbull. Non- and semi-parametric estimation of the receiver operating characteristic curve. Technical Report 1026, school of operations research, Cornell University, 1992.

F. Hsieh and B. Turnbull. Nonparametric and semiparametric estimation of the receiver operating characteristic curve. *The Annals of Statistics*, 24:25–40, 1996.

P. Huber. The behavior of maximum likelihood estimates under nonstandard conditions. *Proceedings of the 5th Berkeley Symposium*, 1:221–233, 1967.

P. Huber. *Robust Statistics.* Wiley, New York, USA, 1974.

R. Hyndman and Y. Fan. Sample quantiles in statistical packages. *The American Statistician*, 50:361–365, 1996.

T. Inglot and A. Janic-Wróblewska. Data driven chi-square test for uniformity with unequal cells. *Journal of Statistical Computation and Simulation*, 73:545–561, 2003.

T. Inglot and T. Ledwina. Towards data driven selection of a penalty function for data driven Neyman tests. *Linear Algebra and its Applications*, 417:579–590, 2006.

T. Inglot, W. Kallenberg, and T. Ledwina. Data driven smooth tests for composite hypotheses. *Annals of Statistics*, 25:1222–1250, 1997.

A. Janic-Wróblewska. Data-driven smooth test for a location-scale family. *Statistics*, 38: 337–355, 2004.

A. Janic-Wróblewska and T. Ledwina. Data driven rank test for two-sample problem. *Scandinavian Journal of Statistics*, 27:281–297, 2000.

A. Janic-Wróblewska and T. Ledwina. Data-driven smooth tests for a location-scale family revisited. *Jourmal of Statistical Theory and Practice*, to appear, 2009.

A. Janssen. Global power functions of goodness of fit tests. *Annals of Statistics*, 29:239–253, 2000.

A. Janssen. Principal component decomposition of non-parametric tests. *Probability Theory and Related Fields*, 101:193–209, 1995.

A. Janssen and T. Pauls. A monte carlo comparison of studentized permutation and bootstrap for heteroscedastic two-sample problems. *Computational Statistics*, 20:369–383, 2005.

H. Javitz. *Generalized Smooth Tests of Goodness of Fit, Independence and Equality of Distributions*. PhD thesis, unpublished thesis, Univ. of Calif., Berkely, USA, 1975.

M. Kac and A. Siegert. An explicit representation of a stationary Gausian process. *Annals of Mathematical Statistics*, 18:438–442, 1947.

M. Kac, J. Kiefer, and J. Wolfowitz. On tests of normality and other tests of goodness-of-fit based on distance methods. *Annals of Mathematical Statistics*, 26:189–211, 1955.

W. Kaigh. EDF and EQF orthogonal component decompositions and tests of uniformity. *Nonparametric Statistics*, 1:313–334, 1992.

W. Kallenberg and T. Ledwina. Consistency and Monte Carlo simulation of a data driven version of smooth goodness-of-fit tests. *Annals of Statistics*, 23:1594–1608, 1995a.

W. Kallenberg and T. Ledwina. On data-driven Neyman's tests. *Probability and Mathematical Statistics*, 15:409–426, 1995b.

W. Kallenberg and T. Ledwina. Data-driven smooth tests when the hypothesis is composite. *Journal of the American Statistical Association*, 92:1094–1104, 1997.

W. Kallenberg, J. Oosterhoff, and B. Schriever. The number of classes in chi-squared goodness-of-fit tests. *Journal of the American Statistical Association*, 80:959–968, 1985.

M. Kaluszka. On the Devroye-Gyorfi methods of correcting density estimators. *Statistics and Probability Letters*, 37:249–257, 1998.

R. Kanwal. *Linear Integral Equations, Theory and Technique*. Academic Press, New York, USA, 1971.

M. Karpenstein-Machan and R. Maschka. Investigations on yield structure and local adaptibility. *Agrobiological Research*, 49:130–143, 1996.

M. Karpenstein-Machen, B. Honermeier, and F. Hartmann. *Produktion Aktuell, Triticale*. DLG Verlag, Frankfurt, Germany, 1994.

J. Kiefer. Deviations between the sample quantile process and the sample DF. In M. Puri, editor, *Proceedings of the Conference on Nonparametric Techniques in Statistical Inference*, pages 299–319, Cambridge, UK, 1970. Cambridge University Press.

B. Kimball. On the choice of plotting positions on probability paper. *JASA*, 55:546–560, 1960.

B. Klar. Diagnostic smooth tests of fit. *Metrika*, 52:237–252, 2000.

M. Knot. The distribution of the Cramér-von Mises statistic for small sample sizes. *Journal of the Royal Statistical Society, Series B*, 36:430–438, 1974.

D. Knuth. *The Art of Computer Programming, Volume 2*. Addison-Wesley, Reading, MA, USA, 1969.

A. Kolmogorov. Sulla determinazione empirica di una legge di distribuzione. *Gior. Ist. Ital. Attuari*, 4:83–91, 1933.

M. Kosorok. *Introduction to Empirical Processes and Semiparametric Inference*. Springer, New York, USA, 2008.

H. Lancaster. *The Chi-Squared Distribution*. Wiley, London, UK, 1969.

R. Larsen, T. Curran, and W. J. Hunt. An air quality data analysis system for interrelating effects, standards, and needed source reductions: Part 6. calculating concentration reductions needed to achieve the new national ozone standard. *Journal of Air Pollution Control Association*, 30:662–669, 1980.

T. Ledwina. Data-driven version of Neyman's smooth test of fit. *Journal of the American Statistical Association*, 89:1000–1005, 1994.

A. Lee. *U-Statistics*. Marcel Dekker, New York, USA, 1990.

J. Lee and N. Tu. A versatile one-dimensional distribution plot: The BLiP plot. *The American Statistician*, 51:353–358, 1997.

E. Lehmann. Consistency and unbiasedness of certain nonparametric tests. *Annals of Mathematical Statistics*, 22:165–179, 1951.

E. Lehmann. The power of rank tests. *Annals of Mathematical Statistics*, 24:23–43, 1953.

E. Lehmann. *Nonparametrics. Statistical Methods Based on Ranks*. Prentice Hall, Upper Saddle River, NJ, USA, 1998.

E. Lehmann. *Elements of Large-Sample Theory*. Springer, New York, USA, 1999.

E. Lehmann and J. Romano. *Testing Statistical Hypotheses (3rd Ed.)*. Springer, New York, USA, 2005.

Y. Lepage. A combination of Wilcoxon's and Ansari-Bradley's statistics. *Biometrika*, 58: 213–217, 1971.

P. Lewis. Distribution of the Anderson-Darling statistic. *Annals of Mathematical Statistics*, 32:1118–1124, 1961.

G. Li, R. Tiwari, and M. Wells. Quantile comparison functions in two-sample problems with applications to comparisons of diagnostic markers. *JASA*, 91:689–698, 1996.

W. Lidicker and F. McCollum. Allozymic variation in California sea otters. *Journal of Mammalogy*, 78:417–425, 1997.

H. Lilliefors. On the Kolmogorov-Smirnov test for normality with mean and variance unknown. *Journal of the American Statistical Association*, 62:399–402, 1967.

C. Lin and S. Sukhatme. Hoeffding type theorem and power comparisons of some two-sample rank tests. *Journal of the Indian Statistical Association*, 31:71–83, 1993.

T. Lumley. Non-transitivity of the Wilcoxon rank sum test. Personal communication, 2009.

D. Mage. An objective graphical method for testing normal distributional assumptions using probability plots. *The American Statistician*, 36:116–120, 1982.

H. Mann and A. Wald. On the choice of the number of class intervals in the application of the chi-square test. *Annals of Mathematical Statistics*, 13:306–317, 1942.

H. Mann and D. Whitney. On a test of whether one of two random variables is stochastically larger than the other. *Annals of Mathematical Statistics*, 18:50–60, 1947.

D. Mason. Weak convergence of the weighted empirical quantile process in $l^2(0,1)$. *Annals of Probability*, 12:243–255, 1984.

F. Massey. The distribution of the maximum deviation between two sample cumulative step functions. *Annals of Mathematical Statistics*, 22:125–128, 1951.

M. Matsui and A. Takemura. Empirical characteristic function approach to goodness-of-fit tests for the Cauchy distribution with parameters estimated by MLE or EISE. *Annals of the Institute of Statistical Mathematics*, 57:183–199, 2005.

R. Mee. Confidence intervals for probabilities and tolerance regions based on a generalisation of the Mann-Whitney statistic. *Journal of the American Statistical Association*, 85:793–800, 1990.

S. Meintanis. Goodness-of-fit tests for the logistic distribution based on empirical transforms. *Sankhya, Series B*, 66:306–326, 2004a.

S. Meintanis. A class of omnibus tests for the Laplace distribution based on the empirical characteristic function. *Communications in Statistics - Theory and Methods*, 33:925–948, 2004b.

S. Meintanis. Consistent tests for symmetry stability with finite mean based on the empirical characteristic function. *Journal of Statistical Planning and Inference*, 128:373–380, 2005.

J. Michael. The stabilized probability plot. *Biometrika*, 70:11–17, 1983.

P. Mielke and K. Berry. *Permutation Tests: A Distance Function Approach.* Springer, New-York, USA, 2001.

J. Mielniczuk. Grade estimation of Kullback-Leibler information number. *Probability and Mathematical Statistics*, 13:139–147, 1992.

A. Mood. On the asymptotic efficiency of certain nonparametric two-sample tests. *Annals of Mathematical Statistics*, 25:514–522, 1954.

D. Moore. *Tests of Chi-Squared Type.* In R. D'Agostino and M. Stephens, editors, *Goodness-of-Fit Techniques*, chapter 3, pages 63–95. Marcel Dekker, New York, USA, 1986.

L. Moses. Rank tests for dispersion. *The Annals of Mathematical Statitstics*, 34:973–983, 1963.

G. Neuhaus. Local asymptotics for linear rank statistics with estimated score functions. *Annals of Statistics*, 15:491–512, 1987.

R. Newcombe. Confidence intervals for an effect size measure based on the Mann-Whitney statistic. part 1: General issues and tail-area-based methods. *Statistics in Medicine*, 25: 543–557, 2006a.

R. Newcombe. Confidence intervals for an effect size measure based on the Mann-Whitney statistic. part 2: Asymptotic methods and evaluation. *Statistics in Medicine*, 25:559–573, 2006b.

J. Neyman. Smooth test for goodness of fit. *Skandinavisk Aktuarietidskrift*, 20:149–199, 1937.

J. Neyman. Contribution to the theory of the $\chi^2$ test. In *Proceedings of the First Berkeley Symposium of Mathematical Statistics and Probability*, pages 239–273, 1949.

J. Oosterhoff. The choice of cells in chi-square tests. *Statistica Neerlandica*, 39:115–128, 1985.

E. Parzen. On estimation of a probability density function and mode. *Annals of Mathematical Statistics*, 33:1065–1076, 1962.

E. Parzen. Nonparametric statistical data modeling (with discussion). *JASA*, 74:105–131, 1979.

E. Parzen. FUN.STAT: Quantile approach to two sample statistical data analysis. Technical report, Texas A& M University, College Station, Texas, USA, 1983.

E. Parzen. Statistical methods, mining, tow sample data analysis, comparison distributions, and quantile limit theorems. In *Proceedings of the International Conference on Asymptotic Methods in Probability and Statistics, 8-13 July 1997, Carleton University, Canada*, 1997.

E. Parzen. *Statistical Methods, Mining, Two Sample Data Analysis, Comparison Distributions, and Quantile Limit Theorems.* In *Asymptotic Methods in Probability and Statistics.* Elsevier, Amsterdam, The Netherlands, 1999.

E. Pearson and H. Hartley. *Biometrika Tables for Statisticians, Vol. 2.* Cambridge University Press, Cambridge, UK, 1972.

E. Pearson and M. Stephens. The goodness-of-fit tests based on $W_N^2$ and $U_N^2$. *Biometrika*, 49:397–402, 1962.

K. Pearson. On the criterion that a given system of deviations from the probable in the case of a correlated system of variables is such that it can be reasonably supposed to have arisen from random sampling. *Philosophical Magazine*, 50:157–175, 1900.

A. Pettitt. A two-sample Anderson-Darling rank statistic. *Biometrika*, 63:161–168, 1976.
H. Piepho. Exact confidence limits for covariate-dependent risk in cultivar trials. *Journal of Agricultural, Biological and Environmental Statistics*, 5:202–213, 2000.
E. Pitman. *Notes on Non-parametric Statistical Inference.* Columbia University, New York, USA, 1948.
R. Potthof. Use of the wilcoxon statistic for a generalized Behrens-Fisher problem. *Annals of Mathematical Statistics*, 34:1596–1599, 1963.
N. Pya. Goodness-of-fit tets for the logistic distribution. *Mathematical Journal*, 4:68–75, 2004.
R. Pyke and G. Shorack. Weak convergences of a two-sample empirical process and a new approach to Chernoff-Savage theorems. *Annals of Mathematical Statistics*, 39:755–771, 1968.
A. Qu, B. Lindsay, and B. Li. Improving generalised estimating functions using quadratic inference functions. *Biometrika*, 87:823–836, 2000.
M. Quine and J. Robinson. Efficiencies of chi-square and likelihood ratio goodness-of-fit tests. *Annals of Statistics*, 13:727–742, 1985.
R Development Core Team. *R: A Language and Environment for Statistical Computing.* R Foundation for Statistical Computing, Vienna, Austria, 2008. URL http://www.R-project.org. ISBN 3-900051-07-0.
R. Randles and R. Hogg. Adaptive distribution-free tests. *Communications in Statistics*, 2:337–356, 1973.
K. Rao and D. Robson. A chi-square statistic for goodness-of-fit tests within the exponential family. *Communications in Statistics*, 3:1139–1153, 1974.
L. Rayleigh. On the problems of random vibrations and flights in one,two, or three dimensions. *Philosophical Magazine*, 37:321–347, 1919.
J.C.W. Rayner and D.J. Best. *A Contingency Table Approach to Nonparametric Testing.* Chapman and Hall, New York, USA, 2001.
J.C.W. Rayner and D.J. Best. Neyman-type smooth tests for location-scale families. *Biometrika*, 73:437–446, 1986.
J.C.W. Rayner and D.J. Best. *Smooth Tests of Goodness-of-Fit.* Oxford University Press, New York, USA, 1989.
J.C.W. Rayner, O. Thas, and B. De Boeck. A generalised Emerson recurrence relation. *Australian and New Zealand Journal of Statistics*, 50:235240, 2008.
J.C.W. Rayner, O. Thas, and D.J. Best. *Smooth Tests of Goodness of Fit: Using R.* Wiley, Singapore, 2009.
T. Read and N. Cressie. *Goodness-of-Fit Statistics for Discrete Multivariate Data.* Springer-Verlag, New York, 1988.
R. Risebrough. Effects of environmental pollutants upon animals other than man. In *Proceedings of the 6th Berkeley Symposium on Mathematics and Statistics*, pages 443–463, Berkeley, 1972. University of California University Press.
J. Romano. A bootstrap revival of some nonparametric distance tests. *Journal of the American Statistical Association*, 83:698–708, 1988.
M. Rosenblatt. Remarks on some nonparametric estimates of a density function. *Annals of Mathematical Statistics*, 27:832–837, 1956.
F. Scholz and M. Stephens. $k$-sample Anderson-Darling tests. *Journal of the American Statistical Association*, 82:918–924, 1987.
G. Schwarz. Estimating the dimension of a model. *Annals of Statistics*, 6:461–464, 1978.
D. Scott. *Multivariate Density Estimation.* Wiley, New York, USA, 1992.
P. Sen. A note on asymptotically distribution-free condence intervals for $\Pr[x < y]$ based on two independent samples. *Sankhya, A.*, 29:95–102, 1967.
J. Shao. Jackknife variance estimators for two sample linear rank statistics. Technical Report 88-61, Department of Statistics, Purdue University, USA, 1988.
J. Shao. Differentiability of statistical functionals and consistency of the jackknife. *The Annals of Statistics*, 21:61–75, 1993.

G. Shorack. *Probability for Statisticians.* Springer-Verlag, New York, USA, 2000.

G. Shorack and J. Wellner. *Empirical Processes with Applications to Statistics.* Wiley, New York, USA, 1986.

S. Siegel and J. Tukey. A nonparametric sum of rank procedure for relative spread in unpaired samples. *Journal of the American Statistical Association*, 55:429–444, 1960.

B. Silverman. *Density Estimation for Statistics and Data Analysis.* Chapman and Hall, London, UK, 1986.

J. Simonoff. *Smoothing Methods in Statistics.* Springer, New York, USA, 1996.

N. Smirnov. Sur les ecarts de la courbe de distribution empirique (in Russian). *Rec. Math.*, 6:3–26, 1939.

T. E. Speed. *Statistical Analysis of Gene Expression Microarray Data.* Chapman and Hall, Boca Raton, Florida, USA, 2003.

M. Stephens. Use of the Kolmogorov-Smirnov, Cramér-von Mises and related statistics without extensive tables. *Journal of the Royal Statistical Society, Series B*, 32:115–122, 1970.

M. Stephens. Asymptotic results for goodness-of-fit statistics when parameters must be estimated. Technical Report 159 and 180, Department of Statistics, Stanford University, 1971.

M. Stephens. EDF statistics for goodness-of-fit and some comparisons. *Journal of the American Statistical Association*, 69:730–737, 1974.

M. Stephens. Asymptotic results for goodness-of-fit statistics with unknown parameters. *Annals of Statistics*, 4:357–369, 1976.

B. Sukhatme. On certain two-sample nonparametric tests for variances. *Annals of Mathematical Statistics*, 28:188–194, 1957.

P. Switzer. Confidence procedures for two-sample problems. *Biometrika*, 63:13–25, 1976.

O. Thas. *Nonparametrical Tests Based on Sample Space Partitions.* PhD thesis, Ghent University, 2001.

O. Thas and J. Ottoy. Goodness-of-fit tests based on sample space partitions: An unifying overview. *Journal of Applied Mathematics and Decision Sciences*, 6:203–212, 2002.

O. Thas and J. Ottoy. Some generalization of the Anderson-Darling statistic. *Statistics and Probability Letters*, 64:255–261, 2003.

O. Thas and J. Ottoy. An extension of the Anderson-Darling k-sample test to arbitrary sample space partition sizes. *Journal of Statistical Computation and Simulation*, 74: 561–666, 2004.

O. Thas and J.C.W. Rayner. Informative statistical analyses using smooth goodness-of-fit tests. *Journal of Statistical Theory and Practice*, to appear, 2009.

H. Thode. *Testing for Normality.* Marcel Dekker, New York, USA, 2002.

M. Tiku. Chi-square approximations for the distributions of goodness-of-fit statistics $U_N^2$ and $W_N^2$. *Biometrika*, 52:630–633, 1965.

A. Tsiatis. *Semiparametric Theory and Missing Data.* Springer, New York, USA, 2006.

J. Tukey. *Exploratory Data Analysis.* Addison-Wesley, Reading, MA, USA, 1977.

V. Tuscher, R. Tibshirani, and G. Chu. Significance analysis of microarrays applied to the ionizing radiation response. *Proceedings of the National Academy of Sciences*, 98: 5115–5121, 2001.

S. Vallender. Calculation of the Wasserstein distance between probability distributions on the line. *Theory of Probability Applications*, 18:785–786, 1973.

M. van de Wiel. The split-up algorithm: A fast symbolic method for computing p values of rank statistics. *Computational Statistics*, 16:519–538, 2001.

A. van der Vaart. *Asymptotic Statistics.* Cambridge University Press, Cambridge, UK, 1998.

A. Van der Vaart and J. Wellner. *Weak Convergence and Empirical Processes.* Springer, New York, USA, 2nd edition, 2000.

B. van der Waerden. Order tests for the two-sample problem and their power. *Indagationes Mathematicae*, 14:453–458, 1952.

B. van der Waerden. Order tests for the two-sample problem and their power. *Indagationes Mathematicae*, 15:303–310, 1953.

C. van Eeden. Note on the consistency of some distribution-free tests for dispersion. *Journal of the American Statistical Association*, 59:105–119, 1964.

R. von Mises. *Wahrscheinlichkeitsrechnung*. Deuticke, Vienna, Austria, 1931.

R. von Mises. On the asymptotic distribution of differentiable statistical functions. *Annals of Mathematical Statistics*, 18:309–348, 1947.

G. Wahba. Data-based optimal smoothing of orthogonal series density estimates. *Annals of Statistics*, 9:146–156, 1958.

L. Wasserstein and J. J. Boyer. Bounds on the power of linear rank tests for scale parameters. *American Statistician*, 45:10–13, 1991.

G. Watson. Goodness-of-fit tests on a circle. *Biometrika*, 48:109–114, 1961.

G. Watson. Density estimation by orthogonal series. *Annals of Mathematical Statistics*, 40:1496–1498, 1969.

F. Wilcoxon. Individual comparisons by ranking methods. *Biometrics Bulletin*, 1:80–83, 1945.

H. Wouters, O. Thas, and J. Ottoy. Data driven smooth tests and a diagnostic tool for lack-of-fit for circular data. *Australian and New Zealand Journal of Statistics*, to appear.

S. Zaremba. A generalisation of Wilcoxon's test. *Monatshefte f ur Mathematik*, 66:359–370, 1962.

J. Zhang. Powerful goodness-of-fit tests based on the likelihood ratio. *Journal of the Royal Statistical Society, Series B*, 64:281–294, 2002.

J. Zhang and Y. Wu. A family of simple distribution functions to approximate complicated distributions. *Journal of Statistical Computing and Simulation*, 70:257–266, 2001.

# Index

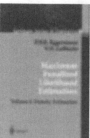